Quality Demystified

Foundations of Quality in Healthcare: Principles, Concepts & Tools

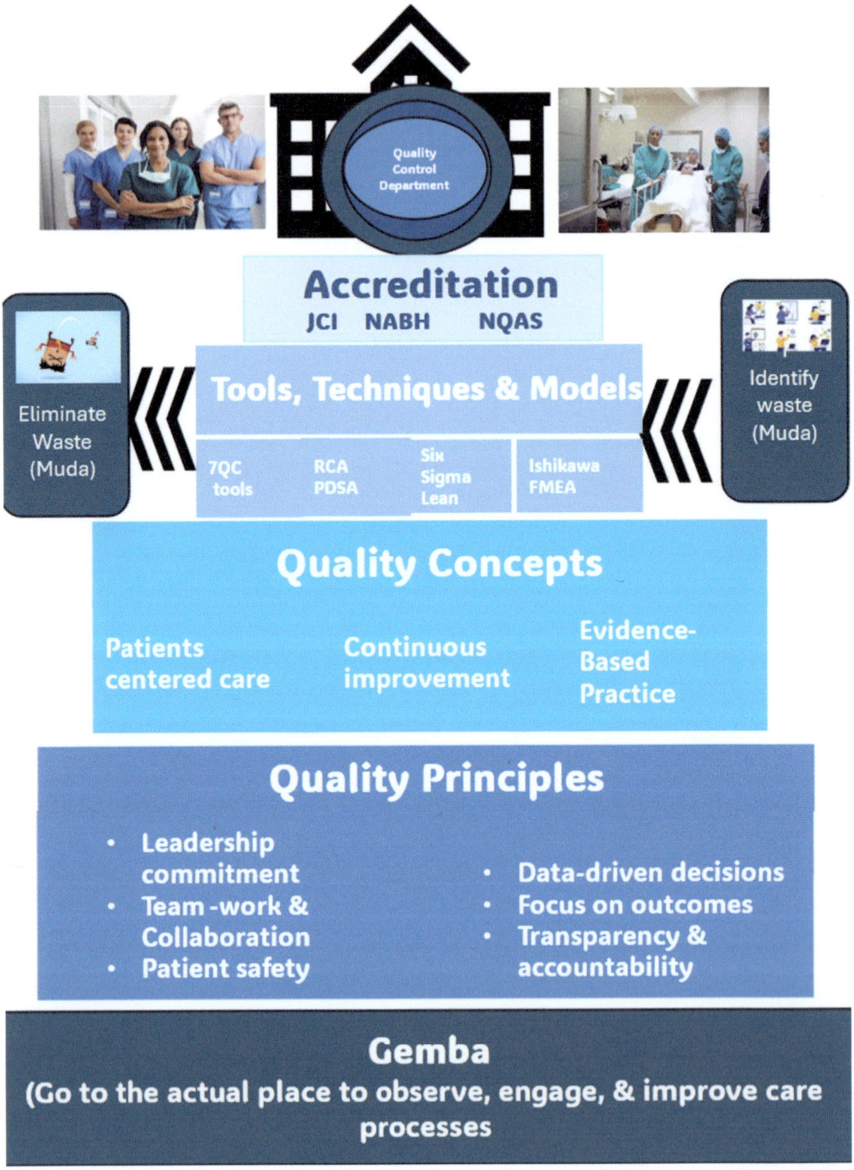

Sangeeta Sharma

Quality Demystified

A Comprehensive Guide to Quality Improvement, Patient Safety, and Clinical Audit in Healthcare

Springer

Sangeeta Sharma
Department of Neuropsychopharmacology
Institute of Human Behavior & Allied Sciences
Delhi, India

ISBN 978-981-95-0338-4 ISBN 978-981-95-0339-1 (eBook)
https://doi.org/10.1007/978-981-95-0339-1

This Springer imprint is published by the registered company Springer Nature Singapore Pte Ltd.
The registered company address is: 152 Beach Road, #21-01/04 Gateway East, Singapore 189721, Singapore

Preface

The journey toward delivering safe, high-quality healthcare is a continuous and evolving one. In a world where healthcare is becoming increasingly complex, ensuring patient safety and maintaining quality standards have never been more critical. This book is designed to be a practical guide for healthcare professionals, administrators, and quality managers, providing the tools, strategies, and insights necessary to link patient safety with quality improvement effectively.

Healthcare in the twenty-first century faces unique challenges—rapid technological advancements, increased patient expectations, and the need to deliver efficient care while maintaining the highest safety standards, especially in diverse healthcare systems. This book addresses these challenges head-on, focusing on practical, day-to-day applications that healthcare professionals can use to create safer environments for patients.

This book is structured to serve as both a learning resource and a reference manual. It begins with an exploration of the foundational principle of patient safety: *First, do no harm*. Common sources of errors, their impact on patient outcomes, and practical methods to mitigate these risks are examined in detail. We also delve into key concepts like medication safety, communication errors, adverse drug reactions, and the role of leadership and clinical governance in fostering a culture of safety. This book also addresses emerging areas like telemedicine, which presents new patient safety challenges and opportunities, further expanding the scope of quality improvement in modern healthcare.

A key feature of this book is its emphasis on quality management tools and techniques, which are essential for driving continuous improvement in healthcare. From flowcharts to control charts, root cause analysis to Six Sigma, the techniques covered here are explained with real-world examples and step-by-step guides for implementation. These tools, often underutilized in healthcare settings, can provide tremendous value when applied correctly, making them a central focus of this book. Additionally, accreditation processes, clinical audits, and performance measurement frameworks are presented in a simple way. As accreditation becomes an essential element of healthcare improvement, this book aims to demystify the process and offer insights into how healthcare organizations can not only achieve but also sustain quality beyond accreditation surges.

In addition to theoretical concepts, this book includes specific audit examples, comprehensive checklists, and a self-test/assessment section to help reflect and reinforce learning through practical application. By showcasing worked-out examples and ready-to-use tools for clinical audits, such as prescription audits, dengue management audits, surgical audit, and nursing audits, the book offers examples of how auditing can be an effective tool for transition from theory to practice for continuous quality improvement in healthcare.

By combining the science of quality management with the art of patient safety, this book aspires to serve as a cornerstone for those committed to advancing healthcare standards. I invite you to explore the tools and methods outlined here and to apply them in your daily practice, contributing to the creation of a healthcare system that not only saves lives but ensures that every life saved is treated with dignity, care, and precision.

Delhi, India Sangeeta Sharma

Acknowledgments

I would like to express my heartfelt gratitude to everyone who supported me in writing this book on quality improvement systems, patient safety, and clinical audits in healthcare.

First, I am deeply thankful to the healthcare professionals who shared their knowledge and experiences, enriching the content with practical insights. My sincere appreciation goes to the experts and researchers in healthcare quality improvement, whose guidance and feedback shaped the book's concepts and ideas.

I also acknowledge the healthcare organizations that allowed me to observe and learn from their quality improvement initiatives, contributing to the practical examples and case studies. I would like to specially thank Chitranjali Gupta and Vineeta Bablani for the artwork.

My heartfelt thanks to my family and friends for their unwavering support and encouragement throughout this journey. Lastly, I am grateful to the publishing team for their professionalism and dedication in bringing this book to fruition.

With profound gratitude,
Dr. Sangeeta Sharma

Competing Interests The author has no competing interests to declare that are relevant to the content of this manuscript.

Contents

Part III Clinical Audit

Part I

Introduction

Introduction 1

Patient safety refers to the prevention of injury and harm during the provision of care, while quality is defined as achieving the best possible outcomes in healthcare. It is crucial for health services worldwide to not only be effective, efficient, equitable, timely, and patient-centered, but also safe. Patient safety is a vital component of healthcare quality and is considered inseparable from it. Ensuring safety is one of the fundamental goals in healthcare settings, aiming to prevent and reduce risks, errors, and harm to patients.

In recent years, there has been a growing sense of dissatisfaction among the public due to the failure of the healthcare delivery system in India to meet expectations. The quality of medical care and hospital services has faced significant scrutiny from consumers and the community. The rising cost and limited resources have added to the complexity of the problem. Access to quality healthcare is not only a basic human need and right but also a matter of concern. Patients have started questioning healthcare providers about the quality and cost of care, and some have sought legal recourse. Factors such as increased consumer awareness, growth in medical tourism, development of the health insurance market, and the involvement of the corporate sector have further fueled the demand for quality benchmarked against global standards.

Improving the quality of health systems is a core component of healthcare. Although access and utilization of health services have improved in the past two decades, the burden of disease and the complexity of healthcare have also increased, resulting in patient harm in healthcare facilities. Unsafe healthcare practices pose a significant public health burden globally and can adversely affect an organization's reputation, staff morale, and public trust in the entire healthcare system. Poor quality of care and underutilization of healthcare services are major barriers to achieving universal health coverage (UHC). Patient safety is also central to the attainment of global goals, particularly Sustainable Development Goal 3, which aims to ensure healthy lives and promote well-being for all. UHC aims to provide essential, quality health services to everyone, from promotion and prevention to treatment,

S. Sharma, *Quality Demystified*, https://doi.org/10.1007/978-981-95-0339-1_1

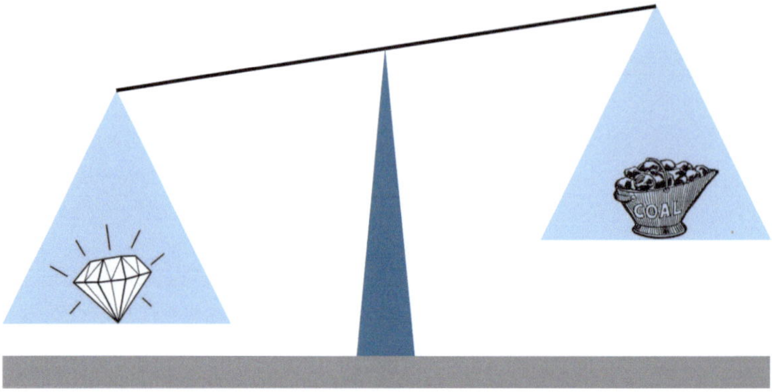

Fig. 1.1 The trade-off between quantity and quality in various contexts

rehabilitation, and palliative care, without causing financial hardship. However, the sustainable development agenda cannot be achieved without ensuring the safety of health services. Patient safety not only reduces costs associated with patient harm but also improves the efficiency of healthcare systems. Safe practices also help build and restore people's trust in healthcare systems. Prioritizing patient safety in health sector policies and programs is crucial in advancing efforts toward achieving UHC. Figure 1.1 illustrates the trade-off between healthcare coverage (quantity) and patient safety (quality). Expanding access increases service reach but may compromise quality, while prioritizing patient safety may limit coverage. The balance highlights the challenge of optimizing both in healthcare policy and practice.

Key Challenges in Healthcare 2

Challenges in healthcare, particularly concerning quality and patient safety, are multifaceted and can vary across different healthcare settings and regions.

Medical Errors Despite advancements in healthcare, medical errors remain a significant challenge, leading to patient harm and even death. Although there is limited data available on the extent of harm caused by unsafe medication practices and medical errors in India, patient safety remains a multidimensional and multifaceted challenge. Addressing this requires robust reporting systems, error analysis, and implementing preventive measures.

Resource Constraints Limited resources, both human and material, can compromise the quality of care. This includes shortages of healthcare professionals, inadequate infrastructure, and insufficient funding.

Communication Barriers Ineffective communication among healthcare teams, as well as between healthcare providers and patients, can lead to misunderstandings, errors, and compromised patient safety.

Healthcare-Associated Infections (HAIs) HAIs are a major concern in healthcare settings, contributing to morbidity, mortality, and increased healthcare costs. Strict adherence to infection control protocols is required to prevent HAIs.

Technology Integration While technology offers numerous benefits, its integration into healthcare systems can be challenging. Issues such as interoperability, data security, and user training need to be addressed to harness the full potential of technology for improving quality and safety.

S. Sharma, *Quality Demystified*, https://doi.org/10.1007/978-981-95-0339-1_2

Cultural and Organizational Barriers Resistance to change, hierarchical structures, and organizational culture can impede efforts to improve quality and safety. Creating a culture of safety and continuous improvement is essential but challenging.

Patient Engagement Engaging patients in their care is crucial for improving outcomes and safety. However, barriers such as limited health literacy, cultural differences, and communication gaps can hinder effective patient engagement.

Regulatory and Compliance Challenges Keeping up with evolving regulations and compliance requirements can be daunting for healthcare organizations. Ensuring adherence while maintaining focus on quality and safety is a constant challenge.

Multisector Challenge Establishing a nationwide institutional structure is particularly challenging, given the country's vast size, diverse infrastructure with a mix of regulated and unregulated public and private sectors, weak regulatory systems, and the decentralized nature of healthcare governance. Each state needs to devise its legislative and administrative structure for implementation. The Indian healthcare delivery system consists of various health institutions with mixed ownership patterns. Tertiary care is dominated by private and/or public-private partnerships, while secondary healthcare is a combination of private and public facilities, and government health systems primarily cater to primary care.

> Money and time spent for training will be ineffective unless inhibitors to good work are removed – **Edwards W. Deming**

Multiple Stakeholders In the delivery of healthcare, involving multiple stakeholders is essential to ensure comprehensive and coordinated care. Each stakeholder, including healthcare providers, patients, organizations, policymakers, insurers, and community organizations, brings unique perspectives, expertise, and contributions to the healthcare ecosystem. However, this diversity can also lead to challenges in communication and coordination. Differences in priorities, goals, and communication styles among stakeholders can create barriers to effective collaboration, potentially compromising the quality, safety, and efficiency of care. Clear communication, mutual respect, and collaborative approaches are crucial to overcoming these challenges, fostering a cohesive healthcare team, and delivering patient-centered care that meets the needs of individuals and communities.

Addressing these challenges requires a multifaceted approach that involves collaboration among healthcare providers, policymakers, patients, and other stakeholders. Continuous education, training, data-driven decision-making, and a commitment to a culture of safety and quality improvement are essential for overcoming these challenges and delivering high-quality, safe care to all patients.

> Quality starts at the top because leaders are responsible for the design of the system – **Mark Graban**

3 Quality and Patient Safety Are Inextricably Linked

The global patient safety framework serves as a foundational blueprint for advancing patient safety across healthcare systems worldwide. It emphasizes the importance of creating a culture of safety, establishing effective systems and processes, and engaging stakeholders at all levels to prevent harm to patients. The framework sets forth patient safety goals that aim to address key areas of concern, such as medication errors, healthcare-associated infections, surgical complications, and diagnostic errors. These goals provide a structured approach to identifying and mitigating risks, improving clinical practices, and enhancing the overall quality of care. By aligning efforts and resources toward achieving these patient safety goals, healthcare organizations can significantly reduce adverse events, improve patient outcomes, and build trust and confidence in healthcare services.

Patient safety goals are crucial guidelines that aim to enhance the safety and quality of healthcare services provided to patients. These goals typically focus on key areas where errors or lapses in care commonly occur, such as medication management, infection control, surgical procedures, and communication among healthcare providers. By setting clear and measurable objectives, patient safety goals help healthcare organizations prioritize efforts to reduce risks, prevent harm, and improve patient outcomes. Adhering to these goals requires a multidisciplinary approach, involving clinicians, administrators, and other stakeholders working collaboratively to implement evidence-based practices, monitor performance, and continuously evaluate and improve care processes.

The six patient safety goals, often highlighted by organizations like The Joint Commission, serve as essential guidelines to improve the quality and safety of healthcare delivery:

1. Identify Patients Correctly: Ensuring that patients are accurately identified using at least two identifiers, such as name and date of birth, helps prevent errors like medication mix-ups or wrong-patient surgeries.

S. Sharma, *Quality Demystified*, https://doi.org/10.1007/978-981-95-0339-1_3

2. Improve Communication: Effective communication among healthcare providers is vital to ensure safe and coordinated care. Goals in this area focus on standardizing communication methods, especially during handoffs between care settings or shifts.
3. Use Medications Safely: Medication errors are a significant cause of patient harm. Goals related to medication safety include ensuring accurate medication reconciliation, proper storage, and administration, as well as patient education about medication use.
4. Prevent Infections: Healthcare-associated infections are a leading cause of morbidity and mortality. Goals aim to reduce the incidence of infections by implementing evidence-based practices in hand hygiene, sterile technique, and infection control protocols.
5. Identify Patient Safety Risks: Recognizing potential risks and implementing preventive measures is crucial. This involves regular risk assessments, fall prevention strategies, and protocols for managing deteriorating patients or emergencies.
6. Prevent Surgical Errors: Surgical procedures carry inherent risks, and the goal is to minimize errors through standardized procedures, surgical site marking, timeouts before procedures, and proper communication among surgical teams.

These goals represent a comprehensive approach to serve as a roadmap for healthcare organizations to foster a culture of safety, accountability and emphasizing the importance of systematic strategies, continuous monitoring, and improvement efforts across various aspects of healthcare delivery.

The World Health Organization (WHO) has been at the forefront of championing patient safety through its global initiatives.[1] The WHO Action Plan for Patient Safety outlines strategic objectives to foster a culture of safety in healthcare systems worldwide. This action plan aims to reduce avoidable harm to patients by half over a five-year period, focusing on key areas such as medication safety, infection prevention and control, surgical safety, and maternal and newborn safety. It emphasizes the importance of leadership, reporting systems, education, and patient engagement in driving improvements. The plan encourages countries to prioritize patient safety, integrate safety measures into health policies, and establish mechanisms for continuous monitoring and evaluation. By setting a global agenda and providing guidance, the WHO Action Plan for Patient Safety serves as a roadmap for countries to strengthen their healthcare systems and ensure the well-being of patients everywhere.

The National Patient Safety Implementation Framework (NPSIF 2018–2025)[2] is a significant development in India, aiming to converge diverse programs and initiatives into a unified framework for successful implementation of patient safety strategies across all levels of healthcare and encompassing the full spectrum of health services. The NPSIF identifies six strategic objectives, emphasizing policy

[1] Source: Global Patient Safety Action Plan 2021–2030, World Health Organization. https://iris.who.int/server/api/core/bitstreams/a28c34c0-089c-4f5d-a0b1-5d9c35a3cd67/content.

[2] National Patient Safety Implementation Framework_0.pdf. https://iris.who.int/server/api/core/bitstreams/a28c34c0-089c-4f5d-a0b1-5d9c35a3cd67/content.

development, leadership capacity, data-driven safety improvements, competent healthcare professionals, and effective patient engagement in their care. Accreditation plays a major role in promoting patient safety, particularly in medication safety, but it remains voluntary and incentive-driven. The lack of knowledge about specific quality tools and their applications in quality assurance activities has been identified as a key obstacle. Other challenges include time constraints, negative attitudes among colleagues, and fear of assessment and criticism. Successful implementation of quality assurance activities requires regular meetings, support for data collection, benchmarking against other practices, establishing local peer review systems, and securing financial support.

Patient Safety

4

Contents

4.1 Key Facts

Approximately 1 in 10 patients experience harm in healthcare settings, and unsafe care contributes to over three million deaths annually. In low- and middle-income countries, up to 4 in 100 individuals may die due to unsafe care. More than 50% of these adverse events are preventable, with a significant portion related to medication errors.

In primary and ambulatory care settings, estimates suggest that up to 40% of patients are harmed, but up to 80% of these incidents could be avoided. Common preventable adverse events include medication errors, unsafe surgical procedures, healthcare-associated infections, diagnostic errors, patient falls, pressure ulcers, misidentification, unsafe blood transfusions, and venous thromboembolism.

The global economic impact of patient harm can reduce economic growth by 0.7% annually, with indirect costs amounting to trillions of US dollars each year. Investing in strategies to reduce patient harm not only has the potential to yield significant financial savings but also enhances patient outcomes. For example, effective patient engagement can decrease the incidence of harm by up to 15%.

Data from various studies and reports indicate that patient safety incidents are a significant concern in India.

S. Sharma, *Quality Demystified*, https://doi.org/10.1007/978-981-95-0339-1_4

4.2 Patient Safety: The Principle of "First Do No Harm"

Patient safety, rooted in the principle of "First Do No Harm," aims to prevent and minimize harm to patients during care. This principle, from the Hippocratic Oath, emphasizes avoiding injury or suffering in medical treatment. It involves implementing practices to reduce risks such as medication errors and infections, fostering a culture of continuous improvement and vigilance. By adhering to this principle, healthcare providers ensure that every aspect of care prioritizes patient well-being and safety.

Patient safety is the "absence of preventable harm to a patient and the minimization of unnecessary risks associated with healthcare to an acceptable level." In the broader context of the health system, patient safety is "a framework of structured activities that fosters a culture, processes, procedures, behaviors, technologies, and environments in healthcare that consistently and sustainably lower risks, reduce avoidable harm, make errors less likely, and mitigate the impact of harm when it does occur."

4.2.1 Common Sources of Errors

Medication Errors Medication-related harm affects 1 in every 30 patients in healthcare, with over a quarter of these incidents being severe or life-threatening. Medications are responsible for half of the avoidable harm in healthcare settings.

Surgical Errors With more than 300 million surgical procedures performed globally each year, surgical errors remain prevalent. They account for 10% of preventable patient harm, with most adverse events occurring before or after surgery.

Healthcare-Associated Infections These infections occur at a global rate of 0.14%, increasing by 0.06% annually. They lead to prolonged hospital stays, long-term disability, increased antimicrobial resistance, and significant financial burdens on patients, families, and healthcare systems.

Sepsis Sepsis, an extreme response to infection causing damage to the body's tissues and organs, affects 23.6% of hospital cases. Approximately 24.4% of these patients die from this condition.

Diagnostic Errors Diagnostic errors occur in 5–20% of physician-patient encounters, with harmful errors found in at least 0.7% of adult admissions. Many people experience diagnostic error in their lifetime.

Patient Falls Patient falls are the most common adverse events in hospitals, occurring at a rate of 3 to 5 per 1000 bed-days. Over one-third of these falls result in injury, impacting clinical outcomes and increasing financial strain.

Venous Thromboembolism Venous thromboembolism is a preventable and significant cause of patient harm, contributing to one-third of hospitalization-related complications.

Pressure Ulcers Pressure ulcers affect more than 1 in 10 adult hospital patients. These ulcers, though highly preventable, significantly impact mental and physical health and quality of life.

Unsafe Transfusion Practices Unsafe transfusion practices and unnecessary transfusions expose patients to serious adverse reactions and infections. On average, there are 12.2 serious adverse reactions per 100,000 blood components distributed across 62 countries.

Patient Misidentification Incorrectly identifying patients can lead to severe adverse events, such as wrong-site surgery. A 2018 Joint Commission report identified 409 sentinel events related to patient misidentification out of 3326 incidents (12.3%) from 2014 to 2017.

Unsafe Injection Practices With 16 billion injections administered globally each year, unsafe practices pose risks of both infectious and non-infectious adverse events. Studies estimate that unsafe injections have led to 1.67 million hepatitis B infections, between 157,592 and 315,120 hepatitis C infections, and between 16,939 and 33,877 HIV infections over a decade.

4.2.2 Factors Leading to Patient Harm

Harm to patient during delivery of healthcare, arising from safety breaches, is a widespread issue that can occur across various settings and levels of care. Several interrelated factors typically contribute to such incidents:

1. **System and Organizational Factors**: Complex medical interventions, flawed processes and procedures, workflow disruptions, and resource limitations, such as insufficient staffing and lack of competency development, can result in patient harm.
2. **Technological Factors**: Problems with health information systems, such as electronic health records or medication administration systems, and improper use of technology can contribute to errors.
3. **Human Factors and Behavior**: Communication failures among healthcare workers, within teams, and with patients and their families, along with ineffective teamwork, fatigue, burnout, and cognitive biases, significantly contribute to patient safety incidents.
4. **Patient-Related Factors**: Inadequate health literacy, insufficient patient engagement, and failure to adhere to treatment plans can also result in negative outcomes.

High-Risk Areas Certain areas within the medication use process are more prone to errors, including prescribing, transcribing, dispensing, administration, and medication reconciliation during transitions of care. Each step presents opportunities for errors to occur, highlighting the need for targeted interventions and strategies to reduce errors.

Vulnerable Populations Certain patient populations, such as the elderly, pediatric patients, and those with multiple comorbidities, are at a higher risk of experiencing medication errors. Factors such as complex medication regimens, polypharmacy, cognitive impairments, and communication challenges contribute to the increased vulnerability.

5. **External Factors**: The absence of clear policies, inconsistent regulations, economic and financial pressures, and environmental challenges can impact patient safety.

Multiple interconnected factors can contribute to patient harm, and typically, more than one factor is involved in any given patient safety incident. These factors often interact, complicating efforts to prevent and address patient harm effectively.

4.3 Definitions: Incidents and Errors

Adverse Event (AE) An "adverse event" is defined as any untoward medical occurrence that may be present during treatment with a medicine, but which does not necessarily have a causal relationship with this treatment.

Adverse Drug Reactions (ADRs) It is a response to a drug that is noxious and unintended and occurs at doses normally used in patients for prophylaxis diagnosis or therapy of a disease (WHO).

Medication Errors (MEs) According to The National Coordinating Council for Medication Error Reporting and Prevention (NCCMERP), "A medication error is any preventable event that may cause or lead to inappropriate medication use or patient harm, while the medication is in the control of the healthcare professional, patient or consumer. Such events may be related to professional practice, healthcare products, procedures and systems, including prescribing, order communication, product labelling, packaging and nomenclature, compounding, dispensing, distribution, administration, education, monitoring and use."

Sentinel Event An occurrence unplanned, not scheduled or anticipated, resulting in death, serious harm, or the risk for physical or psychological harm.

Near Miss An occurrence would have resulted in a sentinel event if chance or intervention had not occurred. In other words, a near miss (also known as close call) is an unplanned event that did not result in injury to the patient. Only a fortunate break in the chain of events prevented injury, e.g. order for IV Octreotide at 50 mg/ml was intercepted by pharmacy and corrected to 50 mcg/ml for a patient with an acute upper GI hemorrhage from esophageal varices.

Any error that leads to an adverse drug event is considered preventable. Examples include:

- Physician omits the "mg" when writing a prescription.
- A pharmacist dispenses the wrong dose of medication.
- A nurse administers medication 4 h late.

No Harm *is* used synonymously with near miss. In the no harm scenario, the *error is not recognized, and the deed is done* but fortunately, for the healthcare professional, the *expected adverse event does not occur*, e.g., patient with an acute myocardial infarction (AMI) and immediately after coronary artery stenting inadvertently began receiving subcutaneous heparin instead of full-dose IV heparin. Error was not recognized for 12 h, but no apparent AE occurred.

4.4 Understanding Relationships Between AEs, MEs & ADRs

The relationship between Adverse Drug Events (ADEs), Adverse Drug Reactions (ADRs), and Medication Errors is integral to understanding medication safety and patient care. ADEs refer to any harm or injury resulting from the use of a drug, encompassing all negative outcomes associated with medication use, whether expected or unexpected, and regardless of whether the medication was used correctly. ADRs are a subset of ADEs and are defined as harmful and unintended reactions that occur at normal doses during normal use of a medication; these reactions are intrinsic to the drug's properties and the patient's unique response, occurring even when the medication is administered correctly. Medication Errors, on the other hand, are preventable events that may cause or lead to inappropriate medication use or patient harm while the medication is under the control of the healthcare professional, patient, or consumer. These errors can occur at any stage of the medication use process—from prescribing and transcribing to dispensing and administration. Some medication errors result in ADEs when they cause harm to the patient, such as administering the wrong dose or the wrong medication. However, not all medication errors lead to harm, and thus, not all result in ADEs. Conversely, not all ADEs are caused by medication errors; some are due to ADRs or other factors like drug interactions or patient-specific responses. Understanding the distinctions and

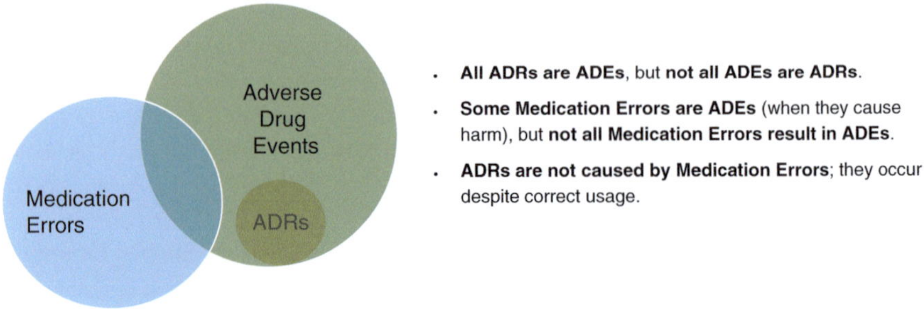

Fig. 4.1 Relationship Between Adverse Drug Events (ADEs), Adverse Drug Reactions (ADRs), and Medication Errors

overlaps among ADEs, ADRs, and medication errors is crucial for healthcare professionals aiming to enhance patient safety by both preventing errors and managing adverse reactions effectively. Figure 4.1 depicts the relationship between Adverse Drug Events (ADEs), Adverse Drug Reactions (ADRs), and Medication Errors.

Medication Safety

5

Contents

S. Sharma, *Quality Demystified*, https://doi.org/10.1007/978-981-95-0339-1_5

Medication errors refer to any preventable event or incident that may lead to inappropriate medication use or patient harm. Medication errors are a significant concern in healthcare and can have serious consequences for patients. The magnitude of medication errors can vary depending on various factors such as the healthcare setting, patient population, and type of medication being prescribed.

Every step-in patient care involves a potential for error and some degree of risk to patient safety. Errors in ambulatory and ICU prescribing are major public health problem. Over one medication error occurs daily on average. The complexity of modern healthcare systems presents a unique challenge in delivering care of the required quality in a safe environment across a large organization. The true essence of patient safety lies in minimizing or completely preventing harm to patients.

5.1 Key Points

Prevalence Medication errors are common, and studies have shown that they occur frequently across different healthcare settings. The exact prevalence rates vary, but research suggests that a significant number of patients are affected by medication errors during their healthcare journey. Prescription errors are considered common in healthcare settings. Studies have reported varying rates of prescription errors, ranging from 5% to 27% of all prescriptions written. The actual prevalence may be higher, as many errors go unreported or undetected.

Harm and Consequences Prescription errors can lead to a range of consequences, from minor inconveniences to severe harm or even death. The severity of harm depends on factors such as the type and dose of the medication involved, the patient's health status, and timely intervention to mitigate the error. Some errors may result in no harm, while others can cause adverse drug reactions, therapeutic failures, or exacerbation of existing conditions.

Impact on Patient Safety Medication errors can result in adverse drug events (ADEs) and harm to patients. ADEs can range from minor side effects to severe complications, hospitalizations, and even fatalities. The severity of harm depends on factors such as the type of medication error, the specific drug involved, and the patient's individual characteristics.

Economic Burden Medication errors not only have a significant impact on patient safety but also impose a substantial economic burden on healthcare systems. The costs associated with treating medication-related complications, hospital readmissions, and extended hospital stays due to medication errors can be substantial.

Setting-Related Factors Prescription errors can occur in various healthcare settings, including hospitals, clinics, long-term care facilities, and community pharmacies. Each setting has its unique challenges and potential for errors. For example, in hospital settings, errors may be more prevalent during care transitions or in high-stress environments like the emergency department.

Underreporting It's important to note that medication errors are often underreported, and the true magnitude of the problem may be higher than reported figures. Healthcare professionals may hesitate to report errors due to fear of retribution, concerns about professional reputation, or lack of awareness about reporting mechanisms.

High-Risk Areas Certain areas within the medication use process are more prone to errors, including prescribing, transcribing, dispensing, administration, and medication reconciliation during transitions of care. Each step presents opportunities for errors to occur, highlighting the need for targeted interventions and strategies to reduce errors.

Vulnerable Populations Certain patient populations, such as the elderly, pediatric patients, and those with multiple comorbidities, are at a higher risk of experiencing medication errors. Factors such as complex medication regimens, polypharmacy, cognitive impairments, and communication challenges contribute to the increased vulnerability.

Preventable Nature Many medication errors are considered preventable, meaning they could have been avoided with appropriate systems, processes, and interventions in place. Enhancing medication safety measures, implementing medication error reporting systems, and promoting a culture of safety are crucial in reducing the occurrence of medication errors.

Medication errors occur at various stages—prescribing, transcribing, dispensing, and administration—posing challenges to patient safety. Prescribing and administration are particularly vulnerable. Figure 5.1 depicts the extent of medication errors at various stages of the medication use process, each contributing differently to the overall incidence of errors. A medication error may or may not result in an actual adverse medicine event. Understanding errors at each stage allows healthcare organizations to implement targeted interventions to improve patient outcomes. The Joint Commission identifies six critical processes for safe medication management, recognizing that MEs can occur at any of these stages and their interfaces.

5.2 Understanding the Stages and Impact of Medication Errors in Healthcare

The six processes are:

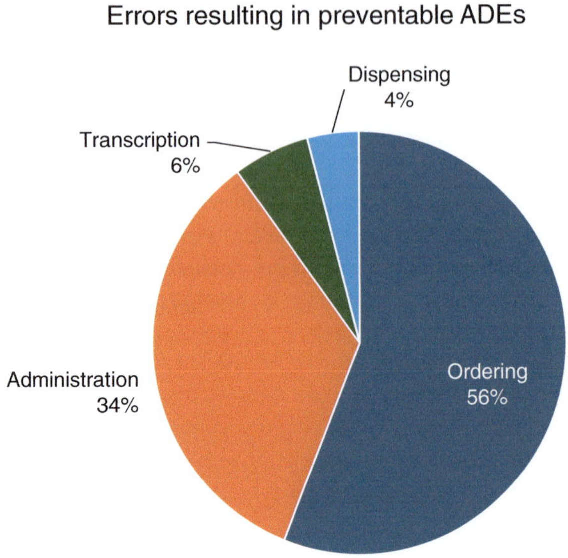

Fig. 5.1 The extent of medication errors at various stages of the medication use process

5.2.1 Selection and Procurement

The process of selecting and acquiring medications must prioritize quality and ensure that they meet patient needs. Medicines should be chosen based on evidence of safety, efficacy, and appropriateness for the intended use. An effective selection process inherently excludes medicines with questionable efficacy or irrational combinations and therapeutic duplication avoiding look-alike dosage forms. Once selected, medications should be procured with stringent quality control measures to ensure that patients receive safe, effective, and high-quality treatments.

5.2.2 Storage

Proper storage of medications is crucial to preserving their efficacy and preventing contamination or degradation. Medicines must be stored in environments that maintain their potency, particularly through temperature & humidity regulation. Without appropriate storage, medications can lose effectiveness, potentially compromising patient outcomes. Whether in healthcare facilities or at home, medicines should be kept in cool, dry places, away from direct sunlight, and stored at a temperature range of 8–25 °C or as specified by the manufacturer. Certain pharmaceutical products, like vaccines, hormones, and enzymes, require colder storage (0–8 °C). Exposure to extreme heat or cold can reduce a medicine's potency over time, and maintaining the appropriate temperature during storage and transportation—especially for cold-chain medicines such as vaccines and sera—remains a challenging yet essential task.

5.2.3 Ordering and Transcribing

Accurately documenting and communicating medication orders is essential to preventing errors in drug administration. A significant number of MEs occur during the ordering and transcribing stages. Medications are prescribed by senior doctors, transcribed onto medication charts by junior doctors or nurses, and indented to the pharmacy by nurses, with all relevant information recorded in patient files. This process involves coordination among multiple individuals, which increases the likelihood of errors.

Medication errors during ordering and transcribing are typically associated with the following five key processes in medication use:

i. Prescribing
ii. Transcribing
iii. Indenting
iv. Documentation
v. Verbal orders

5.2.4 Prescribing

According to the Institute of Medicine (IOM) report *Preventing Medication Errors* (July 2006), every step in the medication process is vulnerable to errors, with prescribing and administering medications being the most error-prone stages. Errors in prescribing can occur when selecting the medication, determining the correct dose, strength, formulation, route of administration, duration of treatment, or when providing usage instructions. Prescription errors encompass mistakes such as irrational prescribing, incorrect drug names or abbreviated or stemmed drug names, dosages, strengths, routes, frequencies, administration rates, and dilution instructions. Mistakes can include selecting the wrong drug, incorrect dosage, inappropriate administration route, or neglecting potential drug interactions. These errors are exacerbated when prescriptions are illegible. Such errors may arise due to poor communication, insufficient patient information, or cognitive lapses. Strategies to prevent prescription errors are described below.

5.2.5 Transcribing

Transcribing involves copying information from medical records or progress notes onto the medication sheet. The purpose of the transcription is to prepare the medication sheet so that the nurse administering the medication can perform the necessary three checks and accurately document the medications given. Errors in transcription, which may be made by doctors, nurses, or medical transcriptionists, are a common cause of medication errors. This stage presents a significant risk for errors due to the potential for miscommunication or incorrect copying of details.

Reducing Transcription Errors

- Avoid transcriptions of medicines or prescription orders as far as possible as it is often associated with errors.
- When it is necessary to transcribe information from one document to another, the original source documents (such as laboratory reports or medication administration records legible and complete in all aspects) should be available with the transcriber.
- Ensure double checks with the original document.
- In case of any confusion due to illegible handwriting always cross-check with the doctors. Do not guess.

5.2.6 Indents

Indents are the written or computerized orders sent to the pharmacy for medication dispensing according to the prescription. While studies show that indent errors have a relatively low incidence (1–2%), their potential impact on patients cannot be overlooked. Common causes of indent errors include illegible prescriptions and confusion with LASA medicines. Additionally, prescribing medications by brand names or abbreviated drug name increases the risk of indenting errors.

5.2.6.1 Methods to Reduce Indent Errors

- Prescriptions ought to be clearly written, ideally using capital letters, and medications should be prescribed by their generic names.
- In case of confusion, always cross-check with the orders written or progress notes or with the concerned doctor.
- List of LASA medicines should be identified and made available to all concerned persons.
- Store particularly high alert medications and LASA medications in separate locations with prominent labeling.
- Highlight the importance of thoroughly reading the label every time a medication is accessed and before it is administered, instead of depending on visual recognition, location, or other vague indicators.
- Ensure that both the generic name and the brand name of the medication are included on medication orders and labels, with the generic name placed near and in a larger font size than the brand name.
- Restricted availability of high alert medicines in different patient care areas. It is especially critical that the availability, access, prescribing, ordering, preparation, distribution, labeling, verification, administration, and monitoring of high alert

medicines be planned in such a way that possible adverse events can be avoided, and, hopefully, be eliminated. Ensuring consistent dosing, measurement units, and terminology is essential for the safe utilization of concentrated electrolyte solutions. Moreover, mix-ups of specific concentrated electrolyte solutions must be avoided (e.g., confusing sodium chloride with potassium chloride). These efforts require special attention, appropriate expertise, interprofessional collaboration, processes of verification, and several forcing functions that would ensure safe use.
- Label all medications, medication containers (e.g., syringes, medicine cups, basins), or other solutions on and off the sterile field. Errors, sometimes tragic, have resulted from medications and other solutions removed from their original containers and placed into unlabeled containers. Medications and other solutions in unlabeled containers are unidentifiable.
- Discard any medications or solutions found unlabeled immediately.
- All original containers from medications or solutions remain available for reference in the preoperative/procedural area until the conclusion of all the procedure.

5.2.7 Documentation

Documentation is the backbone of medication management and safety. An accurate documentation of medication history, disease history, any drug/food allergy, progress history, and administration of medicines can help in detection and prevention of many fatal medication errors. Documentation errors in healthcare can have serious implications for patient safety, treatment outcomes, and legal compliance. These errors occur when information is incorrectly recorded, incomplete, or illegible in-patient records. For example, a nurse may incorrectly document the time of medication administration, leading to a patient receiving the wrong dosage at an inappropriate interval. In another case, a doctor might omit critical details about a patient's allergies, resulting in the administration of a medication that triggers a severe allergic reaction. Additionally, if a healthcare provider fails to update a patient's chart with new diagnoses or lab results, crucial information may be missed by other members of the care team, potentially leading to incorrect treatment decisions. Poor handwriting, transcription errors, and failure to properly record verbal orders also contribute to documentation mistakes, increasing the risk of miscommunication and medical errors. Accurate, clear, and complete documentation is essential to ensure continuity of care and avoid preventable errors. Table 5.1 depicts the various types of documentation errors in healthcare and tips to avoid them.

5.2.7.1 Methods to Reduce Documentation Errors

- The practice of thorough documentation must be instilled in both nurses and doctors, following the principle: "If it's not documented, it didn't happen."
- Frequent training program should be held. The need for and importance of documentation should be emphasized.

Table 5.1 Detailing the types of documentation errors in healthcare and tips to avoid them

Type of documentation error	Description	Tips to avoid
Illegible entries	Handwritten notes or prescriptions that are unclear or difficult to read.	Encourage the use of electronic medical records (EMRs) to eliminate handwriting issues; if writing by hand, ensure legibility and clarity
Incomplete information	Missing important details, such as patient history, allergies, or treatment plan	Use standardized documentation forms with mandatory fields and routinely double-check that all necessary information is filled out
Incorrect entries	Wrong information recorded, such as incorrect dosages, medications, or diagnoses	Verify all information before recording, especially medication names, doses, and patient details; use double-check systems
Omissions	Failing to document key actions, treatments, or patient responses	Implement checklists for critical procedures and ensure real-time documentation to avoid forgetting key details
Miscommunication of verbal orders	Errors in documenting or interpreting spoken instructions from a physician	Always repeat back verbal orders to confirm accuracy; document verbal orders immediately and get written confirmation when possible
Duplicate entries	Recording the same information multiple times, leading to confusion	Standardize documentation processes to prevent unnecessary repetition; use EMRs with integrated systems that link related data points
Time discrepancies	Incorrect or missing documentation of times for medication administration, procedures, etc.	Record information in real-time or as soon as possible; use automated time stamps in electronic systems to reduce manual errors
Unauthorized or unauthenticated entries	Documentation made by an individual not involved in the care or missing a signature	Ensure that only authorized personnel document in the patient's chart; require signatures or electronic authentication for each entry
Inconsistent terminology	Using varying terms for the same condition, procedure, or medication, causing confusion	Standardize medical terminology across the organization; encourage the use of recognized clinical language and abbreviations to ensure consistency
Delayed documentation	Documenting patient information too long after the event, leading to inaccurate recall	Encourage immediate or real-time documentation to ensure accuracy; implement policies for timely record-keeping after patient encounters

- Maintain real-time, accurate, and complete medical records, including notes with a clear date, time, signature and name, partograph, safe birth checklist, safe surgical checklist, etc.
- Document all relevant information, including diagnoses, treatments, and communication with the patient.
- Documentation should be completed immediately after administering medication or receiving any information, without relying on memory.
- Documents should be arranged in chronological order.
- Discharge Procedures. Prepare or countersign discharge summaries, discussing discharge instructions, follow-up, and home care with the patient.
- Do not alter notes or destroy any records. Do not attempt to tamper with the original record.
- If a mistake is discovered later (inaccurate, misleading, or incomplete), insert an additional note as a correction with date. Do not obliterate the original note. For altering cross original words/statements by a single stroke of pen, so that crossed text is still legible & rewrite new one—date & sign both.
- Maintain confidentiality at all times.

5.2.8 Verbal Orders

Verbal orders can significantly contribute to medication errors due to the potential for miscommunication, misunderstanding, or incomplete documentation. These orders, often given in fast-paced or emergency situations, are prone to mistakes such as incorrect drug names, dosages, or instructions being conveyed. Factors like background noise, accents, or unfamiliarity with medical terms can further increase the likelihood of errors. For example, a nurse may mishear a physician's verbal order for "morphine" as "hydromorphone," leading to an administration of the wrong medication. Additionally, verbal orders are sometimes not immediately documented, increasing the risk of incomplete or inaccurate recording, which may cause discrepancies later. A robust verbal order policy minimizes the potential for miscommunication, ensuring patient safety while allowing for necessary flexibility in critical care situations.

A verbal order policy in healthcare sets clear guidelines for issuing, receiving, and documenting verbal instructions, ensuring that communication is accurate and safe, especially in high-stress or emergency situations. Such policies help reduce the risk of medication errors and ensure accountability. Typically, verbal orders are only permitted in situations where written or electronic orders are impractical, such as during emergencies. Policies often require that verbal orders be limited to authorized healthcare professionals, clearly communicated, confirmed, and documented immediately after the order is given.

5.2.8.1 Steps Involved in Verbal Orders

- **Determine Appropriate Scenarios for Using Verbal and Telephone Orders:**

  - Avoid verbal orders to prevent mishaps; establish and adhere to a verbal order policy.
  - Ideally, verbal orders should be avoided, but in emergency situations, they can be given by senior consultants to a doctor or senior nursing staff only.
- Since these orders are given in emergency situations even a slight deviation from the order can result in serious harm and subsequently death of the patients. Verbal order for hazardous and high-alert medicines should be avoided.
- **Issuing the Verbal Order:**
  - The physician or authorized healthcare professional clearly articulates the verbal order, ensuring that the name of the drug, dosage, route, frequency, and any other relevant details are explicitly stated.
  - Use of standardized language is encouraged to avoid confusion, and abbreviations should be avoided where possible.
- **Receiving the Verbal Order**:
  - The receiving healthcare professional (doctor, nurse, pharmacist, or other) listens carefully and writes down the order immediately.
  - The healthcare professional receiving the order must be qualified and authorized to do so. Verbal orders should only be taken by doctors if available, in case not available seniormost qualified and authorized healthcare professionals to ensure accuracy and patient safety. The individuals allowed to take verbal orders typically include Registered Nurses (RNs), pharmacists, and physician assistants. Never assign students to take verbal orders.
  - Order should be enunciated slowly and distinctly and state numbers like pilots—one-five mg for 15 mg, e.g. An order for "NPH insulin 16 units" can easily sound like "NPH insulin 60 units." Always spell out difficult drug names, e.g., "Cerebyx or Celebrex?". Concentration/strength should be specified instead of giving the dose in number of tea spoonsful, tablets, ampoules, or vials. For example, Paracetamol/Iron preparations.
- **Repeating Back the Order**:
  - The receiving professional repeats the entire order back to the prescriber to confirm its accuracy, including the drug name, dose, route, frequency, and other critical details.
  - This "read-back" process ensures clarity and prevents misunderstandings.
  - In unavoidable circumstances, reconfirm instructions to ensure clarity and patient safety.
- **Documenting the Verbal Order**:
  - The verbal order is documented in the patient's medical record immediately, including the date, time, the content of the order, and the names of both the prescriber and the receiver.
  - Notation should indicate that the order was verbal and if required, signed off by the prescriber later.
- **Administering Medication**:
  - The healthcare professional administering the medication double-checks the order to ensure it matches what was prescribed and follows all standard medi-

cation administration procedures (e.g., the "five rights" of medication administration: right patient, right drug, right dose, right route, and right time).

- **Follow-Up and Confirmation**:
  - Whenever possible, verbal orders should be followed by a written or electronic order from the prescriber to ensure proper documentation and verification in the patient's medical record, preferably within twelve (12) hours of issuing order.
  - The prescriber must sign the verbal order as soon as possible to confirm its accuracy and compliance with institutional policies.

5.2.9 Preparing and Dispensing

Ensuring the correct preparation and dispensing of medications involves providing accurate dosage and proper labeling of prescriptions. A dispensing error occurs when there is a discrepancy between the medication prescribed, ordered, and what is actually dispensed. Common dispensing errors include providing the wrong medication, incorrect dosage, wrong dosage form, improper quantity, inaccurate, inadequate, or confusing labeling. These errors may also involve incorrect or unclear instructions for use. At times, pharmacists may substitute a brand without thoroughly verifying its appropriateness, leading to potential errors. Mistakes in this phase can occur due to misinterpreted prescriptions, inadequate checks, or faults in automated dispensing systems, all of which can have serious consequences for patient safety.

5.2.9.1 Dispensing Cycle

Dispensing errors frequently occur due to LASA, and spell-alike medicines. Medications with similar names or packaging can easily be confused, leading to potentially harmful errors and adverse events. LASA issues often arise from changes in brand names or packaging sizes. To prevent such errors, LASA medicines must be regularly identified and updated in both inpatient (IP) and outpatient (OP) pharmacies. This information should be promptly shared with all departments where medications are used, such as hospital floors, ICUs, emergency departments, and operating theatres, to ensure staff awareness and minimize risks. Figure 5.2 illustrates the key stages in the medication dispensing cycle.

Medicine strip cutting and poor storage practices can significantly contribute to medication errors, impacting patient safety and treatment outcomes. When healthcare providers or patients cut medication strips to access individual doses, there is a risk of confusion or misidentification of the medications, especially if the strips contain similar-looking pills. This practice can lead to administering the wrong medication or dosage if not done carefully. Additionally, inadequate storage conditions, such as exposure to heat, humidity, or direct sunlight, can compromise the integrity and potency of medications. For example, improperly stored medications may degrade or lose effectiveness, resulting in suboptimal treatment for patients.

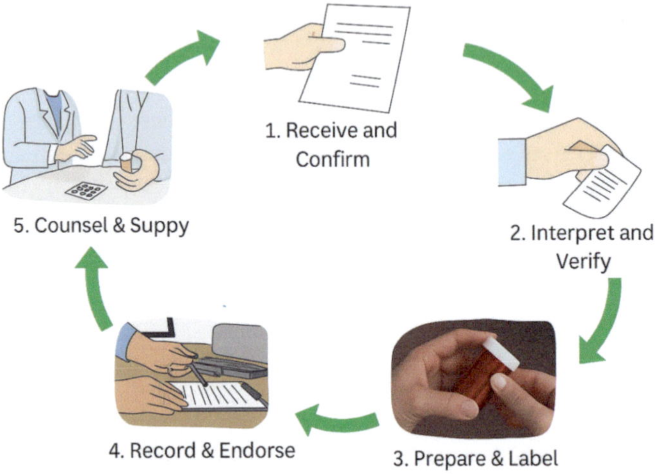

Fig. 5.2 The key stages in the medication dispensing cycle

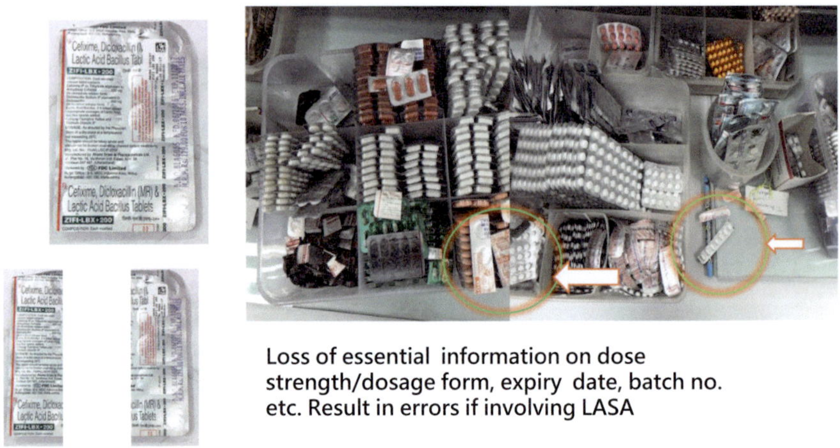

Fig. 5.3 Impact of strip cutting and poor storage increase Look alike and sound alike (LASA) related errors

Ensuring proper storage conditions and avoiding the cutting of medicine strips whenever possible are critical steps in reducing the likelihood of medication errors and ensuring patient safety (Fig. 5.3).

Information on medicine labels is vital for patient safety and effective medication use. Clear labeling provides essential details such as the drug name, dosage instructions, administration route, expiration date, and storage requirements. Strip cutting removes this critical information, increasing the risk of medication errors and potential harm. Accurate labeling helps patients and healthcare providers understand how to take medications correctly, while also including warnings about side effects and drug interactions to enable informed decision-making. Overall, proper labeling promotes adherence, prevents misuse, and enhances the quality of care.

5.2.9.2 Avoid Strip Cutting

To avoid the practice of medicine strip cutting, healthcare facilities and patients should adopt several key strategies. First, implementing a strict "no strip cutting" policy by prescribers is essential, requiring them to indicate the number of tablets or units to be dispensed, particularly for "as needed" (SOS) medications. This policy ensures that patients receive whole units of medication, eliminating the need for cutting strips and reducing the risk of confusion or medication errors. Prescribers can tailor the total dose based on the treatment plan, such as specifying that a medication is to be given three times a day for 7 days, which would warrant dispensing only 20 tablets. Alternatively, they should avoid ambiguous instructions like "three times a day for eight days," which may lead to the need for cutting strips, and instead prescribe a total of either 20 or 30 tablets to ensure the patient receives the correct quantity without the need for modification. Additionally, healthcare providers should prioritize the use of unit-dose packaging whenever possible, which provides individual doses without the need for cutting. Educating healthcare staff and patients about the potential dangers associated with cutting medication strips is crucial, emphasizing the importance of proper medication identification and administration. Pharmacies can further support this by implementing clear labeling and instructions for medications that may require cutting, ensuring patients understand how to manage their medications safely. Finally, healthcare organizations should foster a culture of safety that encourages reporting and addressing any issues related to medication handling and storage, thereby reducing the likelihood of errors associated with cut strips.

5.2.9.3 Auxiliary Labeling

Auxiliary labeling refers to cautionary labels placed on medication containers to provide supplementary information about safe administration, use, and storage. These labels serve as reminders for patients and healthcare providers, reinforcing critical instructions such as "*Take with food*," "*Do not crush*," or "*Keep refrigerated*." Proper auxiliary labeling ensures medication adherence and minimizes errors in handling and administration. Figure 5.4 depicts some sample auxiliary labels.

Color-coded medication labels for anesthesia trolleys enhance patient safety by reducing the risk of drug identification errors. These labels follow standardized color schemes, such as those recommended by ASTM, to differentiate drug classes—*blue for opioids, yellow for induction agents, red for neuromuscular blockers,* and more. By providing a clear visual cue, color-coded labels help anesthesiologists quickly identify medications, especially in high-pressure situations like surgeries. For details visit https://www.asahq.org/standards-and-practice-parameters/statement-on-labeling-of-pharmaceuticals-for-use-in-anesthesiology.

By implementing these measures, including the no strip cutting policy, the risks associated with medicine strip cutting can be significantly minimized, enhancing patient safety and medication efficacy.

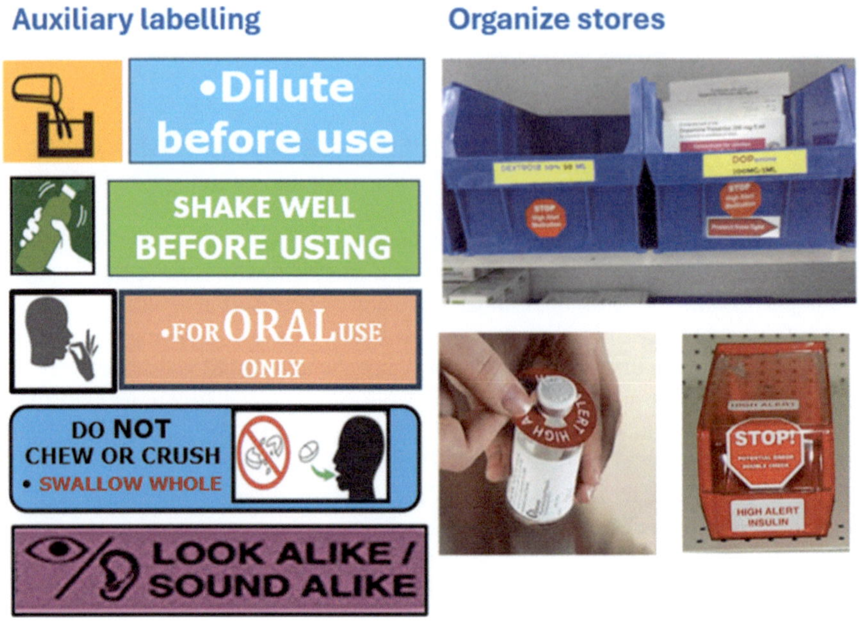

Fig. 5.4 The role of auxiliary labeling and organized storage in preventing medication errors

5.2.9.4 Methods to Reduce Preparing and Dispensing Errors

- The pharmacy should have sufficient storage and dispensing area so that the medicines can be arranged appropriately and retrieved easily as and when required. Medication errors frequently occur during the dispensing process.
- Pharmacists should be trained to adhere to best dispensing practices.
- They must follow all the steps involved in dispensing such as receiving and validating prescription, understanding and interpreting prescription, preparing items for issue, recording action taken and issue medicines to patients with clear instruction and advice.
- They should be aware of LASA and high-risk medicines so that they handle them as per hospital policy.
- Pharmacists should consult with the prescriber before substituting either generic or branded products.
- Avoid strip cutting & procure single dose unit wherever possible.
- Auxiliary labeling of the LASA and high alert medicines.
- Organize medicine stores using organizers, sorters, etc.

5.2.10 Medication Administration

Medication administration is a critical aspect of the medication use process, requiring a meticulous and multi-step approach to ensure that patients receive the correct dose, via the appropriate route, and at the right time. Common administration errors include giving the wrong drug, administering to the wrong patient, missed doses, extra doses, incorrect dosages, improper rates or frequencies of administration, inaccurate dilutions, and administering via the wrong route. Timing errors, such as giving medications too early or too late, are also prevalent in this phase. Studies worldwide indicate that administration errors rank as the most common type of medication error following prescribing errors. For instance, Phillips et al. reported that the most frequent errors included administering an improper dose (40.9%), overdoses (36.4%), wrong medicines (19%), and incorrect routes of administration (9.5%). The leading causes of these errors were identified as performance and knowledge deficits (44%) and communication errors (15.8%).[1]

Medication reconciliation errors occur when the patient's medication list is not accurately updated and reconciled during transitions of care, leading to discrepancies and potential medication errors.

5.2.10.1 Methods to Reduce Administration Errors

- Always follow the *10 Rights* of medication administration to ensure safe practice (Fig. 5.5) and use of identification bands to ensure correct patient identification (Fig. 5.6).
- If there is any uncertainty about the drug's name, dose, dilution, frequency, or route, always cross-check with the prescriber.
- Carefully read the label of each medication before preparing and administering it.
- Label all parenteral medicines immediately after preparation, and before preparing the next medication.
- Identify the patient using at least two identifiers before administering any medication.
- Inspect the general appearance of the medicine before administering it.
- Verify the medication order, ensuring the correct dosage form, strength, dose, frequency, and route, and double-check before administration.
- Label containers for blood or specimen collection at the patient's bedside before taking the sample.

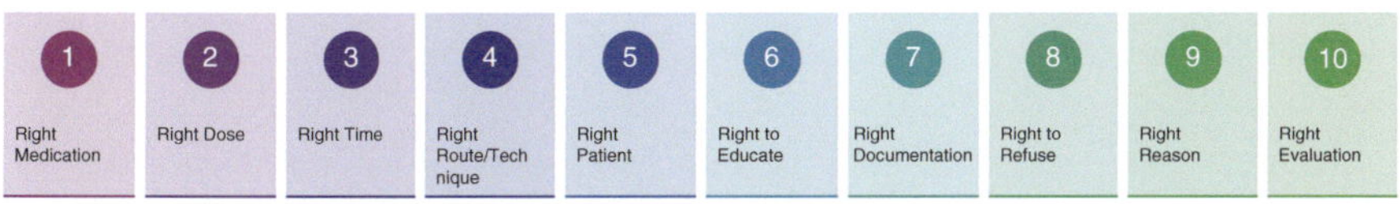

Fig. 5.5 10 Rights of medication administration to ensure patient safety

[1] Phillips J, Beam S, Brinker A, Holquist C, Honig P, Lee LY, Pamer C. Retrospective analysis of mortalities associated with medication errors. Am J Health Syst Pharm. 2001;58(22):2130; Reddy LKV, Modi AG, Chaudhary B, Modi V, Patel M. Medication Errors—A Case Study. Journal of the Academy of Hospital Administration. 2009; 21(1 & 2).

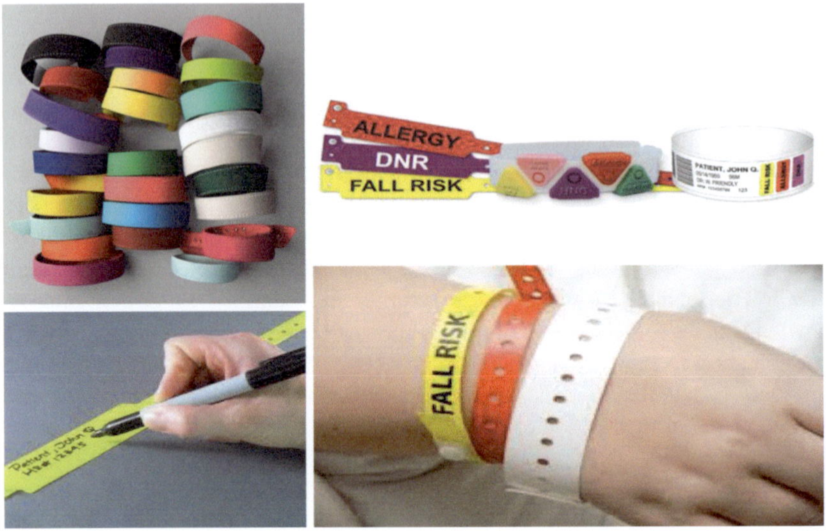

Fig. 5.6 Use of identification bands to ensure correct patient identification

- Keep medicines in their original packaging whenever possible. If removed, label the container with the medicine's name, strength, and route before dispensing or administering.
- Prepare injections at the bedside when necessary.
- For injections, double-check the dilution and rate of administration, and document the time started and stopped.
- Educate patients about their medications to promote safety and adherence.
- Document any drug allergies in bold red capitals on the patient's front sheet. If no allergies are known, write "NO KNOWN HISTORY OF ALLERGY" to avoid leaving this section blank.
- A Drug Allergy Identity Wallet Card provides quick access to critical allergy information in emergencies, helping healthcare providers avoid prescribing or administering harmful medications. It reduces the risk of severe allergic reactions, anaphylaxis, and medication errors, ensuring safer medical treatment across different healthcare settings. Figure 5.7 depicts a sample of drug allergy identity card.

- Exercise Caution During Multi-Disciplinary Care and Transitions:

Ensure medication accuracy especially during transitions in care, such as at admission and across the continuum of care. When multiple clinicians are involved, check for medication duplication or potential drug-drug interactions.

Medication reconciliation should always be performed to prevent therapeutic duplications and drug interactions. Gather comprehensive information on all

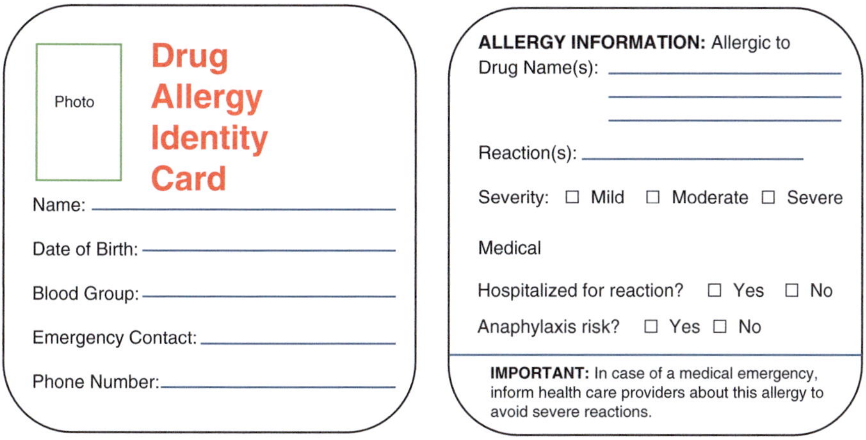

Fig. 5.7 Sample allergy identity card. (Print & Laminate for durability. Keep in Wallet/Purse. Show to Medical Staff in Emergencies)

medications the patient is using, including prescription medicines, over-the-counter (OTC) medications, vitamins, herbal supplements, nutraceuticals, and respiratory therapy-related medicines like inhalers.

5.2.10.2 Three-Step Medication Reconciliation Process

- **Collect** an accurate medication history.
- **Compare** the medication history with physician orders at all transition points, such as:
  - Admission
  - Transfer
  - Discharge
- **Correct** any discrepancies identified.

A complete list of the patient's medications should be communicated to the next care provider when the patient is referred, transferred, or moves to another care setting, service, practitioner, or level of care, both within and outside the organization. Figure 5.8 illustrates the process of medication reconciliation to ensure accurate and safe medication use.

- Develop and make available a "**Do Not Crush" Medication List** at nursing stations for quick reference by healthcare professionals. The *Do Not Crush* medication list includes drugs that should not be broken or crushed due to their formulation, which may alter absorption or effectiveness. Examples include *extended-release tablets* like Metformin ER, *enteric-coated pills* such as Aspirin EC, and *capsules containing beads* like nifedipine, duloxetine. Crushing these medications can reduce efficacy or cause harmful side effects. Educate patients that capsules should not be opened, and only medications with a scoring mark

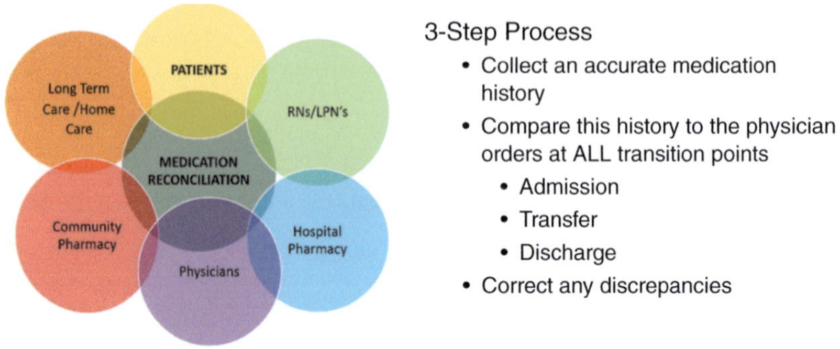

Fig. 5.8 Medication reconciliation process to ensure accurate and safe medication use

should be split. This ensures safe administration of medications, especially through nasogastric tubes, and minimizes the risk of unintended consequences.

- **Effective Crash Cart Management**

- A well-organized crash cart ensures readiness for emergency medical situations. Proper management includes standardizing medication inventory, monitoring expiry dates, labeling of high alert and LASA medicines and conducting regular audits.

1. **Standardized Medication Inventory**
   Maintaining a consistent list of essential medications with predefined quantities prevents shortages during emergencies. For example, *Epinephrine* (adrenaline) must always be stocked in sufficient doses for cardiac arrest and anaphylaxis treatment.
2. **Monitoring Expiry Dates**
   Expired medications can lose potency or become harmful, making regular checks essential. *Adrenaline* has a short shelf life, 12–18 months, and degrades if exposed to light or heat. Ensuring fresh stock is available is crucial, as outdated adrenaline may be less effective in critical resuscitation scenarios. Similarly, *Sodium Bicarbonate*, used for metabolic acidosis, should be checked frequently to maintain efficacy.
3. **Conducting Monthly Audits**
   Routine audits (every month) ensure the crash cart is stocked and compliant with hospital protocols. Checking for missing syringes, verifying *Lidocaine* availability for arrhythmias, expired medicines, and ensuring defibrillator pads are functional are key steps. Monthly audits improve efficiency and emergency preparedness.

A systematic approach to crash cart management enhances patient safety and ensures immediate response capability.

5.2.11 Monitoring

Observing and assessing the effects of medications on patients to identify and address any adverse reactions or efficacy issues. Monitoring errors occur when healthcare providers fail to properly observe and evaluate a patient's response to a medication, leading to adverse outcomes. These errors can stem from inadequate assessments, lack of timely follow-ups, or misinterpretation of vital signs and laboratory results.

For example, if a patient is prescribed a new anticoagulant, but their INR levels are not monitored regularly, this could result in either subtherapeutic levels, increasing the risk of thrombosis, or supratherapeutic levels, raising the risk of bleeding complications. Another instance could involve a diabetic patient whose blood glucose levels are not adequately monitored after insulin administration, potentially leading to hypoglycemia or hyperglycemia.

Monitoring errors highlights the importance of continuous assessment and appropriate interventions in ensuring safe and effective medication use. Inadequate follow-up or lack of patient feedback can contribute to these errors.

Each process and phase require careful attention to prevent errors and ensure safe and effective medication management. These processes are interlinked, and errors in one stage can impact on the others, emphasizing the need for rigorous oversight and coordination to ensure patient safety.

5.3 Moments for Medication Safety

In 2017, the WHO launched the 5 Moments for Medication Safety patient engagement tool to support the implementation of the third WHO Global Patient Safety Challenge: Medication Without Harm. The initiative aims to reduce severe, avoidable medication-related harm by 50% globally over 5 years. This tool identifies five critical moments where patients or caregivers can act to reduce the risk of harm from medication use. Its purpose is to empower patients to actively participate in their own care, working collaboratively with healthcare professionals. Importantly, the tool is designed to stay with the patient, their family, or caregivers throughout their care journey.

5.4 Prescription Errors

Errors in prescribing can occur at various stages, including the selection of medicines, determining the correct dose, strength, pharmaceutical form, route of administration, treatment duration, and usage instructions. Prescription errors encompass all mistakes related to inaccurate medication orders, such as irrational prescribing, incorrect drug names, improper doses, wrong strengths, incorrect routes, improper frequency, incorrect administration rates, and errors in dilution instructions. These errors are compounded when prescriptions are not legible, further increasing the risk of harm.

Case Study 5.1: Prescription Error Due to Illegibility

A 25-year-old male patient experienced generalized itching and was prescribed Tab Fexofenadine (Allegra) 180 mg to be taken once daily at bedtime (HS). However, due to the illegibility of the prescription, the patient misinterpreted the instructions and took the medication three times daily (TID) instead of once daily (OD) for 3 days.

***Type of Error*:** This is a prescription error caused by illegibility, which led to a dosing frequency error.

Why Did This Happen?

The handwritten prescription was unclear, causing the patient to misread the dosage instructions. As a result, the patient took the medication more frequently than prescribed, which could have led to potential side effects or drug overuse.

Case Study 5.2: Therapeutic Duplication Due to Lack of Awareness

A prescription error occurred when a patient was prescribed two medications from the same therapeutic category. The patient received one proton pump inhibitor (PPI) and one H2 receptor blocker, both intended to treat acid-related disorders. This duplication resulted in the administration of a higher daily dose than the recommended maximum. In another instance, the same medication was prescribed under two different brand names—one as a standalone drug and the other as part of a combination therapy.

***Type of Error*:** This is a therapeutic duplication error, where the patient was inadvertently prescribed either the same drug under different names or multiple medicines from the same class.

Why Did This Happen?

The error occurred because the prescriber was unaware of the overlapping effects of different brands and combination medicines or lacked knowledge about the medicines belonging to the same therapeutic category. This oversight resulted in the patient receiving an excessive dosage, which could lead to adverse effects.

5.4.1 People at Risk of Medication Errors and Actionable Patient Safety Solutions

1. Polypharmacy and Irrational Use of Medicines: Polypharmacy is one of the largest risk factors for medication errors, increasing the chances of adverse drug reactions. The complexity of managing multiple medications heightens the risk of mistakes.

   Solution: Aim to prescribe a smaller number of medicines whenever possible and review the necessity of each drug to reduce the risk.
2. Look-Alike, Sound-Alike (LASA) Medications: Confusion between similar-sounding or similar-looking drug names, such as Daonil (Glibenclamide) vs. Diovol (Antacid) or Glynase (Glyburide) vs. Zinase (Serratiopeptidase), can result in errors. Similarly, confusion with LASA vaccines (e.g., TT, Td, Tdap) for both adults and children poses risks.

   Solution: Implement strategies like tall man lettering and ensure proper labeling to differentiate LASA medicines.

3. Use of Abbreviations for Drug Names: Abbreviations like ASA (aspirin), PCM (paracetamol), and CPZ (chlorpromazine) can cause confusion, as they are not standardized across medical practice. This increases the likelihood of errors in prescribing and administering medication.

   Solution: Always write out the full drug name in prescriptions, as it is a legal document, and abbreviations are not universally recognized.
4. Use of Error-Prone Abbreviations, Symbols, and Dose Designations: Common shorthand abbreviations, symbols, and incorrect spacing (e.g., Tegretol300 mg being misread as Tegretol 1300 mg) can lead to significant dosing errors.

   Solution: Avoid using symbols like “>”, “<”, “@”, “cc”, and “μg” for micrograms, and replace them with clear terms. Always place a space between the drug name and its strength.
5. Errors Due to Visual Mix-Ups: Visual similarities between letters and numbers (e.g., “l” and “1,” “O” and “0,” “Z,” and “2”) can lead to misinterpretation. For example, Q1d can be mistaken for QID, leading to a fourfold increase in dosage.

   Solution: Use clear, distinct handwriting or electronic prescribing systems to reduce visual confusion.

   Patients Receiving High-Alert and High-Risk Medications: High-alert medications are those that pose an increased risk of significant harm if used incorrectly. To prevent errors, healthcare facilities can adapt the list provided by the Institute for Safe Medication Practices (ISMP) based on their specific scope and services. Below is a list of high-alert medications as identified by ISMP, risk involved and prevention strategies. By adapting this list to the facility's needs and ensuring adherence to safety protocols, healthcare providers can significantly reduce the risk of medication errors with high-alert medicines. Table 5.2 depicts the risks associated with LASA and high-alert medications and prevention strategies.

Adapting the ISMP List for Facility

Assess Scope: Consider the specific services offered (e.g., pediatrics, oncology, critical care) and the patient population. Each category can be tailored to the specific needs and protocols of the healthcare facility. Ongoing staff education and awareness are essential in all prevention strategies to enhance safety and minimize risks.

Review Usage: Focus on medications commonly used in your facility and adjust protocols for those posing the highest risk.

Customize Protocols: Implement tailored safety protocols for administration, monitoring, and storage based on the adapted high-alert list. Implement extra precautions, such as double-checking and independent verification when prescribing or administering high-alert medications, to minimize risks.

5.4.2 Methods to Reduce Prescription Errors

- Prescriptions should be legible and written in capital letters using the generic name, ensuring they are complete in detail. If necessary, the brand name can be written in parentheses.

Table 5.2 Risks associated with look-alike/sound-alike (LASA) and high-alert medications and prevention strategies

Class/category	Description/risk	Prevention strategy
Look-alike/sound-alike (LASA) medications	Medications that can be confused due to similar names or packaging	Use tall man lettering to distinguish names Implement barcode scanning before administration Conduct staff education
High-alert medications	Medications that have a heightened risk of causing significant harm when used in error	Implement uniform procedures for prescribing, dispensing, and administering Implement independent double-checks
Concentrated electrolytes (e.g., sodium chloride for injection, hypertonic, greater than 0.9% concentration, magnesium sulfate)	High concentrations of electrolytes can cause fatal arrhythmias or imbalances if improperly dosed	Limit access to concentrated electrolytes Use pre-mixed bags whenever possible Dilute concentrated electrolytes and administer via controlled IV infusion provide clear labeling and warnings
Magnesium Sulfate (IV):	Overdosing can cause respiratory depression, cardiac arrest	Administer via controlled IV infusion, and monitor magnesium levels
Opioids (e.g., oral, sublingual, parenteral, transdermal)	Medications that can cause respiratory depression and other significant adverse effects	Use standardized order sets Implement a monitoring protocol for patients receiving opioids Provide staff education on opioid safety
Anticoagulants (e.g., warfarin, low molecular weight heparin, enoxaparin unfractionated heparin)	Medications that affect blood clotting and can lead to serious bleeding if dosed improperly	Use standardized dosing protocols Monitor INR and other relevant parameters closely Provide patient education on risks
Antithrombotic agents: Clopidogrel, dabigatran, Thrombolytics (e.g., Alteplase), etc.	Increased risk of hemorrhage	Carefully monitor coagulation parameters and adjust doses accordingly
Insulin (all forms of insulin subcutaneous & IV)	Medications that can cause severe hypoglycemia if administered incorrectly	Use standardized order sets and protocols Use standardized insulin pens and syringes Implement double-check systems for insulin administration Educate patients on signs of hypoglycemia

(continued)

Table 5.2 (continued)

Class/category	Description/risk	Prevention strategy
Chemotherapy agents (e.g., methotrexate, cisplatin, vincristine, etc.)	Medications used in cancer treatment that can cause significant adverse effects Severe toxicity, narrow therapeutic index	Use dedicated areas for chemotherapy administration Implement strict protocols for handling and disposing of chemotherapy agents Ensure clear communication between departments Educate staff on potential side effects and monitoring
Neuromuscular blockers (e.g., succinylcholine, rocuronium, vecuronium)	Medications that can cause paralysis and respiratory distress if misused	Ensure agents are not administered without appropriate sedation and monitoring Maintain clear labeling and alerts Ensure proper monitoring during and after administration
Antimicrobials	Medications that can contribute to resistance if not used appropriately	Implement antimicrobial stewardship programs Educate prescribers on appropriate use and potential side effects Monitor for adverse effects and efficacy
Adrenergic agonists, IV (e.g., epinephrine, phenylephrine, norepinephrine)	Stimulates adrenergic receptors and can affect cardiovascular function	Use weight-based dosing protocols Monitor vital signs closely during administration Ensure clear labeling and alerts for high-alert status
Adrenergic antagonists, IV (e.g., propranolol, metoprolol, labetalol)	Block adrenergic receptors and can impact heart rate and blood pressure	Monitor heart rate and blood pressure regularly Educate staff on potential side effects and interactions Use standardized protocols for administration
Antiarrhythmics, IV (e.g., lidocaine, amiodarone)	These medicines are used to treat and manage arrhythmias by stabilizing cardiac electrical activity	Use standardized protocols for dosing and administration Monitor cardiac rhythm continuously during administration Educate staff on potential side effects, including bradycardia and hypotension

(continued)

Table 5.2 (continued)

Class/category	Description/risk	Prevention strategy
Inotropic medications (e.g., digoxin, milrinone)	Increase cardiac output by improving the heart's contractility	Monitor for signs of toxicity, such as changes in heart rate and gastrointestinal symptoms Use standardized dosing protocols Educate staff on signs of toxicity and interaction risks
Anesthetic agents, general, inhaled and IV (e.g., propofol, ketamine)	Induce loss of sensation or consciousness	Use weight-based dosing and monitoring protocols Monitor respiratory and cardiovascular status closely Ensure appropriate recovery protocols post-anesthesia
Cardioplegic solutions	Solutions used during cardiac surgery to induce temporary cardiac arrest and protect myocardial tissue	Ensure proper formulation and concentration checks Monitor the patient closely for signs of myocardial recovery Train staff on the proper administration techniques
Dialysis preparations	Solutions used for renal replacement therapy to remove waste products from the blood	Use standardized protocols for preparation and administration Monitor vital signs and laboratory values closely Educate staff on potential complications and management
Liposomal forms of medicines, IV (e.g., liposomal amphotericin B)	Liposomal formulations designed to enhance drug delivery and reduce toxicity	Use standardized protocols for preparation and administration Monitor for infusion-related reactions Educate staff on the differences between liposomal and conventional formulations
Conventional counterparts (e.g., amphotericin B deoxycholate)	Standard formulations may have higher toxicity profiles	Monitor for potential side effects and toxicity Implement clear labeling to differentiate between liposomal and conventional forms Educate staff on appropriate use
Moderate sedation agents, oral (e.g., chloral hydrate, midazolam, ketamine)	Used for sedation in children Chances of overdosage	Use weight-based dosing and thorough pre-sedation assessments Monitor vital signs and consciousness level during sedation Provide education on risks and recovery

(continued)

Table 5.2 (continued)

Class/category	Description/risk	Prevention strategy
Moderate sedation agents, IV (e.g., dexmedetomidine, midazolam, lorazepam)	Used for procedural sedation. Risk of oversedation leading to respiratory failure	Use standardized dosing and monitoring protocols Monitor respiratory and cardiovascular status closely Ensure proper recovery protocols post-sedation
Parenteral nutrition and total parenteral nutrition (TPN)	Improper formulation or administration can lead to severe metabolic imbalances Electrolyte imbalances, infection risk with central lines	Use standardized procedures for mixing and administration, and monitor electrolytes closely Close monitoring of patient's nutritional status and lab values

- **Use Full Drug Names in Prescriptions:** Avoid abbreviating drug names, as prescriptions are medico-legal documents, and abbreviations are not universally recognized. Always write the *full name of the medicine* along with essential details such as its *dosage form* (e.g., tablets, syrup), *strength, dose, frequency, diluents, dilution, duration*, and *purpose* of the prescribed medication. This practice helps prevent errors caused by look-alike or sound-alike drug names, ensuring clarity and patient safety. See Table 5.2 for risks associated with LASA and high-alert medications and prevention strategies.
- Highlight differences in sound-alike medicines using methods like "tall man" (mixed case) lettering (e.g., DOPamine vs. DoBUTamine).
- Avoid certain error-prone abbreviations in prescriptions due to the risk of misinterpretation, which can lead to medication errors and compromise patient safety (Table 5.3). Table 5.3 lists common error-prone abbreviations, their intended meanings, possible misinterpretations, and the recommended terms to use instead. Ensure that *micrograms (mcg)* and *nanograms (ng)* are written in full or use "mcg" instead of "μg," as the (μ) symbol can be misinterpreted as "mg," increasing the risk of a 1000-fold overdose.
- Ensure there is a space between the drug name and its strength to prevent misreading (e.g., "Tegretol 300 mg" instead of "Tegretol300 mg").
- Do NOT use symbols/abbreviations like ">", "<", "D/C" (discontinue), "TCA" (to come again), or "CST" (continue same treatment) as they are not universally recognized. These abbreviations are often used in a local context and may not be standardized across healthcare facilities. Since prescriptions are meant for patients, if they visit another hospital or travel to a different state, they may struggle to understand these abbreviations, potentially leading to confusion or medication errors. Additionally, whenever there is a change in treatment, ensure that **all medicines are written afresh**, especially when using a new prescription page. Listing the full treatment plan each time avoids omissions and clarifies the updated regimen for both patients and healthcare providers, reducing the risk of misunderstanding or missed medications.

Table 5.3 Error-prone abbreviations which should not be written in the prescription

Abbreviation	Intended meaning	Misinterpretation	Recommended term
U	Unit	Mistaken as the number 0 or 4, leading to overdose (e.g., "4 U" seen as "40")	Unit (write out completely)
IU	International unit	Mistaken as IV (intravenous) or the number 10	International unit
Q.D., QD, q.d., QD	Every day	Mistaken as QID (four times a day)	Daily
Q.O.D., QOD, q.o.d., QOD	Every other day	Mistaken as QD or QID, leading to daily dosing	Every other day
D/C	Discharge or discontinue	Misinterpreted as either death certificate, discharge or discontinue, or dilatation and curettage causing confusion	Discharge or discontinue (specify)
MS	Morphine Sulfate	Mistaken for magnesium sulfate ($MgSO_4$)	Morphine Sulfate
MSO_4, $MgSO_4$	Magnesium Sulfate	Mistaken for morphine sulfate (MS)	Magnesium Sulfate
Mg	Microgram	Mistaken as "mg" (milligram), a 1000-fold overdose	mcg or microgram
H.S., HS	Half strength or at bedtime	Mistaken as the opposite term	Half strength or at bedtime (specify)
T.I.W., TIW	Three times a week	Mistaken as "three times a day" or "twice a week"	Three times a week
S.C., S.Q., sub q	Subcutaneous	Mistaken as SL (sublingual) or "every"	Subcut or subcutaneous
A.S., A.D., A.U.	Left ear, right ear, both ears	Mistaken for OS (left eye), OD (right eye), OU (both eyes)	Left ear, right ear, both ears
O.S., O.D., O.U.	Left eye, right eye, both eyes	Mistaken for AS (left ear), AD (right ear), AU (both ears)	Left eye, right eye, both eyes
@	At	Mistaken as the number 2	At (write out completely)
cc	Cubic centimeters	Mistaken as "u" (units)	mL or Milliliters
μ	Micro	Mistaken as "m" (milli), leading to 1000-fold errors	mc (e.g., mcg for microgram)
> and <	Greater than and less than	Mistaken as opposite or as the number 7 or L	Greater than, less than
/ (slash mark)	Separates two doses or indicates "per"	Mistaken as the number 1 or divided by	Per or each (avoid slashes)
Apothecary symbols (e.g., ʒ, Э)	Various measurements	Misunderstood due to unfamiliarity	Use metric units (e.g., mg, mL)

(continued)

Table 5.3 (continued)

Abbreviation	Intended meaning	Misinterpretation	Recommended term
DPT	Demerol-Phenergan-Thorazine	Mistaken for diphtheria-pertussis-tetanus vaccine	Use full drug names
IN	Intranasal	Mistaken as "IM" or "IV"	Intranasal
q.n.	Nightly or at night	Mistaken as "qh" (every hour)	Nightly
q6PM, etc.	Every evening at 6 PM	Misread as every 6 h	6 PM daily
BT	Bedtime	Mistaken as "BID" (twice a day)	Bedtime
UD	As directed ("ut dictum")	Mistaken as a dose unit (e.g., unit dose)	As directed

- Do not abbreviate "units" as "U," since the handwritten abbreviation can easily be misread as "0" or "4". See error-prone abbreviations below.
- Avoid using decimals whenever possible. If unavoidable, use a leading zero (e.g., "0.5" instead of "0.5"). Quantities less than 1 g should be expressed in milligrams (e.g., 500 mg, not 0.5 g), and quantities under 1 mg should be in micrograms (e.g., 100 mcg, not 0.1 mg).
- For "as required (SOS)" prescriptions, specify the minimum dose interval and the maximum dose per day. It's also good practice to state the medication's purpose (e.g., for pain or sleep).
- Avoid abbreviating drug names (e.g., ASA for aspirin, PCM for paracetamol) as they are non-standard and prone to error.
- Identify LASA medications and review this list annually during formulary revisions. Share this list with all staff and display it prominently.
- Control high-alert medications:
  - Include them during formulary additions and annual reviews.
  - Create a list and display it in all clinical care areas.
  - Double-check these medications before dispensing or administration.
  - Avoid verbal orders for high-alert medicines, except in emergencies.
  - Store them separately in pharmacies and patient care units.
  - Carefully control concentrated electrolyte solutions and the use of anticoagulant therapy.
- Always read the label each time a medication is accessed and before administration, avoiding reliance on visual recognition or location cues.
- Create a *DO NOT CRUSH* medication list based on the drugs available within the hospital's scope, ensuring it aligns with the facility's treatment protocols. Hospitals can also refer to the manufacturer's instruction or established guidelines, such as the list developed by ISMP (https://www.ismp.org/sites/default/files/attachments/2018-02/DoNotCrush.pdf) to enhance accuracy. Make this list readily accessible at nursing counters to guide healthcare professionals. Educate patients that *capsules should not be opened* and only medications with a *scoring mark* can be safely split. Clear communication minimizes errors and ensures proper medication administration.

- Utilize technology like computerized physician order entry (CPOE), clinical decision support, and barcoding to reduce errors and enhance oversight. Implementing electronic systems for prescribing medications can help reduce errors caused by illegible handwriting, misinterpretation, and incorrect dosage calculations. CPOE systems provide decision support tools, alerts for potential drug interactions, and dose range checks, improving accuracy and safety.
- Medication Reconciliation: Conducting thorough medication reconciliation processes during care transitions (e.g., hospital admission, discharge, and transfer) helps ensure accurate and complete medication lists. This helps prevent omissions, duplications, or drug interactions that can lead to errors.
- Barcode Medication Administration (BCMA): Using barcode scanning technology at the point of medication administration helps verify the "five rights" (right patient, right medication, right dose, right route, right time). BCMA systems are effective in minimizing administration errors and enhancing patient safety.
- Education and Training: Providing comprehensive education and training to healthcare professionals on medication safety, proper prescribing practices, and error prevention techniques can enhance their knowledge and skills in medication management.
- Standardized Order Sets and Protocols: Developing and implementing standardized order sets and protocols for common conditions and procedures promotes consistent and evidence-based prescribing practices, reducing the risk of errors and variability.
- Double-Checking and Independent Verification: Incorporating independent double-checks for high-risk medications or complex dosing regimens can serve as a safety net to catch potential errors before medication administration.
- Patient and Family Engagement: Involving patients and their families in medication-related discussions, providing clear instructions, and encouraging active participation in their care can help detect and prevent medication errors.
- Medication Safety Culture: Fostering a culture of safety within healthcare organizations encourages open communication, reporting, and learning from medication errors. Creating an environment where healthcare professionals feel comfortable discussing errors and near-misses helps identify system weaknesses and implement improvements. Foster a team culture and encourage Adverse Drug Event (ADE) reporting.
- Continuous Quality Improvement: Establishing robust systems for monitoring and reviewing medication errors, near-misses, and adverse drug events allows for ongoing analysis, identification of root causes, and implementation of corrective measures to prevent future errors.

It's important to note that these strategies should be tailored to the specific needs and context of each healthcare setting. Collaborative efforts among healthcare professionals, administrators, policymakers, and patients are essential for promoting medication safety and reducing medication errors.

5.5 Communication Errors

Communication errors in healthcare occur when information is not accurately conveyed, understood, or documented, leading to potential risks for patients and compromised care. Poor communication accounts for >60% of the root causes of sentinel events reported to the Joint Commission. These errors can arise from various sources, including misunderstandings between healthcare providers, poor handoff processes, and inadequate patient education. Unclear or incomplete medication orders or misinterpretation of medication-related information can contribute to medication errors. For example, if a physician prescribes a medication and verbally communicates it to a nurse, but the nurse mishears the drug name, it could result in administering the wrong medication. Another common scenario is during patient handoffs; if critical information about a patient's allergies or current medications is not clearly communicated, it could lead to adverse drug reactions. Additionally, if patients are not adequately informed about their treatment plans or how to take their medications, they may not adhere to instructions, leading to ineffective treatment or complications.

Another example involves documentation errors, such as a failure to update a patient's medical record with new test results or changes in medication. This can lead to healthcare providers making decisions based on outdated or incorrect information.

Case Scenario

A patient was admitted for hypertension management and prescribed labetalol, hydralazine, and extended-release nifedipine. Due to difficulty swallowing, these medications were crushed and administered via a nasogastric tube. Shortly after administration, the patient experienced profound bradycardia and hypotension, leading to cardiac arrest. Although resuscitation was successful, the same method of crushing and administering the medications was used again the following day.

Key Issues

- **Crushing extended-release medications** can result in dangerously high plasma concentrations and caused the patient to experience profound bradycardia and hypotension, culminating in cardiac arrest.
- **Communication failure** hindered the identification of the underlying issue related to medication administration. Communication failures between the resuscitation team and nursing staff prevented a thorough investigation into the root cause of cardiac arrest. This lack of communication meant that critical information regarding the improper administration of medications was not relayed, allowing the same errors to occur the following day.
- **Lack of knowledge** among nursing staff regarding the proper handling of extended-release medications.

5.5.1 Strategies/Measures to Be Put in Place to Avert Similar Incidents in the Future

1. Strategies to prevent similar errors include creating a "Do Not Crush" medication list to inform staff of medications that should not be altered. Additionally, improving communication through standardized protocols, regular team meetings, and utilizing electronic health records can ensure critical information is shared effectively.
2. Regular training and education sessions can help healthcare staff stay updated on medication safety, proper administration techniques, and the pharmacological properties of medications. This knowledge will empower nurses and other team members to make informed decisions and recognize potential risks associated with medication administration.
3. Need for effective communication within healthcare teams. Effective communication is essential in healthcare to ensure that all team members are aligned and that patients receive safe and appropriate care. Implementing standardized communication protocols and fostering a culture of open dialogue can help mitigate these errors.

Effective communication is a cornerstone of quality healthcare delivery, playing a critical role in enhancing patient safety, improving clinical outcomes, and fostering a collaborative work environment. In healthcare settings, where teamwork is essential, clear and efficient communication can significantly reduce the likelihood of errors and misunderstandings. Figure 5.9 depicts effective communication skills.

Fig. 5.9 Effective communication skills

Key Components of Effective Communication

1. **Clarity and Precision:**
   - Healthcare professionals must convey information clearly and concisely to avoid ambiguity. This includes using standard terminology and avoiding jargon that may not be understood by all team members.
2. **Active Listening:**
   - Team members should practice active listening, which involves fully focusing on the speaker, understanding their message, responding thoughtfully, and remembering key points. This helps in accurately grasping the information being communicated.
3. **Open Dialogue:**
   - Encouraging an open environment where team members feel comfortable sharing ideas, concerns, and feedback fosters trust and collaboration. Regular team meetings and discussions can promote this open dialogue.
4. **Standardized Protocols:**
   - Implementing standardized communication protocols, such as SBAR (Situation, Background, Assessment, Recommendation), can facilitate structured communication, especially during handoffs or critical situations. This ensures that important information is consistently conveyed.
5. **Interdisciplinary Collaboration:**
   - Effective communication between various disciplines (nursing, pharmacy, medicine, etc.) is essential for holistic patient care. Interdisciplinary rounds can help share perspectives and insights that enhance treatment plans.
6. **Utilization of Technology:**
   - Leveraging technology, such as electronic health records (EHRs) and communication platforms, can streamline information sharing and enhance accessibility. These tools can provide real-time updates and ensure that all team members are informed about patient status.
7. **Training and Education:**
   - Ongoing training in communication skills and teamwork can improve the effectiveness of interactions among healthcare professionals. Simulation exercises and role-playing can also prepare staff to handle complex communication scenarios.

Most errors during transitions and hand-offs in healthcare occur due to communication breakdowns and lack of standardized processes. Key errors include:

1. **Incomplete Information**: Critical patient details (e.g., medications, allergies, test results) may be omitted, leading to treatment delays or incorrect interventions.

2. **Miscommunication**: Ambiguities or misunderstandings can arise when information is passed verbally, especially without structured formats, leading to errors in diagnosis or care.
3. **Lack of Documentation**: Failing to update or consult the EHR can result in outdated or missing information, contributing to mistakes in patient management.
4. **Interruptions and Distractions**: Hand-offs often occur in busy environments, and distractions during the process can cause important information to be forgotten or overlooked.
5. **Assumption of Knowledge**: The outgoing provider may assume the incoming team is familiar with the patient, leading to gaps in understanding and follow-up care.
6. **Failure to Prioritize Critical Information**: Not highlighting high-risk patients, urgent test results, or pending tasks can delay critical interventions and increase patient harm.

Addressing these errors requires structured communication, minimizing distractions, and ensuring accurate documentation to improve patient safety during transitions. Effective hand-offs between shifts in healthcare are critical for ensuring continuity of care, minimizing errors, and promoting patient safety. Key strategies for improving hand-offs can significantly improve the quality of hand-offs between shifts, reducing communication errors and enhancing patient safety are depicted in Table 5.4.

5.6 Medication Error Detection

Medication errors can be identified by reviewing patient files for medication charts, doctor's order sheets, progress reports, and by comparing the medication administration record (MAR) with physicians' orders. Errors can also be detected through discharge summaries, patient interviews, direct observation of medication administration, and by verifying indented and dispensed medications.

5.7 Pharmacovigilance and Safety Surveillance in Healthcare

Pharmacovigilance is science and practice focused on detecting, assessing, understanding, and preventing adverse reactions to medicines (i.e., adverse drug reactions or ADRs). In simple words, pharmacovigilance (or drug safety) is the practice of monitoring the effects of medical medicines after they have been licensed for use. India has a national Pharmacovigilance Programme (PvPI). Every hospital should have a special PV cell to monitor and report ADR. Reporting helps in identifying a causal relationship between the drug and the adverse reaction, if one exists. Toll-free helpline number for reporting of ADRs is—*1800 180 3024*. Mail filled ADR reporting form directly to *pvpi@ipcindia.net* or *pvpi.ipcindia@gmail.com*.

Table 5.4 Key strategies for improving hand-offs

Strategy	Description
Structured communication tools	Use standardized tools like SBAR (situation, background, assessment, recommendation) or I-PASS (illness severity, patient summary, action list, situation awareness, and synthesis by receiver) to ensure clear, concise, and complete communication
Face-to-face communication	Whenever possible, conduct hand-offs face-to-face to allow for questions and clarifications. This ensures real-time exchange of critical information and helps clarify uncertainties
Bedside handover	Involve the patient and their family in the hand-off at the bedside to enhance transparency, reduce errors, and engage the patient in their own care. This also allows the incoming team to verify patient conditions and ongoing treatments
Minimize interruptions	Perform hand-offs in a quiet environment, minimizing distractions and interruptions to ensure full attention is on the exchange of information
Use of checklists	Utilize hand-off checklists to ensure all necessary information (e.g., medications, pending tests, clinical status, upcoming tasks) is communicated. This ensures nothing critical is missed
Electronic health record (EHR) integration	Integrate hand-offs with EHR systems to standardize and document the hand-off process. Ensure that the EHR contains updated and accurate information to support effective communication
Critical patient information	Emphasize high-risk patients, critical changes in patient condition, pending diagnostic tests, and required follow-up actions to ensure continuity in critical care aspects
Clear roles and responsibilities	Ensure that both parties (outgoing and incoming) understand their roles in the hand-off process. Define responsibility for pending tasks to avoid confusion
Teach and train handoff skills	Provide training to staff on how to conduct effective hand-offs, emphasizing communication techniques and the importance of relaying accurate, timely information
Double-check for understanding	Encourage the receiving team to summarize and ask questions during the hand-off to ensure mutual understanding and clarify any ambiguities
Standardized timing	Establish consistent timing for hand-offs at the beginning and end of shifts, ensuring all necessary information is exchanged without rushing the process

All suspected adverse reaction to medicines and vaccines (adverse event following immunization [AEFI]) should be reported as soon as possible because over-reporting is always better than under-reporting (*ADR form available at cdsco.nic.in*).

5.7.1 Hemovigilance and Materiovigilance

Hemovigilance and materiovigilance are vital initiatives in India aimed at enhancing patient safety by monitoring adverse events related to blood transfusions and medical devices. Hemovigilance tracks issues such as transfusion reactions and the transmission of infectious diseases. It encourages healthcare professionals to report adverse events through the Hemovigilance Programme of India (HvPI), which is managed by the National Institute of Biologicals (NIB) in collaboration with blood banks and hospitals. Reporting is conducted via the "Blood Transfusion Services" software under the National Blood Transfusion Council (NBTC). Healthcare professionals can report adverse reactions related to blood transfusions by filling out specific forms provided on the platform.

Materiovigilance, launched under the Materiovigilance Programme of India (MvPI) in 2015, aims to monitor and report adverse events linked to medical devices. It is coordinated by the Indian Pharmacopoeia Commission (IPC) in collaboration with medical institutions. Healthcare providers can report issues through the Medical Device Adverse Event (MDAE) reporting form, available on the IPC website or mobile app, Vigiflow.

Both programs emphasize timely reporting to ensure safety interventions and prevent further harm. The process of reporting can also be facilitated through a toll-free number: 1800-180-3024, which provides a direct line for professionals to report adverse events related to either blood transfusion or medical devices. This toll-free service enhances accessibility and encourages the active participation of healthcare workers in promoting patient safety.

All health observed or should report ADRs especially: All ADRs to newly marketed medicines or new medicines added to the hospital formulary, all serious reactions and interactions (expected or not), ADRs which are not clearly stated in the package insert, unusual or interesting ADR, an observed increase in frequency of a given reaction, ADRs in special field of interest such as drug abuse, in pregnancy and during lactation, in children, and ADRs occurring from overdose or medication error or unusual lack of efficacy or when suspected pharmaceutical defects are observed or any product quality problems.

5.7.2 ADR Detection and Surveillance Methods

Key methods for ADR detection include spontaneous reporting systems, where healthcare professionals report suspected ADRs to national or regional databases. This passive surveillance method relies on voluntary reporting and is essential for capturing rare or severe ADRs.

Another approach is active surveillance, which involves systematically monitoring specific populations for ADRs, often through electronic health records or patient registries. This method provides more comprehensive data on drug safety and can identify trends or patterns that spontaneous reporting might miss.

Additionally, signal detection methods are employed to analyze collected data for potential safety issues. Statistical techniques can help identify unusual patterns or associations between medications and adverse events.

Patient self-reporting tools, mobile apps, and health surveys also play a role in detecting ADRs by encouraging patients to report their experiences directly to healthcare providers or regulatory authorities.

Voluntary reporting systems for ADRs has several limitations. One significant drawback is underreporting, where healthcare professionals may not consistently report ADRs due to factors such as time constraints, lack of awareness, or uncertainty about causality. This leads to incomplete data, potentially obscuring the true incidence and nature of ADRs. Additionally, the variability in the quality of reports can affect the analysis, as reports may lack detailed information or include ambiguous descriptions of events.

Intensive monitoring is a proactive approach to the surveillance of ADRs that involves closely observing specific patient populations or particular medications over a defined period. This method is designed to capture a comprehensive and detailed account of ADRs, especially for high-risk medicines, newly approved medications, or specific patient groups. It often requires significant resources, including time, trained personnel, and advanced technology, which may not be feasible in all healthcare settings. Intensive monitoring can also introduce bias, as selected populations or specific drug categories may not represent the broader patient population. Furthermore, the logistics of collecting and analyzing large volumes of data can lead to delays in identifying safety signals, potentially hindering timely interventions.

Intensive monitoring using "triggers" is a targeted approach for identifying ADRs that leverage specific indicators or events that signal potential safety issues. This method involves the systematic use of predefined criteria—known as triggers—that prompt further investigation into a patient's clinical data or medication use.

Intensive monitoring uses "triggers" by reviewing medication charts and medical records of patients admitted in the hospital can actively identify ADRs such as:

- Screening medication orders for abrupt, unexpected discontinuations, significant dosage changes, or prescriptions for specific medications like naloxone, flumazenil, diphenhydramine, 50% glucose, antiemetics, vitamin K, sodium polystyrene sulfonate, corticosteroids, or antidiarrheals may indicate that a practitioner is addressing an ADR.
- Common laboratory tests and reports of changes in laboratory values beyond predetermined thresholds (critical values) can help identify an ADR.
- Laboratory monitoring before and after prescribing medications, such as liver function tests for patients on statins, is essential. For instance, an International Normalized Ratio (INR) greater than 6 would prompt a detailed review for any signs of bleeding to assess whether an ADR has occurred; it's important to note that an elevated INR alone does not constitute an adverse event. Additionally, a new order for a serum drug level may signal the practitioner to investigate whether an ADR is due to drug toxicity or treatment failure.

Daily review of multidisciplinary notes in a patient's chart, e.g., notes pertaining to oversedation, lethargy, and falls may be the sign of an ADR caused by an analgesic, a sedative, or a muscle relaxant. Reports of a rash in a patient's progress notes may suggest an ADR and should be examined for a potential drug-related cause, such as an allergic reaction. Many healthcare systems utilize EHRs and data mining techniques to automatically flag these triggers, allowing healthcare providers to quickly identify patients who may be experiencing ADRs and require closer examination. When a trigger is identified, clinicians can conduct a more detailed review of the patient's medication history and clinical condition to determine whether an ADR has occurred. This focused approach enables more efficient resource use compared to broader monitoring methods. By concentrating efforts on patients with identified triggers, healthcare teams can facilitate early interventions, improve patient outcomes, and reduce the incidence of serious ADRs. Additionally, the data collected from trigger-based monitoring can contribute to ongoing quality improvement initiatives, helping to refine the criteria for triggers and enhance overall patient safety practices.

5.7.3 Prevention of Adverse Drug Reactions

- Avoid polypharmacy by prescribing only the necessary medications.
- Use the few medicines needed to achieve therapeutic goals.
- Ensure the appropriate dose, route, and frequency based on the patient's specific needs.
- Gather a thorough history of drug allergies and allergic diseases, such as asthma.
- Follow correct drug administration techniques; for example, non-steroidal anti-inflammatory drugs (NSAIDs) should be taken after meals, along with appropriate laboratory test monitoring.
- Avoid the inappropriate use of medicines by considering potential drug-drug interactions when adding new medications.

5.7.4 Managing Adverse Drug Reactions

- To manage ADRs effectively, remain vigilant for even minor changes in a patient's clinical status. Pay attention to minor complaints, as they may signal the possibility of future reactions. Consult with a prescriber to consider dose adjustments or rescheduling. Educate patients about potential ADRs they may experience to prevent unnecessary anxiety and ensure they continue their medications.
- Always encourage patients to report any ADRs to their prescriber or nurse immediately.
- If a severe reaction is suspected, withhold the medication and consult with the prescriber.

5.8 Role of Patient Counseling and Education in Patient Safety

Patient counseling and education are vital components of healthcare that significantly enhance patient safety. By providing patients with clear and comprehensive information about their medications, treatment plans, and potential side effects, healthcare professionals empower patients to take an active role in their care. This proactive approach not only improves adherence to treatment regimens but also helps to prevent medication errors and adverse drug reactions.

Case Scenario 1

A 68-year-old male patient with diabetes was admitted to the hospital after experiencing severe hypoglycemia, with a random blood sugar (RBS) level of 24 mg/dL. Upon examination, it was discovered that the patient had been taking oral hypoglycemic agents (OHAs) as part of his diabetes management. However, for the past 2 days, his oral intake had significantly decreased due to a throat infection that made swallowing difficult. Despite this, he continued to take his OHAs because he had not received any information from his treating doctor about the importance of adjusting his medication in light of missing meals.

This case highlights a critical lapse in patient education and communication regarding medication management. The patient's ongoing use of OHAs, despite a significant reduction in food intake, led to dangerously low blood sugar levels. Effective patient counselling would have informed him about the need to adjust his medication when unable to eat, particularly during episodes of illness.

This situation underscores the importance of healthcare providers proactively communicating with patients about medication use, especially in the context of changes in their health status or ability to maintain normal dietary intake. Proper education could have prevented the hypoglycemic episode and emphasized the need for monitoring and adjustments based on the patient's condition. By improving communication and education practices, healthcare providers can enhance patient safety and reduce the risk of adverse events like hypoglycemia.

Case Scenario 2

Consider a patient named Mr. Singh, who has been prescribed anticoagulant medication after a recent surgery. During his discharge, the nurse provides thorough counselling, explaining the importance of taking the medication exactly as directed, the potential side effects (like bleeding), and signs to watch for (such as unusual bruising or blood in urine). The nurse also emphasizes the importance of regular blood tests to monitor his medication levels.

A few weeks later, Mr. Singh experiences mild bruising. However, because he was informed about what to look for and the significance of reporting these changes, he promptly contacts his healthcare provider. Upon evaluation, it is determined that his dosage needed adjustment.

In this scenario, the role of patient counselling and education is evident. Mr. Singh's understanding of his medication, along with clear instructions, enables him to recognize potential issues early. This not only enhances his safety by allowing for timely intervention but also fosters a trusting relationship between him and his healthcare provider. Moreover, the information provided helps mitigate the risk of medication errors, as Mr. Singh is less likely to misuse or misunderstand his medication regimen. In summary, effective patient counselling and education are crucial in promoting patient safety, encouraging adherence, and empowering patients to actively participate in their healthcare journey.

Providing adequate information on medicines helps patients use them properly, safely, and wisely. Studies have shown that when patients are well-informed, they are more likely to adhere to their treatment plans. Conversely, incomplete or insufficient information can result in serious health and economic repercussions.

5.8.1 Minimum Information to Impart During Patient Counseling

During patient counseling, healthcare providers must ensure that patients receive comprehensive information about their medications. After writing a prescription, doctors should explain each drug thoroughly, while pharmacists should provide either verbal or written information at the time of dispensing. The key points to cover are depicted in Table 5.5.

5.9 Reporting and Detecting Medication Errors

Effective reporting and detecting mechanisms for medication errors are essential for improving patient safety, promoting a culture of transparency, and ensuring high-quality healthcare delivery. To promote transparency and accountability, many healthcare institutions encourage voluntary reporting of medication errors, which allows for the collection of data to identify trends, analyze causes, and develop preventive measures. Anonymous reporting systems can further encourage healthcare professionals to report errors without fear of retribution, fostering a culture of safety. In some jurisdictions, mandatory reporting of certain types of medication errors to regulatory bodies may be required, contributing to a broader understanding of the issue at a systemic level.

Detection of medication errors involves several strategies. Clinical surveillance is essential for monitoring patient outcomes and medication use, enabling the identification of potential errors, including tracking ADRs and unexpected patient responses. Medication reconciliation—conducting thorough reviews of patients' medication lists at transitions of care, such as hospital admissions or discharges—can help catch discrepancies and prevent errors. Additionally, implementing technology solutions like CPOE systems, BCMA, and EHRs significantly reduces the risk of medication errors by alerting providers to potential issues, such as drug allergies and interactions. In summary, effective reporting and detecting mechanisms for medication errors are essential for improving patient safety, promoting a culture of transparency, and ensuring high-quality healthcare delivery.

Table 5.5 Minimum Information to be imparted during patient counseling

Information to impart	Description	Strategies to improve compliance
How to take each medicine	Explain the correct dose, timing, frequency, route, and duration of each medication	Use visual aids like charts or diagrams to reinforce instructions
Specific precautions	Inform about precautions regarding meals, concurrent medications, and food interactions	Provide a printed summary of dietary restrictions or interactions
Expected adverse effects	Discuss possible side effects and when to report issues to the doctor	Encourage patients to keep a symptom diary to track side effects and concerns
Storage instructions	Guide on proper storage of medications	Use labeled containers or organizers to help patients remember storage instructions
Missing a dose	Explain what to do if a dose is missed	Provide reminders, such as smartphone alerts, for medication times
Device usage	Instruct on how to use devices like MDIs or Rota halers	Offer demonstrations and practice sessions for using medical devices
Allergy information	Remind patients to inform their doctors about medication allergies and current medications	Create an allergy alert card for patients to carry with them
Medication identification	Encourage learning non-proprietary names of medications as key identifiers	Organize educational sessions on medication identification
Variations in medication	Advise alerting healthcare providers if medications appear different	Foster open communication, encouraging patients to voice concerns during consultations
Medication lists	Ensure a complete list of medications is communicated to the next provider upon referral	Implement standardized checklists for medication reconciliation during transitions
Accessibility strategies	Develop strategies for patients with visual impairments, language barriers, and limited knowledge	Provide materials in multiple languages and accessible formats (e.g., large print)
Discharge instructions	Provide a complete list of medications upon discharge	Schedule a follow-up call or visit to reinforce discharge instructions and address questions

5.10 Safety Event Reporting and Learning

A key approach to learning from medication errors is establishing a voluntary reporting system for adverse events, including sentinel events and near misses. Such systems provide valuable data for enhancing patient safety, but underreporting remains a significant issue, as reported incidents often represent just the tip of the iceberg. Healthcare professionals should be encouraged to self-report errors, but fear of punitive actions can lead to silence, preventing organizations from learning and improving.

Creating a blame-free culture prioritizes transparency over punishment, as the root of patient safety issues lies in flawed systems, not individual practitioners. Blaming does not address underlying risk factors and is ineffective for preventing recurrences, except in rare cases of reckless behavior. To improve reporting, healthcare institutions can implement anonymous reporting systems, emphasize education and training, and recognize individuals for identifying potential errors.

Shifting the focus from blame to accountability and learning fosters a just culture, encouraging openness and continuous improvement. Effective learning systems require analyzing and understanding the causes of errors through critical incident analysis, aiming to identify system failures and the reasons for incidents, ultimately enhancing patient safety.

Reporting near misses is equally crucial, as it helps identify at-risk situations and prevents future incidents and injuries. A near miss, often referred to as a close call, is an unplanned event that does not result in injury, illness, or damage but had the potential to do so. Developing a dashboard to track and analyze errors can further enhance this process, while real-time reporting through automated systems streamlines the identification and resolution of errors.

5.11 Harm Categorization

Harm categorization plays a crucial role in improving patient safety by providing a systematic approach to identifying, analyzing, and addressing medication errors based on their severity. By categorizing errors—ranging from those that cause no harm to those leading to severe injury or death—healthcare organizations can prioritize safety interventions and preventive strategies. This helps in understanding patterns of errors, implementing targeted training, refining clinical practices, and promoting a culture of transparency and accountability. It also facilitates learning from errors, thereby reducing the likelihood of recurrence and enhancing overall quality of care.

The National Coordinating Council for Medication Error Reporting and Prevention (NCC MERP) index is a standardized tool used to categorize medication errors based on the severity of harm caused to patients (Fig. 5.10). It classifies errors from Category A, where no error occurs, to Category I, where the error contributes to or results in patient death. The index helps healthcare professionals assess the impact of errors, from those that cause no harm (Categories B–D) to those causing temporary or permanent harm (Categories E–H), enhancing the focus on patient safety and the prevention of adverse events in clinical practice.

Root cause analysis is another vital component of medication error management in healthcare, focusing on identifying and addressing the fundamental causes of errors rather than just treating their symptoms. Techniques such as the Fishbone (Ishikawa) diagram, Failure Mode and Effects Analysis (FMEA), and the Five Whys method systematically explore contributing factors, including process issues, communication breakdowns, equipment failures, or human errors. By thoroughly understanding these root causes, healthcare providers can implement targeted interventions and quality improvement measures, fostering a culture of continuous learning and improving overall patient care.

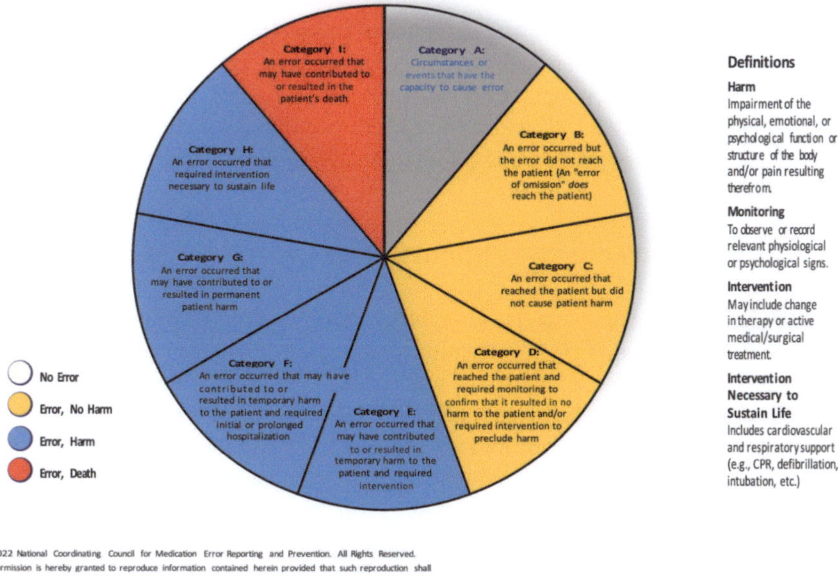

Fig. 5.10 The NCC MERP index for categorizing medication errors based on severity

> Where there is fear you do not get honest figures—**Deming, W. Edwards**
> Fear invites wrong figures. To keep his job, anyone may present to his boss only good news—**Deming, W. Edwards**

5.12 Understanding the Victims of Medical Errors

Harm resulting from adverse events in healthcare extends beyond patients and their families. Increasingly, attention has been drawn to the "second victims"—the healthcare workers directly involved in an incident—who can experience significant emotional and psychological distress. Medical errors can lead to prolonged illness, disability, increased medical costs, and, in severe cases, death, profoundly affecting patients' quality of life and eroding their trust in the healthcare system. Families of affected patients often endure substantial emotional distress, financial burdens, and feelings of betrayal when their loved ones are harmed by preventable errors. Additionally, healthcare providers responsible for these errors, referred to as the "third victims," may suffer intense psychological trauma, including feelings of guilt, shame, anxiety, and burnout, which can impair their professional performance and personal well-being. However, the impact does not stop there. Individuals with indirect exposure to adverse events, such as patient safety professionals involved in incident investigations and improvement activities, can also become third victims.

Recognizing and addressing the multifaceted impact of medical errors on all parties involved is crucial for fostering a culture of safety, accountability, and compassionate support within healthcare settings. By providing comprehensive support systems and promoting open communication, healthcare organizations can mitigate the adverse effects on all victims and work towards preventing future errors, ultimately enhancing overall patient care and professional resilience.

5.13 Telemedicine and Patient Safety

Recent COVID-19 pandemic has brought a paradigm shift in the healthcare sector. Prior to the pandemic, in-person hospital visits and face-to-face consultations were standard practice, while telemedicine was largely overlooked. The global crisis has prompted millions to experience telemedicine for the first time, and this trend is poised for rapid expansion. In India, COVID-19 has significantly accelerated this shift, making digital health a vital enabler in the transformation of the healthcare system. Telemedicine is now recognized as a primary consultation method rather than a secondary option.

With the introduction of the Telemedicine Practice Guidelines in 2020 under the IMC Act of 1956, registered medical practitioners (RMPs) are now legally authorized and protected to provide teleconsultations across various settings. This legal backing ensures that the practice of telemedicine can grow and evolve in a structured manner.

The habits established during this period are likely to become the new norm, moving away from traditional episodic, hospital-based care toward virtual visits facilitated by telemedicine. This innovative approach connects patients with healthcare providers through technology; a trend now embraced across nearly all medical specialties. Telemedicine can be utilized for patient education, monitoring, self-management support, and clinical decision-making for diagnosis and ongoing care. The terms "telemedicine," "telehealth," "eHealth," "mHealth," and "connected health" are often used interchangeably, though "telemedicine" is the most commonly referenced.

Telemedicine has evolved from basic telephone communications to sophisticated, algorithm-driven smartphone applications. Various modalities now include:

1. Telephonic consultations
2. SMS
3. WhatsApp
4. Chat mode
5. Video consultations
6. Comprehensive telemedicine platforms using approved Electronic Medical Records (EMR), which enable storage and retrieval of case records, images, and investigations. Advanced systems may also incorporate remote-controlled peripheral medical devices that provide real-time results.

While telemedicine can help reduce medical errors, there are genuine concerns regarding patient safety, especially with the rapid growth of mobile medical applications. Some issues include inappropriate use of telemedicine services and defining acceptable boundaries for telehealth care. The decision to limit telemedicine in favor of face-to-face consultations rests on clinical judgment, considering the context, professional expertise, and the patient's best interests.

Compliance and adherence to treatment plans pose additional challenges to patient safety. Patients may fail to follow prescribed plans due to miscommunication, misunderstanding, lack of access to necessary facilities, or overly complex regimens that require further explanation.

Moreover, maintaining accurate records and documentation over time is crucial. This includes logs or records of telemedicine interactions—such as phone logs, email exchanges, chat/text records, and video interaction logs. RMPs are required to retain comprehensive patient records, including reports, documents, images, diagnostics, and data (both digital and non-digital) used in telemedicine consultations. In particular, if a prescription is issued, the RMP must maintain prescription records in the same manner as they would for in-person consultations.

Feedback and dissemination of critical incident analyses raise awareness of errors within the system, leading to improved designs that can reduce or eliminate medication errors. Therefore, healthcare organizations and professionals should be encouraged to engage in voluntary reporting systems as a fundamental part of their commitment to patient safety.

5.14 System Approach to Patient Safety

Mistakes leading to patient harm are often not the result of individual or group errors but rather stem from system or process failures that contribute to these mistakes.

The Swiss Cheese Model for medical errors, developed by James Reason, illustrates how errors occur when multiple layers of defense (like slices of Swiss cheese) fail (Fig. 5.11). Each layer is designed to prevent harm, but "holes" (vulnerabilities) in these defenses can align, allowing an error to pass through and cause harm. See section on Tools and Techniques in Quality Management & Quality Management Models.

Key Components

Layers of Defense: Policies, procedures, human interventions, and technology serve as safeguards against errors.

Holes in the Layers: Weaknesses (e.g., incomplete communication, system flaws) that allow errors to slip through.

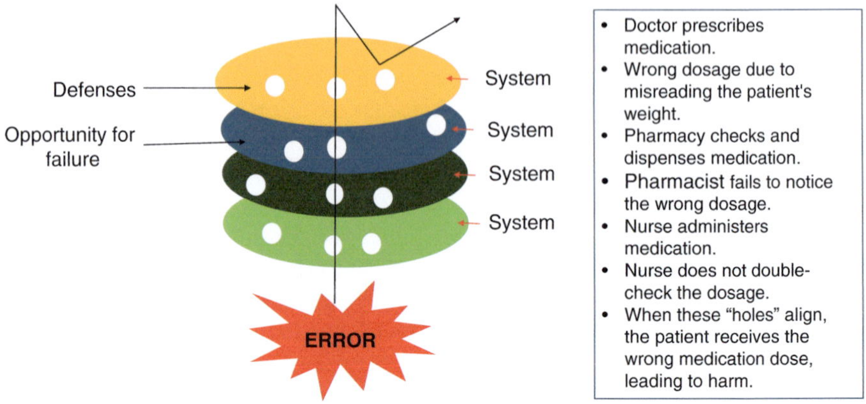

Fig. 5.11 The Latent Failure Model (Swiss Cheese Model) showing contributing factors for medication errors

Example 1: Medication Error

Layer 1: Doctor prescribes medication.
Hole: Incorrect dosage resulting from misinterpreting the patient's weight.
Layer 2: Pharmacy checks and dispenses medication.
Hole: Pharmacist fails to notice the wrong dosage.
Layer 3: Nurse administers medication.
Hole: Nurse does not double-check the dosage.

When these "holes" align, the patient receives the wrong medication dose, leading to harm.

Example 2: Surgical Error

Layer 1: Preoperative checklist to verify the surgical site.
Hole: Incomplete verification due to rushed preparation.
Layer 2: Surgeon confirms the site.
Hole: Miscommunication with the team, leading to a wrong site being marked.
Layer 3: Operating room team verification.
Hole: Team fails to speak up due to hierarchy.

These errors align and culminate, leading to surgery being performed on the wrong site.

> Yes, excessive automation at Tesla was a mistake. To be precise, my mistake. Humans are underrated—**Elon Musk**

The Swiss Cheese Model highlights that errors are often systemic rather than individual failures, stressing the need for multiple, reliable safety barriers to reduce risks. To address errors effectively, a shift from a traditional blame-oriented approach to a system-based perspective is essential. This approach recognizes that errors arise from poorly designed system structures and processes, acknowledging the inherent stress and complexity faced by healthcare workers. While this perspective does not ignore instances of negligence or misconduct, it emphasizes improving systemic factors. (For details see section on Root Cause Analysis).

The System Approach to Patient Safety focuses on designing healthcare systems that prevent errors by addressing underlying processes and systemic weaknesses. It acknowledges that humans are fallible, and errors are inevitable in complex environments, so systems must be designed to anticipate and mitigate those errors. The system approach aims to create safety nets within healthcare to catch errors before they reach the patient, focusing on systemic resilience rather than individual fault.

5.14.1 Key Principles

1. **Designing Safe Systems**: Create environments that minimize error opportunities through standardized protocols, automation, and fail-safes.
2. **Error Prevention**: Use redundant safety mechanisms and processes that catch errors before they reach the patient.
3. **Continuous Improvement**: Regularly analyze errors and near misses to identify system weaknesses and implement improvements.

Error-proofing, or **poka-yoke**, is a key strategy in the systems approach to patient safety, designed to prevent mistakes or catch them early before they cause harm.

Examples

- **Medication Safety**: Instead of blaming a nurse for administering the wrong drug, the system approach would analyze contributing factors such as poor labeling, similar drug packaging, or lack of electronic verification systems. An example of a system solution is to implement BCMA using barcodes on patients' wristbands and medications to ensure the right drug is given to the right patient at the right dose and time, significantly reducing medication errors.
- **Surgical Safety**: If a wrong-site surgery occurs, the system approach looks at issues like communication breakdowns, incomplete surgical checklists, or unclear site markings. A system-based intervention would be the Universal Protocol, which includes preoperative independent verification, site marking, and a time-out before surgery such as the WHO Surgical Safety Checklist, which are used to verify critical details like the patient's identity, surgical site, and procedure, reducing the risk of wrong-site surgeries or miscommunication.

- **Smart infusion pumps** with built-in drug libraries that alert healthcare workers if incorrect dosages are entered, preventing over- or underdosing during IV therapy.
- **Color-coded medication labels** help differentiate between similar-looking medicines, while EHR alerts notify clinicians of potential drug interactions, allergies, or inappropriate dosages, enhancing clinical decision-making.
- **Automatic lockout systems** in devices like PCA pumps prevent overdoses by restricting additional doses once the safe limit has been reached. Use devices incompatible by design (tubings/catheters) that cannot mate with patient IV lines (Fig. 5.12).
- **Unit-dose drug dispensing**, where medications are pre-packaged in single-use doses, reduces handling errors and contamination risks.

Together, these error-proofing measures create safer healthcare systems by integrating safeguards that anticipate and mitigate errors before they affect patient care.

A safe healthcare system undertakes organized efforts to prevent and minimize harm, including:

- **Leadership Commitment**: Ensuring that safety is a top priority and fostering a culture focused on patient safety.
- **Safe Environment**: Creating a secure working environment and maintaining safe procedures and clinical processes.
- **Competency Building**: Enhancing the skills of healthcare workers and improving teamwork and communication.
- **Patient and Family Engagement**: Involving patients and families in policy development, research, and shared decision-making.
- **Incident Reporting**: Establishing systems for reporting safety incidents to facilitate learning and continuous improvement.

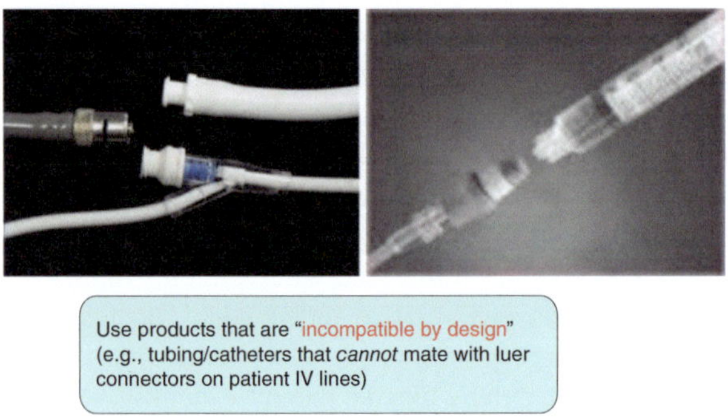

Fig. 5.12 Human factor recommendation to use products incompatible by design

Investing in patient safety measures leads to improved health outcomes, reduced costs associated with patient harm, enhanced system efficiency, and strengthened community trust in healthcare systems.

Any information technology must meet the acid test of supporting people and processes and prove it adds value before it is implemented broadly—**Jeffrey Liker**

Quality in Healthcare

6

Contents

Many audit projects have struggled due to inadequate project design. Common issues include clinical data that is often of poor quality and difficult to access. In some cases, data has been collected for administrative convenience rather than as relevant measures of clinical quality.

Quality assurance in medical practice has a rich historical foundation that acknowledges the contributions of key figures in healthcare but lacks a universal definition. Hippocrates, often referred to as the "father of medicine," emphasized the importance of closely observing and documenting medical practices, including failures and successes, to ensure quality care. Florence Nightingale, a pioneer in modern nursing, advocated for the principle that no harm should be inflicted on patients under care, establishing standards for nursing in 1859.

The concept of quality of care pertains to the level of excellence in the delivery of medical services, whether it meets or surpasses accepted standards. As technology advances, the criteria for quality inevitably change, and expectations from patients evolve accordingly. For instance, in the past, an ECG test might have satisfied a cardiac patient, whereas now they might expect a more advanced Treadmill or ECHO test. Achieving satisfactory quality involves addressing quality needs throughout the entire care process, from diagnosis and treatment to post-treatment follow-up and monitoring.

S. Sharma, *Quality Demystified*, https://doi.org/10.1007/978-981-95-0339-1_6

The Joint Commission on Accreditation of Healthcare Organizations (JCAHO) defines quality of care as the extent to which health services enhance the likelihood of desired health outcomes for individuals and the population while aligning with current professional knowledge.

In the past, there was a prevailing belief among healthcare professionals that higher quality equated to higher costs. Improved quality was associated with the adoption of new technologies, medications, and increased staffing. However, the concept of the "cost of quality" is often misunderstood. It does not refer to the price of creating a quality product or service but rather to the costs incurred when a product or service lacks quality. Whenever work needs to be redone, the cost of quality rises. Examples include re-testing, repeat visits or prolonged hospital stays, and the treatment of complications, adverse events, or medication errors.

Despite efforts to improve quality, challenges such as limited access, affordability, efficiency, and effectiveness of healthcare services persist. Instances of substandard care, inadequate facilities, unnecessary interventions, and insufficient information highlight the need for a closer examination of our healthcare delivery system. In this context, hospital accreditation assumes significant importance. The current situation calls for the efficient and cost-effective management of medical care systems capable of providing adequate healthcare in terms of both quality and quantity, satisfying the needs of consumers. This necessitates proper organization and efficient management of hospitals, optimizing physical facilities, staff, equipment, clinical services, and diagnostic and therapeutic resources.

> Quality and safety are interlinked; one cannot truly exist without the other. – **W. Edwards Deming**
> The essence of quality is the integration of safety; they are two sides of the same coin. – **Philip B. Crosby**
> Safety is a crucial dimension of quality; you cannot have one without the other. – **Joseph M. Juran**

6.1 Cost-Efficiency and Cost-Effectiveness

Cost-efficiency and cost-effectiveness are related but distinct concepts in financial and resource management, especially in sectors like healthcare, business, and public policy. While they may seem similar, they address different aspects of cost management. Cost-efficiency focuses on achieving the desired outcome using the least amount of resources or at the lowest possible cost. Cost-efficiency refers to the optimal use of resources to achieve a specific outcome at the lowest possible cost. It focuses on minimizing waste and reducing costs without compromising the quality or output.

For example, a hospital might streamline administrative processes or reduce unnecessary use of medical supplies to minimize operational expenses. This approach does not necessarily consider the quality or effectiveness of the outcome but rather ensures that resources are used optimally. A cost-efficient hospital, therefore, could be one that provides services at a lower cost by cutting waste, improving workflow, or negotiating better prices for equipment.

In contrast, cost-effectiveness considers both the costs and the benefits of different interventions or strategies. It emphasizes obtaining the maximum possible benefit or outcome for the money spent. In healthcare, for instance, selecting a medication that offers the best health outcome (e.g., higher survival rates) relative to its cost would be a cost-effective choice, even if it isn't the cheapest option. A treatment might be more expensive upfront, but if it leads to quicker recovery times, reduced hospital stays, and fewer side effects, it ultimately provides better value for the cost.

Differentiating between cost-efficiency and cost-effectiveness is important because it guides decision-making based on context and goals. In a hospital setting, focusing solely on cost-efficiency (cutting costs) could lead to using cheaper but less effective treatments, ultimately resulting in poorer patient outcomes and possibly higher long-term costs (Table 6.1). On the other hand, a focus on cost-effectiveness ensures that investments are made in interventions that offer the best outcomes relative to their costs, leading to improved quality of care and long-term savings. Understanding the distinction helps organizations balance cost-saving measures with the need to achieve high-quality results, leading to more informed and strategic resource allocation.

In practice, cost-efficiency and cost-effectiveness are often considered together to ensure that resources are used both wisely and effectively.

6.2 Quality Management System (QMS)

Quality of care should encompass technical standards and patients' expectations. While there is no universal definition of health service quality, the term "quality management system" encompasses planning, control, assurance, and improvement in healthcare. This framework is vital at both organizational and national levels. Effectively balancing these elements is crucial.

Healthcare systems employ four interconnected elements to manage quality effectively:

Quality Planning: Identifying population and user needs through equity assessments and designing tailored interventions.
Quality Control: Measuring system performance against locally set standards using tools like quality dashboards for internal oversight.
Quality Assurance: Systematically gathering evidence to ensure compliance with internal or external standards through audits, incident reports, surveys, and inspections.
Quality Improvement: Employing systematic methods to enhance care quality and patient outcomes continuously.

Table 6.1 Difference between cost-efficiency and cost-effectiveness

Aspect	Cost-efficiency	Cost-effectiveness
Definition	Optimal use of resources (money, time, materials) while maintaining quality at the lowest cost.	Best outcomes achieved for a given cost.
Objective	Minimize costs while maintaining quality.	Maximize value or outcomes for the cost.
Example	Streamlining processes to reduce expenses.	Choosing the treatment that offers the best health outcome for its price.

The terms "quality assurance" and "quality control" are often used interchangeably to refer to methods of ensuring the quality of a service or product. However, these terms have distinct meanings (Table 6.2). Nonetheless, some quality control and quality assurance actions are interconnected and complement each other.

Quality assurance focuses on the planned and systematic activities aimed at fulfilling quality requirements, while quality control involves techniques and activities used to meet those requirements. Both are essential for ensuring high-quality care and can complement each other in improving overall quality.

The concept of a QMS entails a set of policies, processes, and procedures that are necessary for the planning and execution of an organization's core business activities, whether they involve production, development, or services. QMS integrates internal processes within the organization, adopting a process-oriented approach to project execution. Its purpose is to enable organizations to identify, control, and improve their core business processes, leading to enhanced business performance. Additionally, a learning system within this framework is crucial, facilitating evidence creation, evaluation, and knowledge dissemination through networks.

In today's globalized and liberalized economy, quality has become a crucial factor for survival in the competitive world market. The escalating costs of healthcare are evident, emphasizing the need for effective management in hospitals. To achieve efficient management, administrators should adopt a conceptual model of a hospital as a system, with various service areas functioning as subsystems, all working toward the common objective of patient care. In a large hospital, these departments can be functionally grouped into categories such as:

- Clinical and nursing services (e.g., outpatient department, operation theatre, in-patient department, emergency services).
- Supportive services (e.g., X-ray, laboratory, blood bank, records, dietary).
- Utility services (e.g., laundry, central sterile services department, engineering, transport, communication, security).
- Business services (e.g., personnel, finance, materials, administration).

Important features within this model of hospital system, particularly relevant to quality management systems, include:

Table 6.2 Difference between the terms "Quality Assurance" and "Quality Control"

Quality Control vs *Quality Assurance*	
Assurance: The act of instilling confidence, the state of being certain, or the act of ensuring certainty.	***Control***: An assessment to identify necessary corrective measures; the act of managing a process in which variability is attributed to a constant system of chance causes.
Quality assurance: The planned and systematic activities implemented within a quality system to ensure that the requirements for quality in a product or service are met. In other words, quality assurance comprises activities conducted to monitor and enhance performance, ensuring that the care provided is as effective and safe as possible.	***Quality control***: The techniques and activities employed to meet quality requirements through observation and assessment.

- Continuous flow of information: From the moment of patients' admission until after discharge, there should be a seamless flow of information that is essential for decision-making purposes.
- Clear objectives and processes: Each subsystem (such as a ward unit) should have a clear objective and operate with defined processes or methods to achieve the desired output or ultimate objective of the system. These subsystems obtain inputs from both internal and external environments.
- Decision centers: Both doctors and administrators act as decision centers, exercising control over the input and output rates of patients, supplies, and services within the system.
- Feedback system: A well-defined feedback system should be in place to facilitate planning, monitoring, evaluation, and decision-making processes for the benefit of the system's users.
- Role of medical records: Medical records play a vital role in providing essential patient care-related data for both professional and administrative management purposes.

By adopting a comprehensive quality management system and incorporating these salient features, hospitals can strive for improved performance, ensuring the delivery of high-quality healthcare services.

> The first step in improvement is dissatisfaction with the status quo – **Shigeo Shingo**

6.3 Dimensions of Quality in Healthcare

Quality is defined as the extent to which a product or service satisfies or surpasses customer expectations and requirements. It encompasses various attributes, including performance, reliability, durability, and consistency, ensuring that the end product or service is free of defects and performs its intended function satisfactorily. Quality is not just about adherence to specifications but also about providing value and satisfaction to the customer, enhancing their overall experience and trust in the product or service.

Quality in health care can be defined as "doing the right thing right, right away," according to advocates of total quality management. The dimensions of quality can differ between the perspectives of patients and service providers due to their distinct viewpoints and priorities.

From the patient's perspective, quality is often evaluated based on safety, effectiveness, patient-centeredness, timeliness, efficiency, and equity. On the other hand, service providers may prioritize clinical quality, operational efficiency, staff satisfaction, financial performance, compliance, and continuous improvement.

While there is overlap between the two perspectives, the emphasis and specific dimensions can vary based on the role and priorities of each group within the healthcare system.

Quality Dimensions

For patients:

- Safety: Ensuring care doesn't harm patients.
- Effectiveness: Providing evidence-based care with clear benefits.
- Experience:
  - Caring: Staff show compassion, dignity, respect, and involvement in care decisions.
  - Responsive: Services meet patients' needs and empower them as partners.
  - Timeliness: Receiving care without unnecessary delays.
- Efficiency: Perceiving that resources are used wisely without wastage.
- Equity: Receiving consistent care regardless of personal characteristics.

For Service Providers:

- Clinical Quality: Ensuring effective and safe care delivery based on professional standards.
- Operational Efficiency: Streamlining processes to deliver care in a cost-effective manner.
- Staff Satisfaction: Promoting a positive work environment to enhance job satisfaction and retention.
- Financial Performance: Achieving financial sustainability while maintaining quality standards.
- Compliance: Adhering to regulatory and accreditation requirements.
- Continuous Improvement: Focusing on ongoing learning and adaptation to improve care delivery.

Experts recognize several dimensions of quality that hold varying importance depending on the context of quality assurance efforts. The following nine dimensions have been synthesized from the technical literature on quality and insights from QA experts. They provide a helpful framework for health teams to define, analyze, and measure their adherence to program standards in clinical care and management services supporting service delivery. Although all dimensions are relevant, their significance may differ across programs and settings, requiring adaptation to local contexts.

Technical performance: The extent to which health workers and facilities meet expectations of technical quality by adhering to standards.

> Quality is conformance to requirements – **Philip Crosby**
> Quality is a predictable degree of uniformity and dependability, at low cost and suited to the market – **Dr. Edward Deming**
> Quality is fitness for use/purpose – **Dr. Juran**

1. Access to services: The absence of geographic, economic, social, organizational, or linguistic barriers that restrict access to healthcare services.
2. Effectiveness of care: The attainment of desired outcomes or results from care.
3. Efficiency of service delivery: The ratio of service outputs to associated costs, ensuring optimal resource utilization.
4. Interpersonal relations: Trust, respect, confidentiality, courtesy, responsiveness, empathy, effective communication, and active listening between providers and clients.
5. Continuity of services: Consistent delivery of care by the same healthcare provider (when appropriate), along with timely referral and communication between providers.
6. Safety: The minimization of risks related to injury, infection, or other harmful side effects.
7. Physical infrastructure and comfort: The appearance, cleanliness, comfort, privacy, and other aspects of the facility that are important to clients.
8. Choice: Enabling client choice of provider, insurance plan, or treatment when appropriate and feasible.

> Quality planning consists of developing the products and processes required to meet customer's needs. – **Joseph M. Juran**

To improve health services and individual performance, three core quality assurance, also referred as the Quality Assurance Triangle, activities are incorporated (Fig. 6.1):

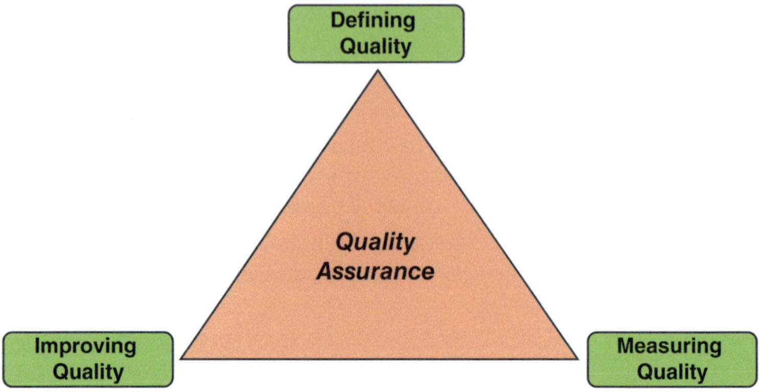

Fig. 6.1 Three core quality assurance activities by defining, measuring, and improving quality

- Measuring quality
- Defining quality
- Improving quality

While the triangle emphasizes these activities and their interplay, each core QA activity encompasses several subactivities. The greatest impact on quality of care can only be achieved when all three activities are implemented in a coordinated manner.

The triangle's shape implies that there is no fixed or optimal sequence for initiating QA activities. The order depends on the healthcare system or facility's capacity and the providers' interests. QA is a cyclical and iterative process, adaptable to specific program needs. It may start with comprehensive standards definition, small-scale quality improvement activities, or monitoring. Some teams may even choose to initiate activities in multiple areas simultaneously (Fig. 6.2). The time and effort required for each step depends on project details and past QA activities, if any.

> Quality in healthcare means doing the right thing, at the right time, in the right way, for the right person—and having the best possible results. - **Donald Berwick**

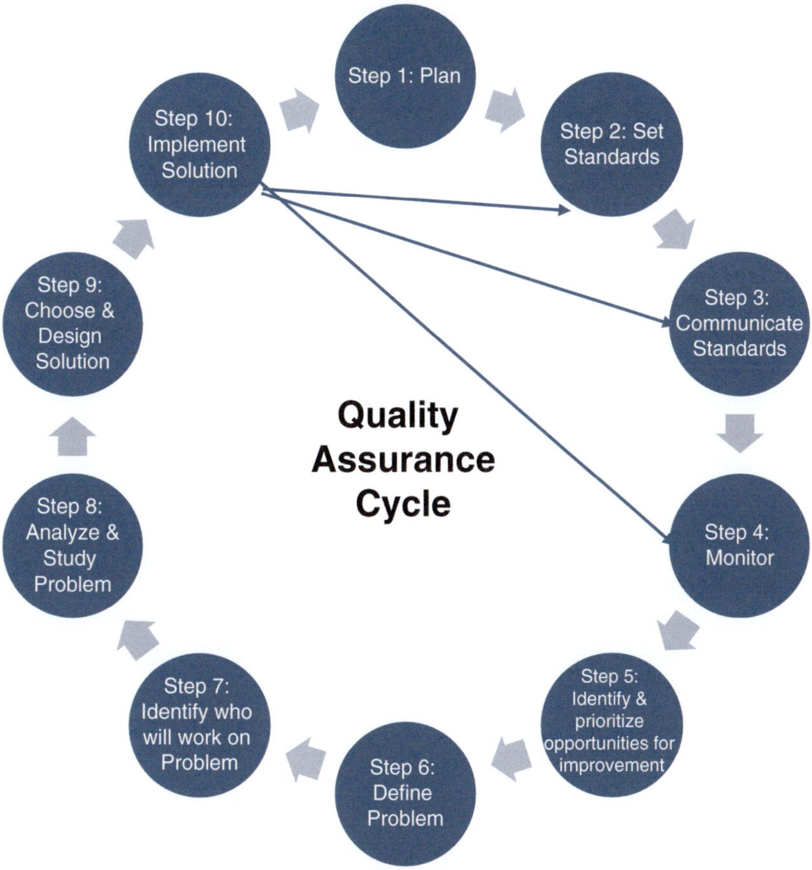

Fig. 6.2 Steps of Quality Assurance Cycle

6.4 Implementing a Quality Management System (QMS)

Implementing a QMS in an organization brings about numerous advantages and positive outcomes such as:

1. Competitive advantage and enhanced reputation: A well-implemented QMS contributes to a competitive edge in the market, improves the organization's image and credibility, and leads to increased customer satisfaction. It also fosters a higher level of commitment among workers, which positively impacts overall performance.
2. Improved management and organizational effectiveness: A QMS provides a systematic framework for better management practices and more effective organizational processes. It promotes streamlined workflows, optimized resource

allocation, and clearer communication channels, resulting in enhanced efficiency and productivity.
3. Market share expansion and positive publicity: Satisfied customers become advocates for the organization, leading to positive word-of-mouth and favorable publicity. This, in turn, helps the organization expand its market share and attract new customers.
4. Increased productivity and profitability: By identifying areas for improvement and implementing quality measures, a QMS reduces operational costs and increases efficiency. This leads to higher productivity and profitability for the organization.
5. Potential for growth: A strong QMS lays the foundation for sustainable growth. It enables organizations to adapt to changing market conditions, meet customer expectations, and seize new opportunities.
6. Baseline information for planning and improvement: A QMS provides valuable data and information that can be used for strategic planning and future service enhancements. It serves as a baseline for measuring progress and identifying areas that require improvement.
7. Safeguarding interests of providers and recipients of care: Implementing a QMS improves the quality of care provided, ensuring better patient outcomes and satisfaction. It also helps in maintaining proper documentation and a robust database, which can be crucial in addressing complaints or legal challenges.

6.5 Importance of Quality in Medical Practice

Improvement initiatives in medical practice offer several benefits, including:

1. Evidence for resource allocation: Audits provide evidence of specific resource needs to enhance quality standards and performance. For example, audits can highlight the need for additional manpower to reduce waiting times or the requirement for improved equipment to enhance patient outcomes.
2. Patients experience assessment: Conducting patient journey surveys and mapping the care process helps identify gaps and areas where the actual care delivery deviates from the intended process. This information guides improvements in patient experience and satisfaction.
3. Feedback for community outreach: Auditing and monitoring patient details can uncover trends in attendance for specific specialties or procedures. Understanding these trends helps community outreach teams address any disparities based on factors such as age, gender, area of residence, or medical status.
4. Assessing the impact of changes: Auditing patient attendance over a significant period enables the evaluation of changes in the quality of care, counselling services, or hospital operational efficiency. It helps identify the effects of interventions and facilitates necessary adjustments.
5. Identification of high-risk groups: By monitoring preoperative and postoperative health assessments, audits can identify high-risk patients prone to complications.

This knowledge aids in preparing operating theatres and ensuring the availability of experienced surgeons to handle any potential complications.

6. Measuring quality of care: Long-term treatment outcomes may be difficult to evaluate, but quality of care can be measured by monitoring complication rates against national or international standards. This allows for continuous improvement and comparison against set benchmarks.

Measurement of Quality: A Key Component of Quality Improvement 7

Contents

Measurement is indeed a key component of quality improvement in healthcare. Without accurate and reliable measurements, it is difficult to assess the current state of quality, track progress over time, and identify areas for improvement. Measurement provides the necessary data and evidence to drive decision-making and prioritize interventions.

By measuring key indicators and outcomes, healthcare organizations can identify variations, detect trends, and compare their performance against established benchmarks or standards. This allows them to pinpoint areas of weakness or opportunities for improvement and implement targeted interventions to enhance quality and patient safety.

Measurement also enables the evaluation of the effectiveness of implemented interventions and interventions, helping organizations determine whether the desired improvements have been achieved. Continuous measurement and

S. Sharma, *Quality Demystified*, https://doi.org/10.1007/978-981-95-0339-1_7

monitoring of quality indicators allow for ongoing assessment and adjustment of improvement efforts, ensuring sustained progress over time.

Moreover, measurement promotes transparency and accountability by providing stakeholders, including patients, with objective data on the quality of care delivered. This facilitates informed decision-making, empowers patients to make choices based on quality metrics, and encourages healthcare providers to strive for excellence.

In summary, measurement is the key that unlocks the potential for quality improvement in healthcare. It provides the foundation for evidence-based decision-making, targeted interventions, evaluation of outcomes, and ultimately, the delivery of high-quality care that improves patient outcomes and safety.

Without data, you're just another person with an opinion. – **W. Edwards Deming**
In God we trust; all others must bring data. – **W. Edwards Deming**
Measurement is the first step that leads to control and eventually to improvement. – **H. James Harrington**
You can't manage what you can't measure. – **Peter Drucker**

7.1 Approach to Measurement of Quality

Quality management approaches work by viewing all work within an organization as interconnected processes and systems. A system consists of various components, including organizations, people, materials, and procedures, working together to achieve specific functions or outcomes. It comprises inputs, processes, and outputs or outcomes. A process, on the other hand, refers to a series of steps that convert inputs from suppliers into outputs for customers. These processes are aimed at achieving a specific goal or outcome that encompasses the entire system.

In healthcare delivery, multiple processes occur simultaneously, involving various professional functions within the organization. Inefficiencies can arise within processes due to problems in execution or the transition between steps. Unnecessary and complex steps can add waste, complexity, and extra work to the system, ultimately diminishing the overall quality of care provided. Tools like flowcharts can assist in understanding the steps involved in a process and identify areas for improvement. QA activities can enhance understanding of processes and systems, identify weaknesses, and implement changes that lead to improved outcomes.

Measuring the quality of care or services presents challenges as it is difficult to establish universal standards for judgment across all activities. Indicators, criteria, and standards are necessary for evaluating services and assessing performance efficiency and community benefit. In healthcare, measuring quality requires considering the entire care process, including available resources, the care setting, the delivery process, and the competence of the care providers. Therefore, qualitative studies encompass the examination of the structure, process, and outcome of care. Studies analyzing factors influencing quality care can be termed "Resource

Utilization Studies," while evaluating the actual performance of care can be referred to as "Professional Activity Studies" aimed at assessing quality care.

Approaches to measuring quality in healthcare encompass a range of methods and tools designed to evaluate and monitor the delivery of care. These approaches help healthcare organizations identify areas for improvement, track progress over time, and ensure that care meets established standards. The measurement of quality in any context requires a systematic approach that involves defining relevant metrics, collecting data, analyzing the results, and using the findings to drive improvement. In the healthcare industry, various approaches can be employed to measure quality. Some commonly used approaches are:

7.1.1 Outcome Measures

Outcome measures focus on the results of healthcare interventions or treatments. These measures assess the impact on patients' health status, functional abilities, or overall well-being, recovery rates, and mortality rates. They provide a tangible assessment of the effectiveness of interventions. Examples include mortality rates, readmission rates, patient satisfaction scores, and disease-specific outcomes. Outcome studies focus on the results of care and assess several aspects:

(i) Quality of work performed: Evaluating the level of excellence in the service provided.
(ii) Quantity of work performed: Assessing whether the amount of work is optimal, minimal, or maximal.
(iii) Expenses related to patient care and the hospital's operational costs.
(iv) Consumer satisfaction with the type of attention received.

Regarding outcomes related to quality care, credit points are awarded for indicators such as the number of lives saved, the number of individuals who have recovered and been rehabilitated successfully. On the other hand, debit points are given for negative outcomes, including deaths, complications resulting from medication or procedural errors, hospital-acquired infections, and so forth. These outcomes provide valuable insights into the effectiveness and safety of the care provided, contributing to the overall assessment of quality.

Even though we've been acclimated to the concept of 'voice of the customer,' companies often believe they know how to satisfy the customer without ever asking what is important to them – **Mark DeLuzio**
Quality can be stabilized, when 'feel' is abandoned in favor of quantitative measures – **Shigeo Shingo**

7.1.2 Process Measures

Process measures assess the adherence to evidence-based guidelines and best practices during the delivery of healthcare services. They evaluate whether appropriate procedures, treatments, or interventions were followed. Process measures can include adherence to clinical protocols, medication protocols, completion of preventive screenings, compliance with surgical safety checklists, timely diagnosis, and treatment initiation, to ensure that care is delivered consistently and efficiently.

"Process" analysis focuses on the practice of delivering care, and it involves various aspects in clinical areas, including direct observation of patient care and studies based on medical records (professional service review). The following elements are examined:

(i) Physician-patient interaction:

- Art of care: Evaluating the quality of the healthcare professional's interaction and communication with the patient.
- Technical management of care: Evaluating the proper utilization of diagnostic and therapeutic facilities to ensure effective treatment.
- Efficiency of care: Examining the overall management of patients' problems while minimizing resource consumption.
- Documentation: Reviewing the adequacy and accuracy of medical documentation.

(ii) Interpersonal relations: Investigating conflicts and the handling procedures for addressing grievances between individuals involved in care delivery.

(iii) Techniques and procedures: Assessing the availability and appropriateness of techniques and procedures used in delivering care.

In areas other than clinical care, such as supportive and utility services, it is important to study the appropriateness and effectiveness of the methods employed to achieve the desired outcomes. These evaluations contribute to identifying areas for improvement and ensuring that processes are optimized to deliver high-quality care.

7.1.3 Structural Measures

Structural measures assess the physical resources, facilities, and organizational characteristics that support the delivery of healthcare services directly or indirectly. These measures focus on the infrastructure, qualifications of healthcare providers, availability of necessary equipment, and adherence to safety standards. It encompasses the following aspects:

- Beds: Evaluating the number of available beds and their utilization to ensure sufficient capacity for patient care.

- Personnel: Evaluating the adequacy of staff in terms of numbers, qualifications, and motivation levels.
- Equipment: Reviewing the sufficiency and condition of equipment, including repairs and replacements.
- Supplies: Ensuring an adequate supply of resources and addressing any issues such as shortages or pilferage.
- Space: Verifying if the physical facilities meet minimum standards and are properly maintained.
- Rules & Procedures: Examining the availability and adherence to established rules and procedures in all areas.
- Technology: Assessing the appropriateness and effectiveness of technology used in the delivery of care.
- Finance: Evaluating the adequacy of financial resources, proper allocation, and utilization to support quality care.

By assessing the structure of the healthcare system, it becomes possible to identify areas that require improvement and allocate resources effectively to ensure the provision of high-quality care.

7.1.4 Patient-Reported Outcome Measures (PROM) and Patient-Reported Experience Measures (PREM)

These are patient-reported measures to capture the patient's perspective on their healthcare experience, including their perception of the quality of care received. These measures assess aspects such as communication with healthcare providers, shared decision-making, access to care, and overall patient satisfaction.

Unlike general satisfaction surveys, PREMs serve as standardized tools that provide objective insights into care encounters, helping healthcare providers identify areas for improvement in *quality, safety,* and *performance*. CAHO (https://www.caho.in/prems) has developed validated PREMs specifically for Indian healthcare settings, addressing areas such as *anesthesia safety, emergency care,* and *medication administration*. These tools assist hospitals in delivering patient-centered care and enhancing service quality.

The main difference between **PREMs** and **PROMs** is what they measure. While PREMs assess the patient's experience of the care they received, PROMs measure the patient's health outcomes, such as symptom severity, quality of life, and functional status. Both PREMs and PROMs are essential for providing a comprehensive view of patient care and identifying areas for improvement.

7.1.5 Safety Measures

Safety measures evaluate the effectiveness of strategies and protocols implemented to prevent adverse events, medical errors, and healthcare-associated infections.

Examples of safety measures include rates of healthcare-associated infections, medication errors, patient falls, and other safety-related incidents help in ensuring patient safety and reducing harm.

7.1.6 Clinical Audits

These are systematic reviews of clinical performance against established standards or guidelines. These systematic reviews evaluate clinical performance against established standards or guidelines. They help to identify gaps in care and areas for improvement.

7.1.7 Cost Measures

Cost measures assess the economic efficiency of healthcare services by evaluating resource utilization, healthcare expenditures, and cost-effectiveness. These measures examine factors such as cost per procedure, length of hospital stay, and cost savings achieved through quality improvement initiatives.

7.1.8 Patient Surveys

Collecting feedback directly from patients about their experiences can provide valuable insights into aspects of care like communication, respect, and satisfaction.

7.1.9 Balanced Scorecards

This comprehensive approach uses a mix of performance indicators across various dimensions of care, including clinical outcomes, patient experience, financial performance, and organizational effectiveness.

7.1.10 Regulatory and Accreditation Measures

Compliance with regulatory standards and accreditation requirements ensures adherence to established guidelines and benchmarks set by governing bodies.

7.1.11 Composite Measures

Composite measures combine multiple indicators and dimensions of quality into a single measure or index. These measures provide a comprehensive assessment of overall quality performance. They are often used to compare and rank healthcare providers or organizations based on their overall quality scores.

7.1.12 Lean and Six Sigma

These quality improvement methodologies focus on reducing waste, improving efficiency, and enhancing the quality of care through data-driven decision-making and process optimization.

When approaching the measurement of quality, it is essential to define clear objectives, select appropriate measures aligned with those objectives, ensure data reliability and validity, and involve relevant stakeholders in the process. Regular monitoring and analysis of quality data allow for continuous improvement efforts and the identification of areas for targeted interventions to enhance the overall quality of healthcare services.

By employing a combination of these measurement approaches, healthcare organizations can gain a comprehensive understanding of the quality of care they provide, identify areas for improvement, and drive meaningful change to enhance patient outcomes and satisfaction.

> The only performance measurement that counts is the customers. – **Philip B. Crosby**

Total Quality Management (TQM)

8

Contents

Total Quality Management (TQM) is a management approach that emerged in the 1950s and gained increasing popularity since the early 1980s. It embodies a company's culture, attitude, and organizational structure, all aimed at delivering products and services that meet customer needs.

TQM aims to unify all functions within an organization—such as marketing, finance, design, engineering, production, and customer service—with the goal of meeting customer requirements and achieving organizational objectives. It views an organization as a collection of interconnected processes and emphasizes the continuous improvement of these processes by leveraging the knowledge and experiences of employees. The primary aim of TQM is to "do things correctly from the start, every time."

The principles of TQM are as follows:

1. Management Commitment (Plan-Do-Check-Act cycle):
    - **Plan**: Identify opportunities for improvement and devise a plan (drive, direct).
    - **Do**: Implement the change on a small scale (deploy, support, participate).
    - **Check**: Analyze data to evaluate the impact of the change (review).
    - **Act**: Based on the results, implement the change more broadly or revise the plan accordingly (recognize, communicate, revise).

S. Sharma, *Quality Demystified*, https://doi.org/10.1007/978-981-95-0339-1_8

2. Employee Empowerment:
   - Training: Equip employees with the essential knowledge and skills.
   - Suggestion scheme: Encourage and reward employee suggestions for improvement.
   - Measurement and recognition: Recognize and reward employees based on performance metrics.
   - Excellence teams: Foster cross-functional teams dedicated to achieving excellence in specific areas.
3. Fact-Based Decision-Making:
   - Utilize tools such as ICOR (Input, Control, Output, Resource), process mapping, cause and effect diagrams, root cause analysis, Pareto charts, PDCA cycles, control charts, statistical process control, failure modes and effects analysis, risk priority number, Gantt charts, network analysis, queuing techniques, and the seven statistical tools.
   - Apply techniques like FORD 8D (Team-Oriented Problem Solving Eight Disciplines tool) to address complex issues.
4. Continuous Improvement:
   - Systematically measure and focus on continuous quality improvement and CONQ (Cost of Non-Quality). This metric represents the total cost incurred by an organization due to poor quality products or services. It includes costs related to rework, scrap, inspection, warranties, customer complaints, and lost sales due to dissatisfied customers.
   - Form excellence teams to drive improvement initiatives.
   - Implement cross-functional process management.
   - Establish, maintain, and improve standards.
5. Customer Focus:
   - Develop partnerships with suppliers to enhance quality throughout the supply chain.
   - Foster a service-oriented relationship with internal customers.
   - Never compromise on quality.
   - Establish customer-driven standards.

By adhering to these principles, organizations can embrace TQM and strive for continuous improvement, customer satisfaction, and organizational excellence.

> Empowerment is not giving people power; it's giving people the freedom to use their power. – **John C. Maxwell**

The concept of Continuous Quality Improvement (CQI) within the framework of TQM is based on the philosophy of ongoing improvement, similar to the gradual ascent of a mountain climber. It emphasizes the importance of understanding the process of improvement, exploring different approaches (learning by trial), and

integrating successful practices into everyday routines to maintain progress. CQI encompasses four key elements: problem-solving discipline, interpersonal skills, teamwork, and a quality improvement process.
In essence, TQM provides an overarching framework and philosophy for quality management within an organization, while CQI represents the practical application of TQM principles through systematic, iterative improvement initiatives. Together, they form a symbiotic relationship where TQM sets the foundation and guiding principles for quality management, while CQI serves as the operational mechanism for implementing those principles in day-to-day activities. Ultimately, both TQM and CQI aim to drive continuous improvement, enhance organizational effectiveness, and deliver value to customers and stakeholders.

> There is no magic method. Rather, a total management system is needed … to best enhance creativity and fruitfulness, to utilize facilities and machines well, and to eliminate all waste – **Taiichi Ohno**

Early efforts in Quality Improvement (QI) focused on the notion that adding more resources, such as machines, procedures, training, or supplies, would automatically lead to improved quality. However, it was discovered that simply increasing resources does not guarantee efficient utilization or quality enhancements. In fact, a valuable lesson learned was that improvements in quality can often be achieved by optimizing healthcare systems without necessarily increasing resources. By enhancing healthcare processes, better outcomes can be attained while simultaneously reducing costs through the elimination of waste, unnecessary work, and rework.

In the past, inspecting activities or processes was a common approach used by managers to identify and solve problems. This approach aimed to increase control over staff and often placed blame on individuals for mistakes. However, this limited perspective on improving quality yielded minimal success as it failed to identify barriers to improvement and lacked the support of workers who felt resistant to evaluation. In contrast, current QI approaches focus on examining how activities can be modified to enable employees to perform their work more effectively. For instance, poor employee performance may be attributed to factors such as inadequate supplies, inefficient processes, or a lack of training or coaching, rather than solely blaming the workers.

Quality improvement models are structured approaches used by organizations to enhance processes, systems, and outcomes. These models provide systematic frameworks for identifying areas needing improvement, implementing changes, and evaluating their effectiveness. Some common models for quality improvement include Six Sigma, Lean, the PDCA cycle, TQM, etc. Each model offers unique methodologies and tools tailored to specific contexts, but all share the common goal of driving continuous improvement and achieving higher levels of quality and

Fig. 8.1 The KAIZEN approach for continuous, incremental improvement

efficiency. By adopting these models, organizations can streamline operations, reduce errors, enhance customer satisfaction, and ultimately deliver better outcomes.

Kaizen, often referred to as the "Continuous Improvement" model, is a key quality improvement approach originating from Japan. It emphasizes making small, incremental changes over time to improve processes, products, or services. Kaizen focuses on involving employees at all levels of an organization in identifying and implementing improvements, fostering a culture of continuous learning and improvement. Its key principles include continuous improvement, respect for people, Gemba (going to the source), standardization, waste elimination, Kaizen events, and the PDCA cycle. By adopting Kaizen, organizations cultivate a culture of continuous improvement and excellence, paving the way for sustained success. Figure 8.1 represents the KAIZEN approach for continuous, incremental improvement.

Improving quality entails employing appropriate methods to bridge the gap between current and desired levels of quality as defined by established standards. TQM emphasizes continuous improvement across all aspects of work, ranging from high-level strategic planning and decision-making to the precise execution of tasks. It stems from the belief that mistakes can be avoided, and defects can be prevented through ongoing efforts to enhance quality.

Kaizen is a healthy dissatisfaction with the status quo – **Mark DeLuzio**
Kaizen dictates that even though we will always strive for perfection, we will never reach a perfect state – **Mark DeLuzio**
Kaizen without a target would resemble a trip without a destination – **Masaki Imai**
We will never be experts at everything, so we all have to be willing and able to learn constantly – **Tracey & Ernie Richardson**

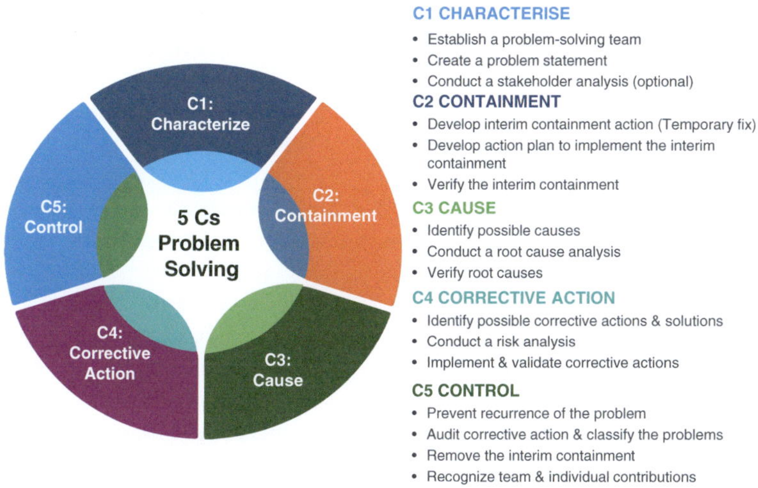

Fig. 8.2 The 5Cs of problem-solving combined with the KAIZEN approach for continuous improvement

The 5Cs of problem-solving—Characterize, Contain, Cause Analysis, Corrective Action, and Control—combined with the Kaizen approach emphasize continuous improvement by systematically identifying and resolving issues while fostering a culture of ongoing efficiency, quality enhancement, and proactive problem prevention (Fig. 8.2).

8.1 Continuous Improvement vs. Continual Improvement

Continuous improvement and continual improvement are often used interchangeably but have distinct meanings. Continuous improvement is an uninterrupted, ongoing effort to enhance processes and systems. It involves consistently making small, incremental changes over time. Continual improvement, on the other hand, involves periodic assessments and improvements, allowing for strategic pauses to evaluate progress and implement more significant changes. Both approaches aim to achieve higher efficiency and effectiveness, but they differ in their implementation rhythm and scope.

Continuous improvement encompasses not only enhancing current results but also developing capabilities to achieve better outcomes in the future. There are five major areas of focus for capability improvement: demand generation, supply generation, technology, operations, and people capability.

> Continuous improvement is better than delayed perfection – **Mark Twain**
> Don't look for the big, quick improvement. Seek the small improvement one day at a time. That's the only way it happens - when it happens, it lasts – **John Wooden**
> Continual improvement is an unending journey – **Lloyd Dobyns**

TQM recognizes that while people may make mistakes, most errors are caused or permitted by faulty systems and processes. By identifying and eliminating the root causes of mistakes, changes can be made to prevent their recurrence. Within TQM, three essential mechanisms for prevention are employed. The mechanisms of prevention include:

1. Preventing Mistakes: Proactively designing processes to avoid defects. Implementing measures to make mistakes or defects impossible or highly unlikely (mistake-proofing).
2. Early Detection: Identifying mistakes at an early stage to prevent their progression through the value chain (inspection at the source or by the next operation). Ensuring staff are skilled in identifying mistakes and are engaged in quality practices.
3. Process Correction: Halting production when recurring mistakes are identified until the underlying process can be corrected, thus preventing the production of additional defects (stop in time).

The biggest danger is that we wind up blaming people for a badly designed process" – **Tracey & Ernie Richardson**

To successfully implement TQM, organizations should focus on eight key elements:

1. **Ethics:** Ethics is the discipline that deals with distinguishing between good and bad in any situation. It encompasses both organizational ethics, which involve establishing a business code of conduct that all employees must follow, and individual ethics, which pertain to personal morals and values.
2. **Integrity**: Upholding honesty, morals, values, fairness, and adherence to facts, fostering trust between stakeholders. People view the opposite of integrity as duplicity, and TQM cannot function in an environment of duplicity.
3. **Trust:** Trust stems from integrity and ethical conduct. Without trust, the foundation of TQM cannot be established. Trust is built through integrity and ethical behavior, promoting full participation, appropriate decision-making, and customer satisfaction.
4. **Trust fosters full participation of all members**: It enables decision-making at appropriate levels within the organization, encourages individual risk-taking for continuous improvement, and ensures that measurements focus on process improvement rather than being used to evaluate individuals. Trust is crucial for ensuring customer satisfaction.
5. **Training**: Providing comprehensive training to employees to enhance productivity, including interpersonal skills, teamwork, problem-solving, decision-making, and technical expertise. Supervisors bear the primary responsibility for implementing TQM within their departments and educating their employees on its principles. Training for these employees should encompass interpersonal skills,

teamwork, problem-solving, decision-making, job management, performance analysis and improvement, business economics, and technical skills.

6. **Teamwork:** Encouraging collaborative efforts to find quicker and better solutions to problems, utilizing quality improvement teams, problem-solving teams, and natural work teams. In teams, people feel more comfortable bringing up problems that may occur and can get help from other workers to find a solution and put it into place.

In teams, individuals feel more at ease bringing up potential issues and can seek assistance from colleagues to find and implement solutions. TQM organizations typically adopt three types of teams:

Quality Improvement Teams (Excellence Teams): Temporary teams established to address specific recurring problems. They usually operate for three to twelve months.

Problem-Solving Teams: Temporary teams formed to tackle particular issues and identify and eliminate their causes. Their duration typically ranges from one week to three months.

Natural Work Teams: Small groups of skilled workers who share tasks and responsibilities. These teams employ concepts like employee involvement teams, self-managing teams, and quality circles, usually meeting for one to two hours per week.

These team structures facilitate open communication and collaborative problem-solving, essential for continuous improvement within TQM frameworks.

7. **Leadership**: It is possibly the most crucial element in TQM, permeating every part of the organization. It involves inspiring a vision, setting strategic directions, and instilling values that guide subordinates. TQM must be introduced and championed by top management, who must demonstrate commitment and personal involvement. This includes establishing clear quality values and goals aligned with the company's objectives, as well as developing and implementing well-defined systems, methods, and performance measures to achieve these goals.
8. **Communication**: Establishing effective communication channels within the organization, promoting a common understanding of ideas among all stakeholders. It binds everything together. From the foundation to the roof of the TQM house, everything is held together by the strong mortar of communication. It serves as a crucial link between all elements of TQM. Communication ensures a shared understanding of ideas between the sender and the receiver. The success of TQM relies on effective communication with and among all organization members, suppliers, and customers.
9. **Recognition:** Recognition is the final element in the system, crucial for acknowledging contributions and achievements of both teams and individuals. It fosters motivation, productivity, and quality. Recognition should be given for both suggestions and accomplishments. Employees aim to receive recognition for themselves and their teams. Identifying and recognizing contributors is a key task for supervisors. Recognition can significantly boost self-esteem, productivity,

quality, and effort. The best recognition is immediately following an employee's action. It can come in various ways, places, and times:

Ways: Personal letters from top management, award banquets, plaques, trophies, etc.

Places: In front of departments, on performance boards, in front of top management.

Time: During staff meetings, annual award banquets, etc.

Effective recognition ensures continuous improvement and a positive work environment.

> The voice of the customer must be driven through every process from design to manufacturing – **Jeffrey Liker**

Achieving sustained improvements often requires a change in attitude and a sense of ownership for the quality of services provided. Integration of Quality Assurance (QA) into the organization's structure and function relies on several supporting factors, known as essential elements of institutionalization. These elements can be divided into three categories:

1. Internal Enabling Environment: Creating a supportive internal environment that values and prioritizes quality, including leadership commitment, culture, and employee engagement.
2. Structure: Establishing the necessary structure to support quality improvement efforts, such as clear roles and responsibilities, defined processes, and accountability mechanisms.
3. Support Functions: Providing the necessary support functions, including training, resources, tools, and continuous monitoring and evaluation, to enable effective quality management and improvement.

> Leadership is not about being in charge. It's about taking care of those in your charge. – **Simon Sinek**
> The best leaders are those most interested in surrounding themselves with assistants and associates smarter than they are. – **John C. Maxwell**

By focusing on these essential elements, organizations can institutionalize quality assurance practices and foster a culture of continuous improvement. To achieve sustained improvements and successfully integrate Quality Assurance (QA) into an organization, a change in attitude and a sense of ownership for service quality are crucial. Several supporting factors, known as essential elements of institutionalization, are required. These elements can be categorized into three main areas:

1. Internal Enabling Environment: Creating an internal environment that fosters and supports QA initiatives. This includes cultivating a culture of quality, emphasizing the importance of service excellence, and promoting employee engagement and commitment to quality improvement.
2. Structure: Establishing a well-defined structure that enables effective implementation of QA practices. This involves defining roles and responsibilities, clarifying reporting lines, and ensuring accountability for quality outcomes. Streamlining operational activities and addressing conflicts between line and staff functions are essential for efficient QA implementation.
3. Support Functions: Providing the necessary support functions to facilitate QA efforts. This includes developing policies and guidelines that align with quality objectives, ensuring strategic leadership and commitment from top-level management, allocating sufficient financial resources for QA initiatives, and empowering employees through involvement in decision-making and process improvement.

In the context of implementing TQM in a hospital setting, the following factors are critical:

1. Policy: Establishing clear policies that outline the organization's commitment to quality and define the strategic direction for TQM implementation.
2. Strategic Leadership: Demonstrating strong leadership at all levels of the organization, with a focus on driving TQM initiatives, setting quality goals, and fostering a culture of continuous improvement.
3. Financial Resources: Allocating adequate financial resources to support TQM implementation, including investment in training, technology, and infrastructure necessary for quality enhancement.
4. Employee Involvement and Empowerment: Encouraging active participation of employees in TQM activities, empowering them to contribute ideas, take ownership of quality improvement, and fostering a sense of responsibility for delivering high-quality care.
5. Streamlining Operational Activities: Optimizing operational processes and workflows to eliminate bottlenecks, reduce inefficiencies, and enhance the overall quality of healthcare delivery.
6. Commitment at the Top-level Management: Ensuring commitment and support from top-level management to drive TQM efforts, allocate resources, and promote a culture of quality across the organization.
7. Organizational Structure: Adapting the organizational structure to minimize conflicts between line and staff functions, enabling smooth coordination and collaboration for effective TQM implementation.
8. Work Culture for Continuous Learning and Improvement: Creating a work culture that values continuous learning, innovation, and improvement, encouraging employees to embrace change, share knowledge, and seek opportunities for professional growth within a TQM framework.

Quality is the result of a carefully constructed cultural environment. It has to be the fabric of the organization, not part of the fabric. – **Philip B Crosby**
Healthcare quality is not a department. It's a culture. – **Unknown**

Key attributes and behaviors of a quality culture include:

1. Leadership Commitment: Leadership demonstrates a strong commitment to quality by setting clear expectations, allocating resources, and actively participating in quality improvement initiatives.
2. Continuous Improvement: There is a culture of continuous improvement where employees are encouraged to identify opportunities for enhancement and implement changes to processes and systems.
3. Customer Focus: The organization prioritizes and places a strong emphasis on understanding and fulfilling the needs and expectations of both internal and external customers, aiming to ensure their satisfaction and loyalty.
4. Employee Involvement: Employees are actively engaged in quality initiatives, encouraged to contribute ideas, and empowered to take ownership of quality improvement efforts.
5. Accountability: Clear roles and responsibilities are established, and individuals are held accountable for their actions and contributions to quality outcomes.
6. Data-Driven Decision-Making: Decision-making is informed by data and evidence, with robust systems in place for collecting, analyzing, and interpreting data to drive improvements.
7. Transparency and Open Communication: There is open communication at all levels of the organization, with a focus on sharing information, feedback, and lessons learned to foster trust and collaboration.
8. Risk Management: The organization proactively identifies, assesses, and manages risks to quality, taking preventive measures to mitigate potential issues and ensure compliance with standards and regulations.
9. Training and Development: Employees receive adequate training and development opportunities to build skills and competencies relevant to quality improvement and are supported in their professional growth.
10. Recognition and Reward: Achievements and contributions to quality are recognized and rewarded, reinforcing desired behaviors and motivating employees to maintain high standards.

These attributes and behaviors collectively contribute to a strong quality culture within an organization, driving sustained performance excellence and delivering value to stakeholders.

Quality in a service or product is not what you put into it. It is what the customer gets out of it. – **Peter Drucker**
Your most unhappy customers are your greatest source of learning. – **Bill Gates**

8.2 Quality Circles

Quality circles play a significant role in implementing CQI. These circles are specifically formed for a particular service area, such as ambulatory care, and typically include the heads of departments or managers within that area. The quality circles report to the CEO or Medical Superintendent, and their purpose is to promote CQI by following these key steps:

1. Identify Activities and Process Flow: The quality circle examines the major activities and process flow within their respective departments to gain a comprehensive understanding of the current operations.
2. Establish Objectives: The circle develops relevant, understandable, measurable, and achievable objectives that align with the department's goals and contribute to overall quality improvement.
3. Define Quality Standards: Quality circles establish quality standards for various activities performed within their departments, ensuring consistency and excellence in service delivery.
4. Identify Bottlenecks: The circle identifies bottlenecks and obstacles that hinder the smooth functioning of the departments or impede the achievement of set objectives.
5. Implement Remedial Solutions: The circle collaboratively finds remedial solutions to address the identified bottlenecks and implements them effectively.
6. Monitor Implementation and Improvement: Continuous monitoring of the implemented solutions and assessing their impact on the operations is a crucial step. The quality circle keeps track of progress and identifies areas for further improvement.

It is impossible to improve any process until it is standardized. – **Masaki Imai**
Without standards, there can be no improvement. – **Taiichi Ohno**
Standards are the foundation of improvement. – **Genichi Taguchi**

In the context of ambulatory care, patients often face issues related to waiting times for sample collection or report delivery, which negatively impacts their overall satisfaction with hospital services. Implementing quality circles brings several benefits, including:

1. Resolution of Interdepartmental Issues: Quality circles enable the resolution of common problems encountered in the ambulatory care department, such as poor interdepartmental coordination and cooperation.
2. Cross-Functional Problem Solving: The cross-functional approach adopted by quality circles allows all members to contribute their ideas and perspectives, increasing the likelihood of finding the best possible solutions to problems. This approach minimizes the chance of missing out on valuable insights.
3. Employee Satisfaction and Motivation: By involving employees from different departments, quality circles provide a platform for them to express their ideas, participate in discussions, and consider the pros and cons of different approaches. This involvement fosters employee satisfaction, motivation, and prevents demotivation, dissatisfaction, and frustration.
4. Cost-Effective Solution: Implementing quality circles does not require substantial financial investments. It leverages the existing expertise and resources within the organization while promoting a collaborative and cordial work environment.

By implementing quality circles, organizations can address challenges, improve interdepartmental cooperation, encourage employee engagement, and achieve significant advancements in service quality within the ambulatory care setting.

The quality control circle is a device of human relations. – **Kaoru Ishikawa**
Quality is everybody's job, but because it is everybody's job, it can become nobody's job without the proper leadership and organization – **Armand Feigenbum**

8.3 Leadership

Leadership is the capacity to influence others through a dynamic, reciprocal covenant aimed toward identifying and accomplishing collective purposes – **Parisi-Carew, 2000**

Leadership in healthcare plays a crucial role in guiding organizations, inspiring teams, and driving initiatives that improve patient outcomes and operational efficiency. There are various types of leaders within healthcare settings, each contributing uniquely to the complex healthcare environment. Clinical leaders, such as department heads or chief medical officers, bring expertise in medical practice and patient care, ensuring clinical excellence and adherence to best practices. Administrative leaders, like hospital administrators or healthcare executives, focus on strategic planning, financial management, and regulatory compliance to maintain operational stability and growth. Nursing leaders, such as nurse managers or directors of nursing, oversee nursing staff, patient care protocols, and ensure patient safety. Additionally, transformational leaders emphasize innovation and change management, driving initiatives like quality improvement, patient experience enhancement, and adoption of new technologies. Effective leadership in healthcare requires collaboration, empathy, and a commitment to patient-centered care, ultimately fostering a culture of excellence and continuous improvement across the organization.

Leadership styles in any organization, including healthcare, vary based on the approach and behaviors of leaders in guiding their teams. Several prominent leadership styles are commonly observed (Fig. 8.3). **Transformational leadership** inspires and motivates teams by setting a compelling vision and empowering them to achieve goals through encouragement and support. **Transactional leadership** focuses on structure and order, rewarding compliance with established rules and procedures while addressing deviations. **Servant leadership** emphasizes humility, empathy, and a focus on the needs of others, enabling leaders to serve their teams and patients effectively. **Democratic leadership** involves decision-making through consensus, encouraging team participation and fostering a sense of ownership among staff. **Autocratic leadership**, in contrast, involves centralized decision-making and strict control over tasks and processes, ensuring efficiency but potentially stifling innovation and team morale. Each style has its strengths and suitability depending on the situation, organizational culture, and the nature of healthcare delivery, impacting how effectively leaders guide their teams towards achieving healthcare goals while ensuring patient-centered care.

> Change before you have to – **Jack Welch**
> Every success story is a tale of constant adaption, revision, and change – **Richard Branson**

Fig. 8.3 The seven styles of leadership

Leadership behaviors encompass effective communication, decisiveness, and empathy. Strong leaders inspire and motivate their teams, fostering collaboration and innovation, while also demonstrating integrity and accountability in their actions. Lean leadership behaviors focus on three core aspects—continuous improvement, waste reduction, and empowering team members. These leaders encourage a culture of problem-solving, prioritize value for the customer, and support their teams in achieving efficiency and quality in all processes. Together, these behaviors help create a culture of excellence and drive sustainable success. Lean leadership behaviors include:

1. Teaches and Engages Workgroups:
   - Leaders regularly visit and audit processes, encouraging team members to challenge themselves and be creative in improving them.
   - Tailor information to meet the specific needs of team members.
   - Actively participate in group activities and engage with physicians during rounds, including senior team members like the CFO and COO.
2. Respect for People:
   - Treat others as you would expect to be treated, fostering an environment of mutual trust and respect.
   - Embrace open communication and encourage diverse opinions.
   - View Errors as Learning Opportunities and treat mistakes as valuable lessons to foster a culture of continuous improvement.
3. Process Focus:
   - Prioritize tackling process issues over blaming individuals when investigating problems.
   - Foster a deep understanding of problems through effective root cause analysis.
   - Emphasize making improvements to processes and systems, while accepting that mistakes happen as long as they lead to learning.
4. Support and Recognition:
   - Establish a clear escalation process to demonstrate that work is valued, and leadership is attentive to challenges.
   - Be available to support team members, especially during critical times.
   - Recognize individual and team efforts and provide training to support their development.
5. Leadership Commitment—Lead by Example:
   - Embrace the "Go, Look, See" philosophy by quickly addressing problems at their source.
   - Demonstrate integrity, openness, and honesty in your own actions.
   - Follow through on commitments and admit mistakes when necessary.
6. Deploy Policy and Objectives:
   - Utilize and regularly review master schedules, keeping them visible and easily accessible to team members.
   - Guide and lead your team to understand their role in achieving program objectives.

7. Commitment to Standards:
   - Use standards to define normal and abnormal conditions, implementing clear, visual controls.
   - Maintain personal discipline and coach others to adhere to standards and procedures.
   - Investigate and address deviations from standards promptly.
8. Understand Lean Vision and Principles:
   - Promote Lean concepts and principles through your actions and decisions.
   - Encourage the visualization of information and processes.
   - Challenge existing care practices and ensure alignment with the organization's vision.
9. Support the Change Process:
   - Actively participate in and lead change initiatives, removing obstacles and ensuring all necessary factors for change are in place.
   - Inspire and motivate to encourage and energize others to adopt and pursue the future state vision.
   - Communicate openly about inhibitors and work collaboratively to overcome them, following the Plan-Do-Check-Act cycle.

Through the adoption of Lean leadership principles outlined by the Nine Leadership Behaviors, healthcare organizations can go beyond superficial implementation and achieve true Lean transformation. The personal changes in leaders' behaviors can be influential and inspiring, leading to rapid and remarkable improvements.

A leader is a coach, not a judge – **Deming, W. Edwards**
Sustainable results that combat flatlining can only be achieved via an enterprise focus that has hands-on involvement from senior leadership – **Mark DeLuzio**

By focusing on all stakeholders, the value stream, and the actual work environment (Gemba), Lean leaders can create a transformative impact in healthcare organizations. The inverted pyramid approach places workers at the forefront, supported by other stakeholders. The role of the Nursing team leader becomes pivotal in embodying Lean leadership behaviors and bridging the gap between frontline realities and management expectations.

The greatest danger in times of turbulence is not the turbulence - it is to act with yesterday's logic – **Peter Drucker**

In a typical healthcare organization, there is a diverse range of stakeholders, including patients, doctors, nurses, consultants, suppliers, housekeeping staff, accounting personnel, and support staff. Lean methodology places a strong emphasis on the value stream and the Gemba, which is the actual place where the work is done. The patient care unit is where most of the value is added, primarily by nurses and doctors.

Traditionally, stakeholders are depicted in a pyramid structure resembling an organizational chart, with the director at the top and the workers at the base. This hierarchical model often entails top-down command-and-control communication. However, in the Lean philosophy, the pyramid is inverted, placing the workers at the top and supported by other stakeholders. This shift recognizes the importance of frontline employees and promotes a collaborative approach. Figure 8.4 highlights human-centered behavior in Lean practices for improved efficiency and collaboration.

The role of the Nursing team leader becomes pivotal as they navigate between the realities of the patient care unit and management. They play a crucial role in convincing their team members that Lean is not merely another management fad but a meaningful approach to improve healthcare. Their ability to embody Lean leadership behaviors becomes critical in this context.

Lean has the potential to bring about radical improvements in healthcare organizations, but its impact has often been limited. Recognizing the significance of Lean leadership is key to achieving a broader and more substantial impact. Leaders serve as examples to go beyond superficial implementation by embracing the Leadership Behaviors framework, which fosters the development of true Lean leadership. Through a pragmatic approach, these principles can be implemented rapidly, with personal behavioral changes acting as a catalyst for infectious and astonishing transformations.

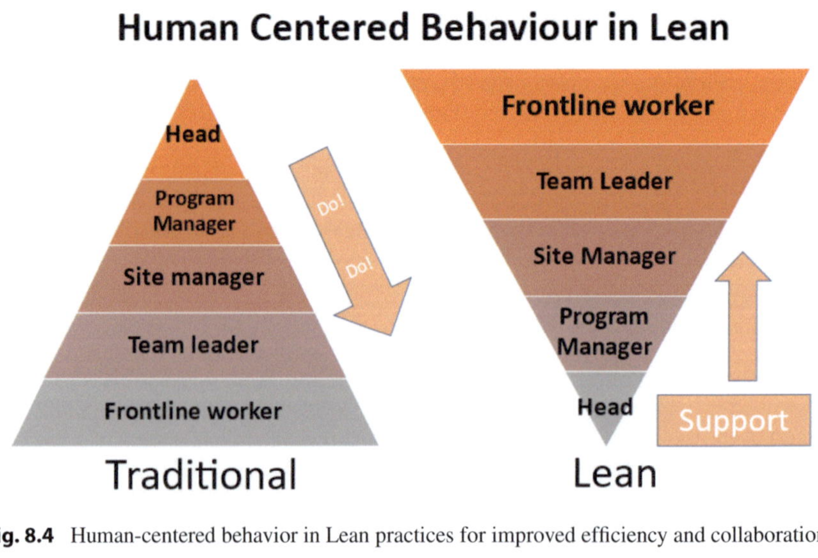

Fig. 8.4 Human-centered behavior in Lean practices for improved efficiency and collaboration

True Lean leadership lies in focusing on stakeholders, understanding the value stream, and creating an environment where the work is done effectively and collaboratively. By adopting Lean leadership principles, healthcare organizations can unlock their potential for meaningful change. Figure 8.5 depicts the difference between traditional and lean leadership.

"Leadership Behaviors" and "leadership styles" are linked and talk about how these behaviors can be applied within different leadership styles (Fig. 8.6) to enhance effectiveness and achieve organizational goals as below:

1. **Enhancing Leadership Effectiveness:** The Leadership Behaviors provide specific actions and strategies that contribute to effective leadership. Regardless of the overarching leadership style a leader adopts (e.g., transformational, transactional, servant), practicing these behaviors can strengthen their ability to inspire, motivate, and guide their teams towards success.
2. **Alignment with Leadership Styles:** Each leadership style incorporates different approaches to decision-making, motivation, and relationship-building. For instance:
   - **Transformational Leadership:** Leaders who exhibit behaviors such as "Inspiring a Shared Vision" and "Modeling the Way" align well with transformational leadership by motivating teams through a compelling vision and setting an example through their actions.
   - **Transactional Leadership:** Leaders can apply behaviors like "Setting Clear Expectations" and "Providing Supportive Feedback" to reinforce structure, clarity, and accountability within transactional leadership frameworks.
   - **Servant Leadership:** Behaviors such as "Building Trust" and "Encouraging the Heart" resonate with servant leadership principles of prioritizing the needs of others, fostering a supportive environment, and nurturing trust and collaboration.

Traditional Leadership	Lean Leadership
Leader plans	Direction setter
Staff meets goals set by leader	Ensures team goals support vision
Leader produces metrics and feeds back when not met	Monitors and audits team's metrics
Rigid enforcement of rules and regulations information controller	Sets expectations
Sole problem solver	Information conduit
Technical expert	Facilitates 'root cause' analysis
Assigner of work	Technical resource
	Provider of forward workloads
	Appraises team performance to team goals/plans

Fig. 8.5 Difference between traditional and lean leadership

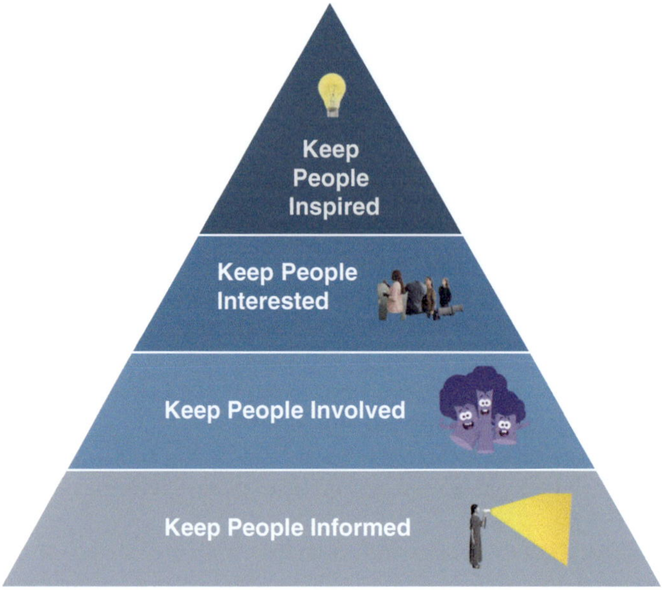

Fig. 8.6 Link between "leadership behaviors" and "leadership styles"

3. **Adaptability Across Contexts:** The Nine Leadership Behaviors are adaptable across different organizational contexts and situations. Leaders may blend these behaviors to suit dynamic challenges, cultural nuances, and the specific needs of their teams and stakeholders.
4. **Comprehensive Leadership Approach:** By integrating these behaviors into their leadership styles, leaders can create a comprehensive approach that enhances their ability to inspire commitment, drive performance, and foster a positive organizational culture.

While leadership styles define overall approaches to leadership, nine leadership behaviors outline specific actions and strategies that leaders can employ to enhance their effectiveness and impact within their chosen style.

- **Focus:** Nine leadership behaviors emphasize specific actions and behaviors that effective leaders exhibit to inspire and guide their teams.
- **Comprehensiveness:** Leadership styles are broader classifications that encompass overall approaches to leadership, encompassing how leaders behave, make decisions, and interact with their teams.
- **Application:** Nine leadership behaviors can be seen as practical strategies or behaviors that contribute to effective leadership within any style adopted by a leader. They provide a framework for leaders to enhance their effectiveness regardless of their primary leadership style.

8.4 Improving Processes and Change Management Go Hand-in-Hand

Broken, outdated, and inefficient processes can drain your team and business of valuable time and resources, leading to increased costs and reduced productivity. It's crucial to regularly review and streamline processes to ensure they are efficient and effective. Unfortunately, process improvement often falls to the bottom of a leader's priority list because motivating employees to engage in continuous improvement can be challenging. It's natural for individuals, teams, and organizations to become comfortable with predictability, leading to complacency and resistance to change. However, for teams and organizations to thrive, they must focus on improving processes. Embracing streamlined and efficient processes leads to numerous benefits: increased efficiency, improved customer experience, reduced waste and errors, and enhanced productivity and consistency. These advantages collectively contribute to a more effective and successful organization.

Enhancing processes inherently involves change; the two are inseparable. Even when process improvements are clearly beneficial and align with the best interests of the team or organization, not everyone will readily embrace change.

> The aim of leadership is … to remove the causes of failure: to help people to do a better job with less effort – **Deming, W. Edwards**

9 Clinical Governance

Clinical governance is strategically placed after quality principles, ensuring that these principles are embedded into the organization's culture and practices. It ensures that the organization adheres to high standards of care, utilizes quality concepts, and employs the appropriate tools and techniques. Clinical governance provides a structured approach for integrating all aspects of quality improvement, including ethical, confidentiality, data protection, and legal issues, thereby ensuring comprehensive oversight and continuous enhancement of healthcare quality and safety.

Clinical governance was implemented as a strategy to improve the quality of healthcare by establishing the necessary components to deliver high-quality care in a culture of learning, supported by strong leadership, and a value-driven environment where staff are respected and collaborate with patients. Initially, the focus was on providing well-trained staff, good facilities, and equipment, which were believed to be indicators of high standards. Over time, more systematic approaches such as medical and clinical audits were introduced. Despite the expertise of healthcare providers, human errors can still occur, leading to incidents like patient misidentification, surgical instruments left inside patients, medication errors, and hospital-acquired infections. These failures in care and resulting harm to patients have raised significant public and professional concerns and eroded confidence in the healthcare system. In the worst cases, they can even lead to legal liabilities for healthcare facilities. Sometimes, hospitals themselves contribute to poor outcomes by implementing procedures with inadequate oversight or inappropriate guidelines for dealing with hazardous situations. This highlights the urgent need to redesign systems to enhance patient safety and improve operational efficiency.

Clinical governance can be viewed as an overarching framework that integrates various quality improvement concepts, principles, and tools. It ensures that quality and safety are maintained and continuously improved across all healthcare services. Clinical governance focuses on ensuring and continuously improving the quality of clinical services across all healthcare organizations. It establishes a supportive

S. Sharma, *Quality Demystified*, https://doi.org/10.1007/978-981-95-0339-1_9

environment that fosters excellence. At its core, clinical governance aims to guarantee that everyone who receives healthcare services is well-cared for. It is a system that enables staff to learn from their experiences and perform at the highest standards possible.

The pillars or components of clinical governance include:

1. Leadership engagement and commitment: Strong leadership involvement and dedication to quality improvement.
2. Staffing and manpower planning: Effective management of staffing needs, including recruitment and retention strategies.
3. Clinical risk assessment and management: Identification and management of risks associated with clinical practice.
4. Evidence-based medicine and effectiveness: Applying evidence-based practices and clinical standards or treatment guidelines.
5. Clinical audit: Evaluating individual and service-level performance against established standards.
6. Continuing professional development and lifelong learning: Encouraging ongoing education and skill enhancement for healthcare professionals to ensure they remain current with the latest advancements and best practices in their field.
7. Patient and caregiver experience and involvement: Ensuring that patients and their caregivers are engaged and actively involved in their care, and their experiences are valued.

These pillars form the foundation of clinical governance, aiming to improve the quality of care and ensure the highest standards are consistently met.

Leadership Engagement and Commitment Clinical governance thrives in organizations which create an open and participative working environment, share ideas and good practice, value education and research, and where blame is used. Leadership engagement and commitment are crucial for the successful implementation of clinical governance. Organizations that foster an open and participative working environment, encourage the sharing of ideas and best practices, value education and research, and promote a blame-free culture are more likely to thrive in clinical governance. However, transforming the organizational culture to embrace these principles and challenge traditional practices requires effective leadership.

To promote ownership of clinical governance at the local level, leaders must lead by example and create an environment that encourages open participation and constructive feedback. They should empower individuals to question established norms and contribute to the continuous improvement of service delivery. Management should allocate the necessary resources to support the implementation of clinical governance initiatives. It is important to adopt a top-down approach that involves leaders at all levels, including heads of departments and nursing staff, working in collaboration with top executives to establish mechanisms for achieving this cultural transformation.

Clinical leaders should be accountable for driving clinical governance as change agents within the organization. They have the responsibility to cultivate a sense of ownership and engagement in every staff member, ensuring that they understand the importance of clinical governance and their role in its implementation. By fostering this ownership mentality, clinical leaders can facilitate a culture of continuous improvement and excellence in the delivery of quality services.

Clinical Risk Assessment and Management In the existing healthcare delivery system, there are inherent clinical risks that need to be proactively identified and managed to ensure patient safety. This involves the following steps:

Progress cannot be generated when we are satisfied with existing situations – **Taiichi Ohno**
To improve is to change; to be perfect is to change often – **Winston Churchill**

Accreditation 10

Contents

Accreditation is a process that establishes the technical competence of healthcare organizations based on specific standards for delivering services. It is a voluntary process that recognizes healthcare organizations that meet these standards, which require continuous improvement in structures, processes, and outcomes. Accreditation focuses on learning, self-development, improved performance, and risk reduction, while promoting optimum standards, professional accountability, and a commitment to ongoing excellence.

Accreditation plays a pivotal role in ensuring quality and patient safety within healthcare organizations. Accreditation is essential for:

Standardized Care Accreditation establishes standardized protocols and guidelines that healthcare providers must adhere to. This standardization ensures that all patients receive consistent, high-quality care, regardless of where they seek treatment.

Continuous Improvement Accreditation encourages healthcare organizations to adopt a culture of continuous improvement. Through regular assessments and feedback, organizations identify areas for improvement and implement changes to enhance patient care and safety.

S. Sharma, *Quality Demystified*, https://doi.org/10.1007/978-981-95-0339-1_10

Accountability Accreditation holds healthcare organizations accountable to both patients and regulatory bodies. It demonstrates a commitment to meeting and maintaining specific quality and safety standards, fostering trust among patients and stakeholders.

Risk Management Accreditation emphasizes risk management strategies, helping healthcare organizations identify and mitigate potential risks before they escalate into more significant issues. This proactive approach minimizes errors, reduces adverse events, and enhances patient safety.

Patient-Centered Care Accreditation standards often prioritize patient-centered care, focusing on patient rights, communication, and involvement in care decisions. This ensures that care is tailored to individual patient needs, preferences, and values.

Competitive Advantage Accreditation can serve as a competitive differentiator for healthcare organizations. Achieving accreditation signals a dedication to excellence, drawing in patients who value high standards of quality and safety in their healthcare decisions.

Regulatory Compliance Accreditation aligns healthcare organizations with regulatory requirements and industry best practices. It helps organizations navigate complex regulatory landscapes, reducing the risk of non-compliance and associated penalties.

> Improvement usually means doing something that we have never done before – **Shigeo Shingo**

Investing in compliances and quality are two critical aspects for healthcare organizations, each serving distinct yet interconnected purposes. A comparative analysis of the two is presented in Table 10.1.

While compliance and quality are distinct areas, they are interconnected and crucial for the overall success and reputation of healthcare organizations. Investing in compliance provides a foundational layer of legal protection and trust-building, whereas investing in quality drives continuous improvement, innovation, and patient satisfaction. Balancing investments in both areas is essential for healthcare organizations to thrive in a competitive landscape while ensuring patient safety and satisfaction. While compliance with regulations is a critical component of healthcare quality and safety, it should be viewed as a baseline rather than an endpoint. To truly ensure quality and safety, healthcare organizations must adopt a holistic approach that goes beyond mere compliance to embrace continuous improvement, patient-centered care, staff engagement, and proactive risk management. By doing so, orga-

Table 10.1 A comparative analysis of the compliances and quality

Compliances	Quality
Purpose	
Compliances ensure that healthcare organizations adhere to legal and regulatory standards set by governmental bodies. These standards cover areas such as patient care, data privacy, billing practices, and safety protocols.	Quality focuses on improving patient outcomes, enhancing patient satisfaction, and optimizing operational efficiency. It encompasses various dimensions such as patient care, clinical outcomes, patient experience, and staff training.
Focus	
Investing in compliance typically involves resources directed toward ensuring documentation, staff training, audits, and infrastructure upgrades to meet regulatory requirements.	Investing in quality involves resources directed toward staff training, technology adoption, process improvements, and patient-centered initiatives.
Scope	
Limited scope: Regulations often set minimum requirements that healthcare organizations must meet to operate legally. However, these requirements may not encompass all aspects of quality care, leaving gaps that could compromise patient safety and satisfaction.	**Individualize care**: Quality lays importance of individualized care tailored to each patient's unique needs, preferences, and circumstances.
Nature	
Static nature: Regulations are typically static and may not evolve quickly enough to address emerging healthcare challenges, technological advancements, or changing patient needs. This can lead to outdated practices that do not reflect the latest evidence-based guidelines or best practices.	Adopts a proactive strategy to ensure quality and safety. Quality-oriented organizations anticipate and mitigate risks before they escalate, and address issues before they occur, minimizing adverse events and patient harm.
	Patient-centered care and staff engagement in quality improvement initiatives and utilize patient feedback in shaping care delivery.
Benefits	
Legal protection: Ensures protection against legal repercussions and penalties. Trust building: Demonstrates commitment to transparency and accountability to stakeholders. Operational efficiency: Streamlines processes by providing standardized guidelines.	Improved patient care: Ensures better patient outcomes and safety. Competitive advantage: Enhances reputation and attracts more patients. Cost-efficiency: Long-term cost savings due to fewer errors and readmissions
Challenges	
Costly: Can be expensive due to continuous updates and requirements. Time-consuming: Requires ongoing monitoring and adjustments to stay compliant. Potential red tape: Strict adherence might sometimes hinder innovation and flexibility.	Initial investment: High initial costs for training, technology, and process changes. Complexity: Quality improvement initiatives can be complex and time-consuming. Measuring success: Sometimes challenging to measure the direct impact on patient outcomes.

nizations can create a culture of excellence that prioritizes the well-being and satisfaction of both patients and healthcare providers.

Typically, accreditation involves a multidisciplinary team of health professionals

> Cease dependence on inspection to achieve quality – **Deming, W. Edwards**

assessing healthcare organizations against published standards for the clinical care environment. These standards are usually derived from a combination of national statutes, governmental guidance, independent reports, overseas accreditation standards, and biomedical and health services research.

Various accreditation schemes such as ISO 9000, NABH (National Accreditation Board for Hospitals & Healthcare Providers), NQAS (National Quality Assurance Standards), and JCI (Joint Commission International) play crucial roles in assessing and enhancing healthcare quality worldwide. Each scheme offers distinct frameworks, criteria, and methodologies tailored to evaluate different aspects of healthcare organizations, promoting adherence to standards and continuous improvement.

ISO 9000 is an internationally recognized set of standards that outlines the criteria for quality management systems (QMS). These standards are designed to help organizations ensure that they meet the needs of customers and other stakeholders while complying with statutory and regulatory requirements related to a product or service.

Clause 3.3.1 of ISO 9000 specifically defines and categorizes improvement activities within an organization. These levels of improvement ensure a structured approach to enhancing processes, products, and services. ISO 9000 provides a foundational approach but lacks specific healthcare criteria compared to specialized schemes like NABH and JCI. NABH and JCI offer comprehensive frameworks tailored to healthcare settings, promoting standardized practices, patient safety, and continuous improvement.

In India, the journey of quality accreditation began in 2006 with the establishment of the National Accreditation Board for Healthcare Facilities Standards (NABH) under the Quality Council of India (QCI). The NABH standards are specifically designed to suit the Indian ethos and working environment, with a focus on patient, employee, visitor, and environmental safety. QCI has introduced various accreditation programs such as the NABH Accreditation for Blood Banks, NABH Accreditation for Primary Health Centres, and NABH Accreditation for Ayurveda Hospitals, among others. These accreditations have incentivized healthcare providers to adopt standardized protocols, ensure patient safety measures, and implement quality improvement initiatives. As a result, there have been significant improvements in the safety and outcomes of blood transfusions, primary healthcare delivery, and traditional medicine practices.

The focus of the Quality Accreditation Council (QAI) is to ensure that organizations meet high standards of quality and performance. It aims to enhance the quality of education, healthcare, and other services by setting benchmarks, conducting evaluations, and providing certifications. The council also emphasizes continuous improvement, accountability, and excellence in service delivery to ensure the best outcomes for stakeholders.

Public healthcare delivery system in India operates at three levels: primary, secondary, and tertiary. Together, these levels aim to ensure comprehensive healthcare coverage and address the varying needs of the population efficiently. The National Rural Health Mission (NRHM), launched in 2005, aimed to improve the functioning of community health centers (CHCs) and district hospitals. However, patients' perception of the quality of services provided was not satisfactory. To address this, the Ministry of Health & Family Welfare introduced the NQAS in 2013 to ensure sustainable quality of services in public health facilities. Both NABH and NQAS accreditation programs align with international standards specified by ISQua (International Society for Quality in Health Care Inc.) and provide a framework for quality assurance and improvement, emphasizing patient safety and quality of care. The patient-centric standards and the organization management standards cover a diverse range of issues varying from access to care, continuity of care, patient rights, medication management and use to issues like involvement of top management and the Board in leading the organization, quality improvement, reduction of risks for the patients and family, staff qualification and education, security, and integrity of information such as accuracy, reliability, and security of information about patient as well as organizational.

While JCI is an international accreditation body, several healthcare organizations in India have pursued JCI accreditation to demonstrate their commitment to quality and patient safety. JCI accreditation has encouraged hospitals to implement robust patient safety programs, enhance infection control measures, improve medication management systems, and develop a culture of continuous quality improvement. This accreditation has led to improved patient outcomes, reduced healthcare-associated infections, and better patient experiences.

Despite variations in specific criteria and focus areas, the overarching goal of improving healthcare outcomes and ensuring high standards of care unites these accreditation schemes, fostering a culture of quality improvement across healthcare settings globally. The cardinal principles of accreditation evaluation include:

1. Sound System-Based Organization: Hospital operations are based on transparent and objective principles of system-based organization.
2. Institutionalized Standards: Accreditation standards are implemented and integrated into the functioning of the hospital.
3. Patient Safety and Quality of Care: Patient safety and quality of care are established as core values, owned by management and staff at all levels and in all functions.
4. Structured Quality Improvement Program: A structured quality improvement program is in place, which includes continuous monitoring and feedback on patient care services.

At their core, these schemes advocate for structured processes that aim to identify, address, and prevent deficiencies while promoting organizational excellence.

1. **Correction**: At the lowest level is correction, which involves taking action to address a detected nonconformity. Corrections typically involve activities such as repair, rework, return, regrade, redo, or other immediate responses to rectify the nonconformity and manage its consequences. However, correction does not delve into identifying and eliminating the root causes of the nonconformity. Therefore, there remains a risk that the same issue may recur in the future.
2. **Corrective Action**: Corrective action is actions taken to eliminate the causes of detected nonconformities or other undesirable situations and to prevent recurrence. Both correction and corrective action are reactive approaches. It addresses issues after they have been identified and aims to prevent the recurrence of problems. Example includes revising procedures and retraining staff following an incident of medication error to prevent future occurrences. Analyzing root causes of patient falls and implementing targeted interventions such as improved lighting, staff training, and patient education.
3. **Preventive Improvement**: Actions taken to eliminate causes of potential nonconformities or other undesirable situations. It is a proactive approach focusing on risk identification and mitigation and involves analyzing data and trends to anticipate issues. Conducting regular risk assessments and updating safety protocols to prevent hospital-acquired infections is an example of this approach.
4. **Continual Improvement**: Ongoing efforts to enhance products, services, or processes through incremental changes. Small, regular improvements involve all levels of the organization and are often driven by feedback and performance data. For example, regularly updating clinical protocols based on the latest evidence and feedback from healthcare professionals and patients. Introducing regular patient satisfaction surveys and using the feedback to make small, ongoing enhancements in patient care and services.
5. **Breakthrough Change/Innovation**: Significant changes that lead to substantial improvements in performance. Innovation goes beyond incremental improvements and involves the creation or adoption of new ideas, processes, technologies, or practices that revolutionize existing norms. It includes radical or transformative changes, typically requires significant resources and planning, and is often initiated in response to major problems or opportunities. For example, implementing a new HER system that drastically improves data management and patient care coordination. Launching a new telemedicine platform that significantly expands access to healthcare services, especially in remote areas.
6. **Reorganization**: It signifies substantial changes in organizational structure, processes, or strategies aimed at achieving significant enhancements in quality and performance. This level transcends incremental improvements and focuses on systemic optimization to better align with organizational goals and stakeholder expectations. For instance, a healthcare network may undertake a re-organization by merging multiple hospitals into a unified system. This consolidation allows for centralized management, streamlined operations, and optimized

resource allocation, leading to improved patient care coordination, enhanced clinical outcomes, and increased operational efficiency. Re-organization at this level requires meticulous planning, stakeholder engagement, and effective change management to successfully integrate new structures and processes while fostering continuous improvement and long-term sustainability in healthcare delivery.

Without continual growth and progress, such words as improvement, achievement, and success have no meaning. – **Benjamin Franklin**

Healthcare professionals should maintain high standards consistently, even when not being monitored. This integrity ensures patient safety and builds a culture of trust and excellence within the organization.

10.1 Key Drivers of Hospital Accreditation in India

1. Consumer Awareness: Increasing consumer awareness and a focus on quality, driven by rising affluence, have led to greater emphasis on benchmarking healthcare services against global standards.
2. Consumer Protection Act: The Consumer Protection Act has made it essential for hospitals to ensure the highest quality standards in their organizations, necessitating the introduction of accreditation systems.
3. Competition: The entry of major corporate players and overseas healthcare providers in India has intensified competition, prompting healthcare providers to reevaluate and improve the quality of their services to attract patients.
4. Health Insurance Penetration: The development of the health insurance market in India, along with tie-ups between insurance companies and healthcare providers, has influenced accreditation decisions.
5. Medical Tourism: Accreditation of hospitals in India has also been driven by the influx of medical tourists. To attract patients from developed countries, it became necessary to benchmark the quality of services provided against global standards.

Accredited establishments have grown into a robust community recognized by the Government of India and various healthcare schemes. Accreditation entails assessment by an external agency based on predefined standards, covering structures, processes, and outcomes. Compliance with these standards ensures improved care delivery, patient experience, and outcomes. The process involves defining and meeting standards, monitoring adherence, and continuous improvement efforts. Accreditation emphasizes safety practices, data-driven processes, and corrective actions, leading to enhanced care quality and patient satisfaction. Patient education, community engagement, and outcome transparency further strengthen the hospital's reputation and community confidence. This includes developing a robust system for identifying indicators, collecting and analyzing data, evolving solutions, and implementing them to improve the level of care.

When quality and safety are a priority, excellence becomes a habit. – **Unknown**
If you think education is expensive, try ignorance. – **Derek Bok**

10.2 Carrying the Accreditation Process Forward

The accreditation agency, NABH, sets the standards for healthcare providers. However, benchmarks can presently be established by the care provider based on their capabilities and process robustness, aiming to enhance performance. These benchmarks should be reassessed once initial goals are met. Additionally, the standards themselves are periodically revised, necessitating ongoing efforts for accreditation maintenance. This continuous evolution of the accreditation process fosters improved performance among care providers. However, achieving and maintaining accreditation can be resource-intensive, requiring significant investments in time, finances, and staff training. Challenges include varying interpretations of standards, scalability issues for smaller facilities, and the need for ongoing updates to reflect evolving healthcare practices and technologies.

While each accreditation scheme serves to enhance healthcare quality and patient safety, organizations must choose based on their specific needs, resources, and strategic goals. Effective implementation requires commitment from leadership, engagement of healthcare professionals, and a culture of continuous improvement to ensure sustainable quality outcomes.

Accreditation of hospitals in India has so far been confined to the metros and the urban areas and has gained currency with the corporate players and larger private healthcare providers. NQAS in India, while a commendable initiative to enhance healthcare quality, faces several limitations. These include challenges in raising awareness and understanding among healthcare providers, resource constraints that hinder compliance, particularly in rural areas, and variability in implementation across regions. Additionally, monitoring and enforcement mechanisms may be inadequate, and the evolving nature of healthcare necessitates periodic updates.

In India, the challenge of accreditation in healthcare settings is multifaceted, with several limitations hindering its widespread implementation. One major obstacle is the lack of awareness and understanding among healthcare providers about the accreditation process and its benefits. Additionally, the financial burden associated with accreditation, including the costs of compliance and assessment, poses a significant challenge, particularly for smaller healthcare facilities and those in rural areas. Furthermore, the variability in healthcare infrastructure and resources across different regions presents challenges in achieving uniform standards of quality care necessary for accreditation. Limited regulatory oversight and enforcement mechanisms also contribute to inconsistencies in accreditation processes and outcomes. Addressing these challenges requires concerted efforts from policymakers, healthcare providers, and accreditation bodies to streamline processes, improve accessibility, and enhance support mechanisms for accreditation initiatives in India.

10.3 Grading of Hospitals in India

In conjunction with the evolution of hospital accreditation, there is a growing need for grading hospitals. Hospital grading has become increasingly important alongside the development of hospital accreditation in India. While accreditation focuses on ensuring compliance with a specific set of standards and results in a binary decision of accreditation or non-accreditation, grading goes beyond compliance and evaluates hospitals for excellence, innovation, and their ability to deliver high-quality care. Grading involves comparing hospitals against optimal standards and assigning them a grade, which allows for differentiation among institutions based on their performance.

The introduction of hospital grading provides essential assurance to patients regarding the services and facilities offered by a hospital. It helps patients make informed decisions about where to seek care by considering the hospital's grade. Additionally, there is a growing demand for hospital quality assessments as part of managed healthcare processes, further emphasizing the need for hospital grading.

While accreditation ensures a baseline level of quality, grading allows for a more nuanced assessment and enables value versus price analysis. The grading scale serves as a quality capability ladder, encouraging hospitals to strive for higher grades by continuously improving and developing capabilities beyond the minimum requirements set by accreditation standards. This focus on continuous improvement and exceeding benchmarks promotes excellence within the industry, ensuring that hospitals adapt to evolving standards and deliver the best possible care.

10.4 Pitfalls of Grading of Hospitals

Grading hospitals is a complex process that aims to provide transparency and assist consumers in making informed decisions about their healthcare choices. While it can serve as a useful tool for patients, there are several considerations and potential pitfalls to be aware of when critically evaluating hospital grades.

Firstly, the criteria used for grading hospitals can vary widely between different rating systems. Some may prioritize clinical outcomes, while others focus on patient experience, safety measures, or hospital amenities. This variability can lead to inconsistencies in ratings and make it challenging for patients to compare hospitals effectively.

Additionally, the grading process often relies heavily on quantitative data, such as mortality rates, readmission rates, or infection rates. While these metrics provide valuable insights, they may not capture the full scope of a hospital's performance, or the quality of care provided. Factors like patient satisfaction, staff-to-patient ratios, and access to specialized services can also significantly impact on a hospital's overall quality but may not be adequately reflected in grading systems.

Moreover, the grading process can create undue pressure on hospitals to prioritize metrics that directly influence their grade, potentially at the expense of other essential aspects of patient care. This focus on meeting specific criteria may discourage innovation and experimentation in healthcare delivery, limiting opportunities for improvement and advancement.

It's also crucial to consider the potential for grading systems to perpetuate disparities in healthcare access and quality. Hospitals serving underserved or marginalized communities may face challenges in achieving high grades due to factors beyond their control, such as socioeconomic barriers, lack of resources, or patient populations with complex health needs. Grading systems that fail to account for these disparities can inadvertently exacerbate existing healthcare inequities.

While hospital grading can offer valuable insights into healthcare quality, it's essential to approach these ratings critically and consider them as just one of many tools available for evaluating hospital performance. Patients should be encouraged to use multiple sources of information, including personal recommendations, qualitative reviews, and their own experiences, to make well-informed healthcare decisions.

Part II

Tools and Techniques in Quality Management & Quality Management Models

Tools and Techniques in Quality Management and Quality Management Models

11

Contents

S. Sharma, *Quality Demystified*, https://doi.org/10.1007/978-981-95-0339-1_11

Quality management employs a range of tools, techniques, and models to ensure that products, processes, and services consistently meet or exceed customer expectations. Quality improvement models provide comprehensive frameworks for achieving organizational goals and fostering a culture of continuous improvement. Tools and techniques, on the other hand, are specific methods employed within these frameworks to address particular issues or analyze processes. These tools and techniques help in identifying, analyzing, and resolving quality issues, improving efficiency, and driving continuous improvement. Tools and techniques are often implemented at operational levels to address specific issues, improve efficiency, reduce defects, enhance customer satisfaction, and achieve measurable results. Models offer strategic guidance and principles, while tools and techniques provide practical methods and steps for implementation. Tools and techniques offer practical methods and instruments to support the implementation of these strategies for improvement goals and achieve measurable outcomes. Both are essential components of a comprehensive approach to quality improvement within organizations. However, models like Six Sigma and Lean are applicable across various industries and organizational functions, while tools and techniques may vary in their applicability depending on the specific improvement goals.

Lean principles apply regardless of what kind of business you are in – **Mark DeLuzio**
In God we trust; all others bring data. – **W. Edwards Deming**

11.1 Tools and Techniques in Quality Management

Some commonly used tools and techniques in quality management:

1. **Flowcharts and Process Maps**: These graphical representations illustrate the sequence of steps in a process, helping to identify bottlenecks, inefficiencies, and areas for improvement.
2. **Cause and Effect Diagrams (Ishikawa/Fishbone Diagrams)**: This tool visually maps out the potential causes leading to a specific problem or effect, enabling teams to identify and categorize possible root causes for analysis and corrective actions.
3. **Pareto Analysis**: A graphical tool for ranking causes based on their significance, with the name derived from economist Vilfredo Pareto. It highlights that a majority of effects stem from a minority of causes, typically 80% of effects originating from 20% of the possible causes. This technique prioritizes problems or causes based on their frequency or impact. It helps identify the vital few issues that contribute most significantly to quality problems, enabling focused improvement efforts.
4. **Histograms**: A visual representation of variation within a dataset. Histograms present data in a graphical format, displaying the frequency distribution of a specific characteristic or variable. They provide insights into the distribution pattern, central tendency, and variation in data, aiding in analysis and decision-making. Histograms enable the identification of patterns that may not be easily discernible in a simple table of numbers.
5. **Control Charts**: Control charts monitor process performance over time, comparing measured values against upper and lower control limits. They provide a visual representation of process stability and identify when the process is out of

control or exhibits unusual variation. It displays upper and lower control limits on which values from a series of samples are plotted. A central line in the control chart helps to detect trends toward the control limits. This central line represents the process average and aids in identifying shifts or trends that could indicate potential issues in the process, ensuring timely corrective actions to maintain quality standards.

6. **Scatter Diagram**: A graphical technique for analyzing the relationship between two variables. It involves plotting two sets of data as dots on a graph, allowing for the identification of patterns and determining if the variables are related.
7. **Statistical Process Control (SPC)**: SPC involves using statistical methods to provide quantitative insights into process performance and identifying sources of variation through control charts and statistical analysis and taking corrective actions to maintain process stability and improve quality. This approach ensures that processes are consistently producing at their optimal level, leading to better overall performance and quality outcomes.
8. **RCA (Root Cause Analysis)**: RCA is a systematic method used to identify the underlying causes of problems or failures. Common techniques employed in RCA include the 5 Whys, Fault Tree Analysis, and Failure Mode and Effects Analysis (FMEA). These tools help organizations uncover the root causes of issues, enabling them to implement effective solutions and prevent recurrence.
9. **Quality Function Deployment (QFD)**: QFD is a customer-driven approach that translates customer needs and requirements into specific design and process characteristics. It ensures that customer expectations are incorporated throughout the product development or process improvement stages.
10. **Six Sigma**: Six Sigma is a data-driven methodology focused on reducing defects and process variation. It employs statistical tools and techniques to measure, analyze, improve, and control processes, aiming for high levels of performance and customer satisfaction.
11. **Lean/Lean Six Sigma**: Lean principles focus on eliminating waste and improving process efficiency. Lean Six Sigma combines these Lean tools with Six Sigma methodologies to optimize processes and minimize defects. This hybrid approach leverages the strengths of both systems to enhance overall performance and quality.

> The fishbone diagram is the systematic tool for understanding the cause-and-effect relationship. – **Kaoru Ishikawa**

Quality tools are valuable resources that assist individuals in understanding and enhancing processes to support quality improvement initiatives. While certain quality tools are specific to particular domains or industries, many tools can be applied across various fields. The expertise of quality professionals lies in their ability to adapt and apply these tools to specific situations in different contexts. These quality tools aid in identifying and comprehending underlying causes, analyzing and interpreting data, generating ideas, managing projects, and making informed decisions for continuous improvement efforts.

1. Flowchart/Process Map: These graphical tools visually represent the steps in a process, documenting the inputs and outputs of each step, aiding in process understanding.
2. Check Sheet: A simple structured data-recording form used for collecting and interpreting data efficiently.
3. Cause-Effect Diagram: Also known as an "Ishikawa diagram" or "fishbone diagram," it analyzes a process by illustrating the main causes and sub-causes leading to an effect or symptom, categorizing ideas effectively.
4. Pareto Chart:
5. Histogram: By leveraging these seven quality tools, organizations can generate ideas, analyze and develop processes, evaluate data, and effectively address process-related challenges to drive continuous improvement and achieve performance excellence.

> For quality procedures to be effective, they must be simple and practical enough to be used every day by the people doing the work – **Jeffrey Liker**

11.2 Flowchart (Also Called Process Mapping/Flow Diagram)

Flowcharts are fundamental, simple, and long-standing tools used for process improvement in various fields. They depict a sequence of events that occur either sequentially or in parallel. Flowcharts help analyze complex processes by revealing the relationships and dependencies between events. Additionally, they provide insights into the critical path of a process, highlighting bottlenecks or breakdowns and the events involved in the critical path.

Process mapping and flowcharts are often used interchangeably, but there is a slight difference between the terms. Process mapping refers to the actual creation of a diagram, while the diagram itself is called a flowchart. Both process maps and flowcharts represent the activities that occur during a process and their sequence. However, process maps tend to be more detailed and show a timeline.

Process maps and flowcharts are crucial initial steps in a Six Sigma approach. Their purpose is to illustrate how a process works and identify areas for improvement to eliminate waste.

Flowcharting a process provides a visual representation of its structure and helps identify issues within the process. It distinguishes how work is currently done from how it should be done and clarifies the functions a system should perform compared to how it is designed to perform them.

Flowcharts are also effective for showcasing process changes resulting from improvements or introducing new workflow processes. They can depict the sequence of actions, inputs and outputs (materials or services), decision points, people involved at each step, time considerations, and process measurements.

> The people doing the work are in the best position to improve the standardized work – **Jeffrey Liker**
> A simple approach to improvement is by making a process visible

Flowcharts are versatile tools that can be adapted for a wide range of purposes, including analyzing administrative or service processes, developing project plans, and more.

The uses of flowcharts include:

- Developing a shared understanding of a process.
- Documenting a process.
- Analyzing processes to identify issues, barriers/blockers, inefficiencies, and other factors.
- Standardizing processes by ensuring everyone follows the same steps.
- Studying a process for improvement.
- Communicating how a process is currently executed or how a new process will be implemented.
- Facilitating better communication among individuals involved in the same process.
- Planning new projects.

Flowcharts employ a standard set of symbols, making them easily understandable and accessible to anyone (Fig. 11.1). Figure 11.2 depicts the process flow template.

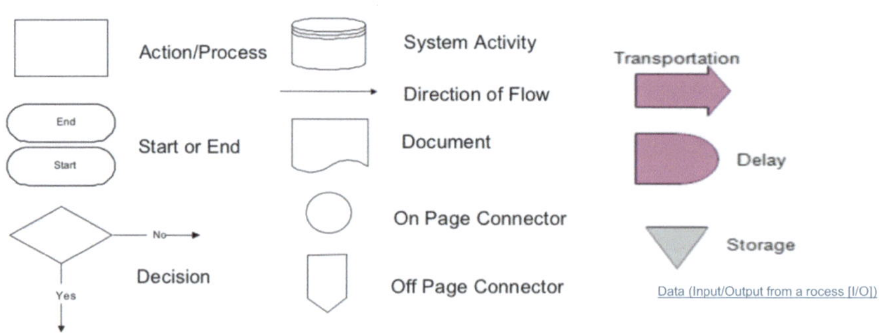

Fig. 11.1 Standard set of symbols use in flow chart to make the flow easily understandable and accessible

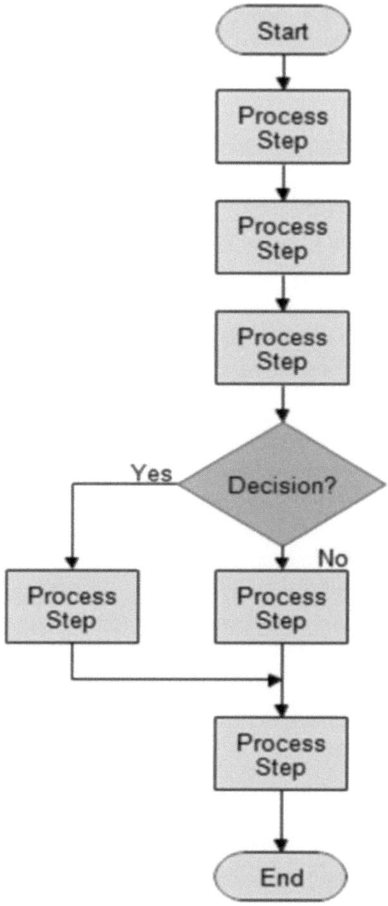

Fig. 11.2 Process flow template

11.2.1 Process Flow Template

Decision/Conditional—The decision shape in a flowchart is represented by a Diamond. It has a distinct feature of two arrows emanating from it, indicating a question being asked and the answer determining the direction of the flow. One arrow emerges from the bottom point, representing "Yes" or "True," while another arrow originates from either the right or left point, representing "No" or "False." It is essential to label the arrows clearly to ensure clarity in the process flow.

Stored Data—he stored data symbol is a general representation of a data storage object within the process flow. It denotes data being stored within the flowchart itself, as opposed to data that may be stored on physical devices such as a hard drive, magnetic tape, memory card, or any other storage medium.

Ensure that every feedback loop in the flowchart is closed, meaning that each path leads either back to a previous step or forward to another step.

11.2.1.1 How to Use a Flowchart

1. Assemble a team consisting of employees involved in carrying out the process for analysis.
2. Identify and list down all the steps involved in the process, starting from the beginning and ending at the final step.
3. Write down the process steps and rearrange them as you identify the flow and logical sequence.
4. Determine the order of the steps. When creating the flowchart, arrange the steps in a logical sequence. Represent the flow from left to right or from top to bottom to ensure clarity and ease of understanding.
5. Connect the flowchart shapes using arrows to indicate the flow and direction.

During the analysis, pay attention to the following aspects:

- Material handling
- Automation opportunities
- Redundancy
- Duplication
- Decision points
- Delays
- Rework loops
- Differentiating between non-value-added steps and value-added steps

Value-Added Steps:

- These steps involve a physical change in the product or service.
- The customer is willing to pay for these steps.
- First-time yield is 100%, meaning they are done correctly the first time.

It is important to review steps that are taken to meet compliance or regulatory criteria, such as testing and documentation. Look for opportunities to streamline these "regulatory" or "business-mandated" processes. Do not simply accept them as requirements. The team should critically assess these steps, challenge the rules, and ensure they are still valid. Seek evidence to validate the rules. It is possible that the current method contains non-value-adding activities, waste, or fails to meet the requirements altogether.

Non-Value-Added Steps:

- Set-ups
- Preparation
- Meetings
- Any of the 7 Wastes (as mentioned earlier)
- Inspections that are not business or regulatory requirements

- Non-Value Adding Steps: Steps that do not contribute to the product or service's overall value
- Customer-Resistant Steps: Steps for which customers are unwilling to pay, as they do not see their benefit or value

An example of withdrawing cash from a machine is presented below. During the process mapping, it was discovered that individuals occasionally left their card in the machine after withdrawing cash. By simply reversing the order of these two steps, the issue was resolved. Now, people are required to remove their card before receiving their money (Fig. 11.3).

11.2.2 Advantages of Flowcharts

Flowcharts offer several advantages that can greatly benefit your operation. These advantages include:

1. Communication: Flowcharts serve as a simple and effective means of communication for conveying processes and potential issues to management. They facilitate discussions about process improvements or the implementation of new equipment and procedures.
2. Analysis: Flowcharts assist in analyzing floor-level issues in a more efficient manner. They highlight areas of waste, cost reduction, and time allocation, allowing for the removal, combination, or rearrangement of steps. This leads to

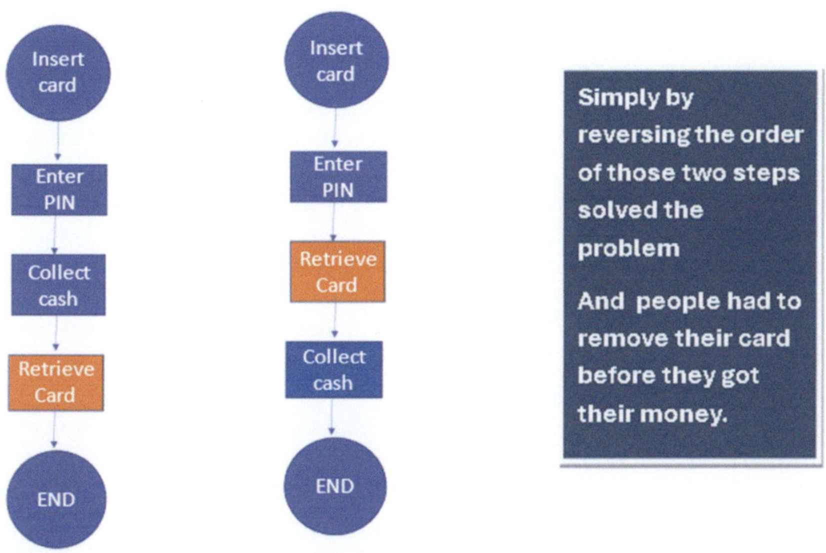

Fig. 11.3 Identification of non-value adding steps in the process giving example of changes made in the withdrawal of cash from machine

reduced waste, increased output per hour, and streamlined work processes. Flowcharts also shed light on delays, duplications, hand-offs, approvals, uncertainties, and errors, promoting efficiency and standardization while fostering consensus among stakeholders.
3. Documentation: Flowcharts provide adequate documentation of floor-level processes, enhancing overall understanding. They serve as a valuable resource that can be utilized across various departments within an organization.

11.2.3 Disadvantages of Flowcharts

Despite their benefits, flowcharts also have a few disadvantages, including:

1. Oversimplification: Flowcharts may oversimplify complex processes, making it challenging for management to fully comprehend specific areas of waste. However, simplifying or combining steps can help address this issue.
2. Logic Complexity: Complex processes can make flowcharts appear messy and cumbersome, particularly when trying to identify areas of waste. This complexity may hinder management's understanding. However, simplifying or combining steps can alleviate this problem.
3. Reproduction/Alteration: Modifying a flowchart can be time-consuming and labor-intensive, especially since flowchart symbols cannot be easily typed. However, using flowchart software available in the market can mitigate this challenge.

11.2.4 Flowchart Basic Procedure to Create a Flowchart, Follow These Basic Steps

1. Define the process to be diagrammed and write its title at the top of the work surface.
2. Clearly define the starting and ending points of the process. Discuss and decide on the level of detail to be included in the flowchart to ensure that it is both comprehensive and easy to understand. This step ensures that all relevant steps are captured without unnecessary complexity.
3. Brainstorm the activities or steps involved in the process and write each one on a card or sticky note. The sequence is not crucial at this stage.
4. Arrange the activities in the proper sequence.
5. After including all the activities and agreeing on the sequence, use arrows to illustrate the flow of the process. These arrows will help visualize the direction and order of each step, making the flowchart easy to follow and understand. This step ensures that the process is clearly mapped out from start to finish.
6. Review the flowchart with all stakeholders involved in the process (workers, supervisors, suppliers, customers) to ensure consensus and accuracy.

11.2.5 High-Level Flowchart

A high-level flowchart provides an overview of the process from a macro perspective. It allows one to focus on the big picture, similar to looking down from an airplane cruising at 30,000 feet. This initial view helps identify the general layout, major components, and key features of the process before delving into the details. In contrast, low-level process mapping at 500 feet delves into detailed, step-by-step workflows, capturing granular actions, responsibilities, and potential inefficiencies for process optimization. Figure 11.4 illustrates the contrast between high-level process mapping at 30,000 feet and low-level process mapping at 500 feet.

11.2.5.1 To Create a High-Level Flowchart

1. Begin by drawing the 30,000-foot view, limiting the map to only 4-7 steps. Avoid adding excessive detail, as it's better to maintain a concise overview.
2. Remember that those directly involved in the process are usually the best individuals to map and improve it.
3. Once the high-level view is established, descend into the details (500-foot view). Imagine descending from the plane to a closer perspective, where you can see houses, streets, and intersections. This level of detail helps understand specific pathways, like reaching a railway station or crossing a river. However, avoid going too deep into detail, which can be overwhelming. Instead of crashing into the minute details like a forest, maintain clarity.
4. While the initial overview can be mapped in a room, detailed mapping benefits from walking the process and speaking with those involved. The goal is to capture the current process, not an idealized version. Collect all insights, then gather the team to map it out on a wall using Post-it notes. Break down the process into

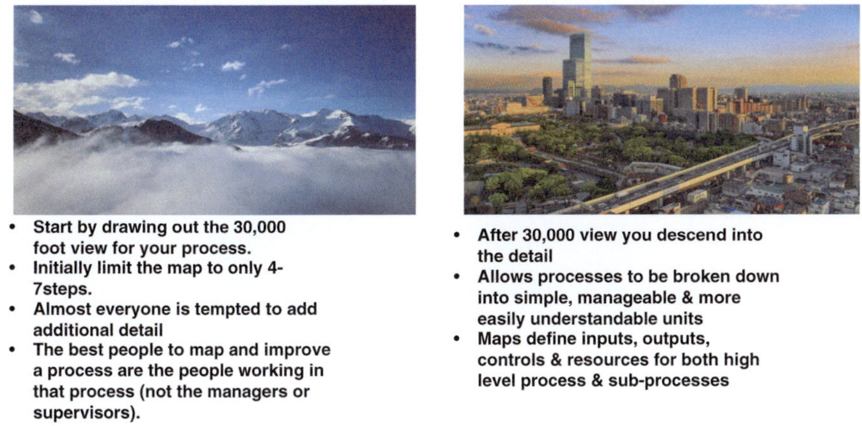

Fig. 11.4 Process Mapping contrast between high-level process mapping at 30,000 feet and low-level process mapping at 500 feet

manageable units, define inputs, outputs, controls, and resources for both high-level processes and sub-processes.

5. Add timing information to the flowchart and mark value-added and waste steps with distinct colors (green and red, respectively).
6. Define metrics for each process, preferably using key performance indicators (KPIs).

Processes inherently vary due to factors like weather, traffic, and road conditions, leading to variable outcomes. Capturing this variation on process maps by recording times for each step reveals improvement opportunities. For instance, document approval might take either 1 hour or 3 days, highlighting a need for improvement. Analyzing these variations can lead to optimizing processes, such as reducing the time for urgent blood tests in an Accident and Emergency ward. Highlighting the delivery time variation of blood samples, ranging from 18 to 128 minutes, on the process map reveals improvement opportunities. It's recommended to display such time variations on all process maps. Ensure accuracy by capturing actual times from the work area, not idealized ones from SOPs. Engage with the staff, explain the process, and ask them to record timings. Adding up individual times from the detailed "500-foot view" can easily incorporate these into the broader "30,000-foot view" flowchart (Figs. 11.5, 11.6 and 11.7).

Each step on the process map should be labelled as either value-added (green) or waste (red), following the 7 Wastes principle from the Toyota Production System. A step is considered value-added if it meets these criteria: the customer is willing to pay for it, the item changes physically, and it's done correctly the first time. Highlighting waste steps in red on the process map identifies areas for improvement, such as reducing inefficiencies in the urgent hospital blood process.

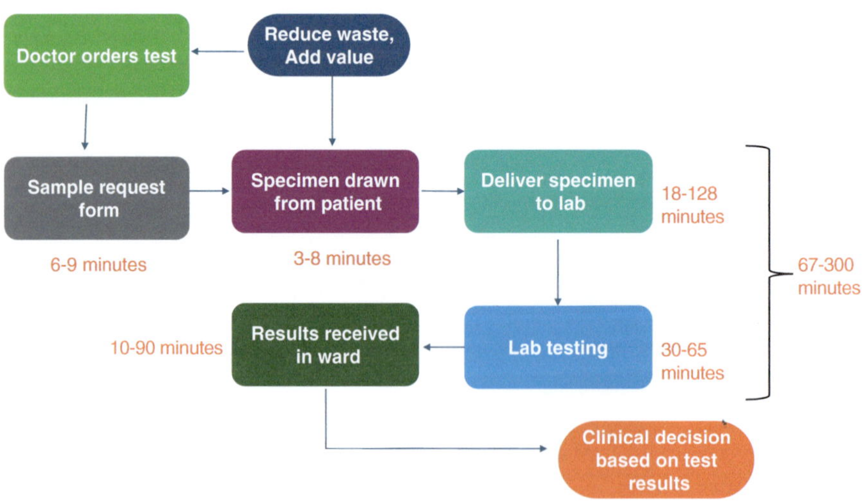

Fig. 11.5 30,000-Foot view flowchart for urgent blood testing

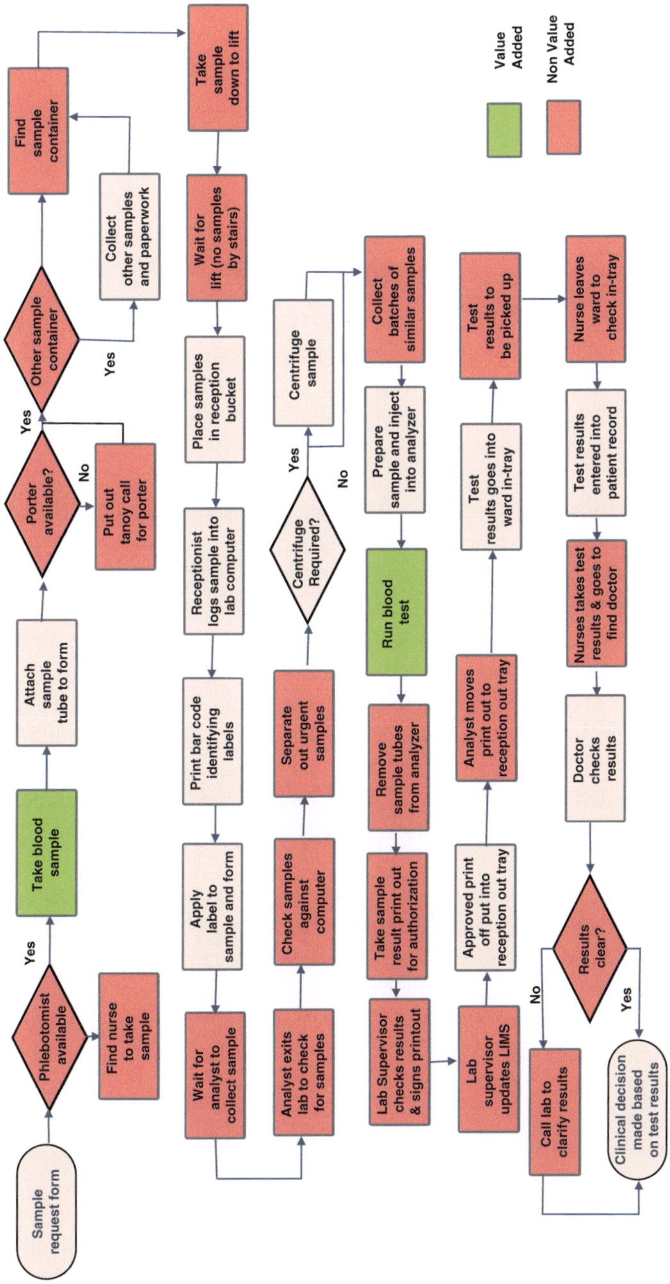

Fig. 11.6 500-Foot view flowchart for urgent blood testing

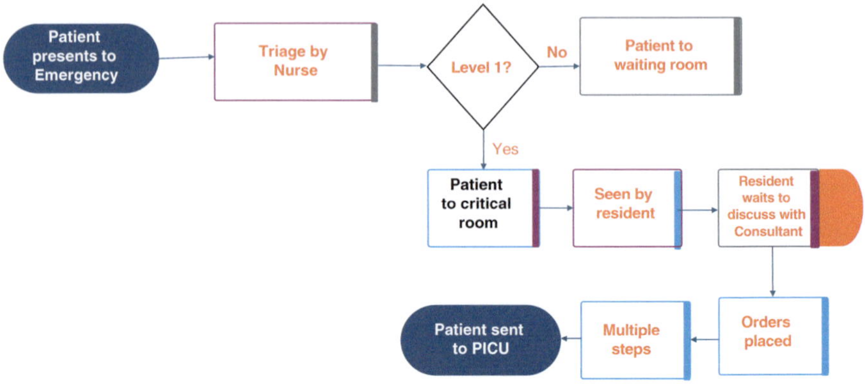

Fig. 11.7 Process mapping of the triaging in the emergency room

When prioritizing process mapping, it is recommended to consider the following factors:

1. Statutory compliances: Give priority to processes that are directly related to meeting statutory requirements and regulations.
2. Customer concerns: Identify and prioritize processes that are currently causing issues or dissatisfaction among customers. By addressing these areas, you can significantly enhance customer satisfaction and loyalty, leading to a more positive overall experience and long-term success for your organization.
3. Safety-related processes: Prioritize processes that involve safety considerations to ensure the well-being of employees, customers, or other stakeholders.
4. Critical and repetitive activities: Identify and map processes that involve crucial and repetitive activities that directly impact the final outcome. Examples include sample collection, testing processes, medication administration, and conducting procedures.
5. Potential for improvement: Consider processes that have the potential for significant improvement, such as streamlining procurement or optimizing the storage of medications and materials. These improvements can lead to cost savings, efficiency gains, and enhanced productivity.
6. Processes of low priority: Lastly, focus on mapping processes that are of lower priority, such as back-office operations that are not directly involved in patient or visitor interactions.

Conducting Process Mapping: When conducting a process mapping exercise, it is important to consider the following considerations and guidelines:

1. Identify and involve key stakeholders: Include all relevant individuals involved in the process, including those who perform the work, suppliers, customers, and supervisors. Engage them in the flowcharting process by conducting interviews, sharing the evolving flowchart, and seeking their feedback.

2. Avoid relying solely on technical experts: It is crucial to involve the individuals who actually perform the process in the flowcharting process. Technical experts may provide guidance and support, but the primary participants should be the people who have hands-on experience with the process.
3. Ensure clarity and simplicity: Strive for clarity and simplicity in the flowchart. Use clear symbols and labels to represent each step or activity, making it easily understandable for all stakeholders.
4. Keep it up-to-date: Regularly review and update the flowchart to ensure it accurately reflects the current state of the process. As processes evolve or new information becomes available, make the necessary revisions to maintain the relevance and effectiveness of the flowchart.
5. Seek feedback and validation: Share the flowchart with all relevant stakeholders and obtain their feedback and validation. This ensures that the flowchart accurately represents the process and captures their insights and perspectives.

By following these considerations and adhering to the do's and don'ts of flowcharting (Table 11.1), one can create comprehensive and valuable visual representations of the processes.

Simple and visual process maps are effective tools for driving improvement, as they make it easier for everyone to understand and identify areas for enhancement. Starting with a basic map that everyone can grasp allows for the addition of more detailed layers over time. For instance, value stream maps not only outline process steps but also depict material and information flow, along with the required personnel at each stage. Jumping to a complex map, especially when many might not comprehend value stream maps, can miss the chance to engage everyone and leverage their improvement suggestions. In most cases, keeping process maps straightforward yields the most valuable improvement insights.

Leadership teams should start by mapping the process at a high level (30,000 feet) and then have teams detail it at 500 feet. This simplified view makes reviewing and identifying improvement areas easier, leading to quick benefits. A small-time investment can yield significant results. Avoid diving into excessive detail too soon; start with a high-level view in 4-7 steps and ensure agreement before proceeding. Ensure the process map reflects the current reality by consulting those involved;

Table 11.1 Do's and don'ts of flowcharting

Do's	Don'ts
Observe and ask questions. Map the process as it actually happens. Think about the process across the entire organization. Engage people who are involved in the process. Establish the start and finish of the process prior to commencing. Map the process at a high level.	Mapping without a clear purpose Outline the process as you perceive it occurs or as you believe it should occur. Lost in the details. Failure to finalize mapping. Not verifying the facts Hidden bias or agenda Not focusing on customers' needs
Process mapping is NOT the sanitized version of Standard Operating Procedures (SOPs)	

mapping from a meeting room is insufficient for detailed views. Use caution with Post-it notes as they can fall off; take photos to document progress. Make maps accessible; electronic versions hidden on computers won't drive improvement. Finally, use the maps to identify and act on areas needing improvement to ensure the effort is worthwhile.

11.3 Check Sheet (Also Called: Defect Concentration Diagram)

Check Sheets These are tools used to collect and update data by marking the occurrence of specific events. They help by systematically recording information during processes or inspections. In a checklist, important checkpoints or events are listed in a tabular or metrics format. As these checkpoints or events occur, their status is updated or marked. This method ensures that critical steps are not missed and provides a clear record of process performance or task completion. By using check sheets and checklists, organizations can enhance accuracy and efficiency in data collection and process monitoring.

11.3.1 When to Use a Check Sheet

- When data can be repeatedly observed and collected by the same person or at the same location.
- Check Sheets and Checklists are highly useful tools for collecting data on the frequency or patterns of events, problems, defects, defect locations, defect causes, or similar issues. They help ensure that every significant occurrence is recorded accurately and consistently, facilitating thorough analysis and informed decision-making. Figures 11.8 and 11.9 illustrate sample check sheets on human resource questions and recording defects/events.
- When collecting data from a production process.

11.3.2 Examples of Check Sheet Applications

- Service Delivery Assurance: Checking the availability of reproductive and child health (RCH) services, using checklists such as the Indian Public Health System (IPHS) checklist for service availability, medicines, and diagnostic tests.
- Mistake-proofing: Utilizing checklists for surgical safety or safe birth procedures.
- Quality Assurance: Conducting housekeeping inspections using checklists.
- Medical Record Completion: Using checklists to ensure the completion of medical records.
- Check sheets can be either observational or quantitative, such as scoring sheets.

11.3.3 Check Sheet Procedure

1. Decide on the event or problem to be observed.
   - Develop operational definitions.
   - Identify the required information type.
   - Before starting data collection, it is essential to clearly define the purpose and objectives. Understanding the reason for gathering data ensures that the information collected will be relevant and useful.
   - Specify the information to be collected.
   - Determine the data collection location and timing.
   - Assign the responsible person for data collection.
   - Define the measurement method.
   - Determine the necessary amount of data to be collected.
2. Design the check sheet in a way that allows data to be recorded simply by making check marks, X's, or similar symbols. Avoid the need for data recopying during analysis.
   - Label all the spaces on the check sheet.
   - Test the check sheet during a trial period to ensure it collects the appropriate data and is easy to use.
   - Record data on the check sheet each time the targeted event or problem occurs.
3. Pre-test the check sheets and ensure the relevance of all rows and columns. Revise the checklists based on feedback received. The Figs. 11.8 and 11.9 show an example of a check sheet used to collect data human resource questions and recording defects respectively. Adding tick marks as data is collected over several weeks allows for a clear visual representation of the number of defects each day. By tracking these defects on the check sheet, one can easily analyze patterns and trends. This method helps in identifying specific days with higher defect occurrences, enabling targeted actions to improve process quality and reduce

	Monday	Tuesday	Wednesday	Thursday	Friday	**Total**
Health Insurance	卌 IIII 9	卌 7	卌 6	卌 8	卌 5	**35**
Disability Insurance	卌 2 卌	卌 5	卌 4 卌	卌 3	卌 2	**16**
Sick Time	8	卌 4	6	3	4	**25**
Paid Time off	7	4	8	9	8	**36**
Tuition Reimbursement	1	卌 7	1	4	卌 3	**16**
Payroll Error	1	5	2	3	7	**18**
Total	**28**	**32**	**27**	**30**	**29**	**146**

Fig. 11.8 Sample of check sheet on human resource questions

Project Name								
Name of the data collector			**Location**			**Date**		
Defect Types/ Event Occurrence	Sunday	Monday	Tuesday	Wednesday	Thursday	Friday	Saturday	Total
Defect 1	3	1	5	3	4	2	1	19
Defect 2	1	1	2	6	5	3	2	20
Defect 3	1	3	2	6	4	2	1	19
Defect 4	2	3	4	2	1	3	5	20
Defect 5	4	5	2	3	1	1	2	18
Defect 6	3	1	2	1	2	4	4	17
Defect 7	3	2	1	1	4	3	1	15
Defect 8	1	4	4	2	3	2	2	18
Defect 9	5	2	2	1	2	1	3	16
Defect 10	1	4	3	2	1	2	1	14
Total	**24**	**26**	**27**	**27**	**27**	**23**	**22**	**176**

Fig. 11.9 Sample check sheet to record defects/events

defect rates over time. Consistent data tracking and analysis are key to maintaining high standards of quality and efficiency.

11.3.4 Check Sheet Example

11.3.4.1 Histogram

Histograms and bar diagrams (bar charts) are both graphical representations used to visualize data, but they serve different purposes and are suitable for different types of data analysis (Fig. 11.10). A histogram is used to display the distribution of continuous data, such as measurements or variables that can take on any value within a range (e.g., age, blood pressure readings). It consists of bars that represent the frequency or count of data points falling within specific intervals, or bins, along the horizontal axis (X-axis), with the height of each bar indicating the frequency of data points in that interval. Histograms help to visualize the shape, central tendency, and variability of data distributions, making them useful for understanding patterns and identifying outliers or unusual data points within a dataset.

On the other hand, a bar diagram or bar chart is used to compare categorical data by displaying the frequency or count of different categories. Each category is represented by a separate bar, where the length or height of the bar corresponds to the frequency or count of data points in that category. Bar charts typically have distinct categories along the X-axis and numerical values (frequencies or counts) along the Y-axis. They are effective for visualizing comparisons between different groups or categories, such as comparing sales figures across different months or patient satisfaction scores across different departments.

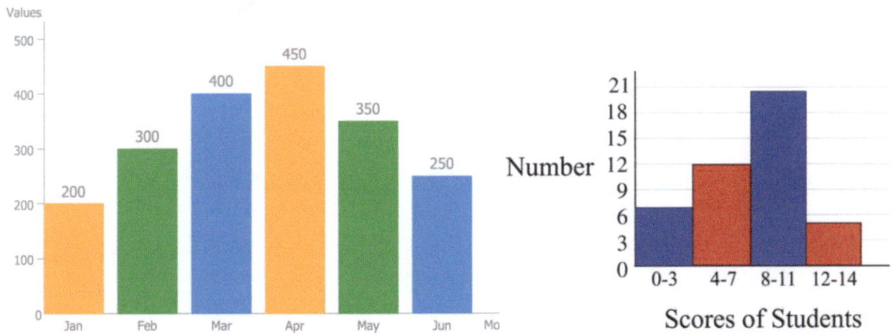

Fig. 11.10 Histogram and bar diagram

Histograms are ideal for understanding data distributions and patterns in continuous data, while bar charts are effective for comparing categorical data and visualizing comparisons between different groups or categories. As one of the fundamental quality tools, the histogram transforms raw data into visual insights, fostering a culture of continuous improvement within businesses. It plays a pivotal role in quality planning and control, swiftly communicating results and guiding preventive and corrective actions. The columns within a histogram depict distribution based on the mean, showcasing a bell-shaped curve for normal frequency distributions and varying shapes for abnormal ones, contingent upon distribution characteristics. Histograms involve measuring at least two variables against each other.

11.3.4.2 Uses of Histograms

1. When analyzing and visualizing the distribution of continuous data, especially when one needs to understand the shape of the data distribution, detect outliers, or assess variability.
2. Simplifying Complexity: They streamline complex data sets, making them accessible and understandable across all organizational levels.
3. Identifying Trends and Variations: Crucial for process improvements, histograms pinpoint outliers and variations, exemplified by a customer service department analyzing daily complaint volumes to flag systemic issues.
4. Predicting Process Performance.
5. Studying Data Density: They elucidate frequently recurring factors or data points within a given distribution.
6. Prioritizing Areas Requiring Attention.
7. Driving Data-Driven Decisions: Empowering teams to base decisions on empirical evidence, histograms aid marketing teams in evaluating campaign success and customer engagement across diverse channels.

11.3.4.3 When to Use a Histogram

- Determining the distribution of numerical data, such as patient distribution, procedure frequencies, test results, surgery rates, infection rates, waiting times, etc.
- Analyzing the shape of the data's distribution, especially to assess if it approximates a normal distribution.
- Evaluating whether a process meets customer requirements.
- Demonstrating process changes over time.
- Analyzing and comparing the outputs of different processes to reveal valuable insights into their efficiency and effectiveness.
- Use a bar chart when comparing categorical data or when you want to visualize frequencies or counts across distinct categories or groups. It is useful for presenting comparisons in a straightforward and easily interpretable manner.

11.3.4.4 Creating a Histogram

1. Collect a minimum of 50 consecutive data points from a process and categorize them.
2. Identify the independent set and the dependent set in the data.
3. The characteristic by which you grouped the data becomes the independent variable.
4. The frequency of each category becomes the dependent variable.
5. Ensure equal increments on both axes.
6. For each category, plot horizontal marks to represent the corresponding frequency.

11.3.4.5 Histogram Shapes

- Normal Distribution: This is the typical bell-shaped curve where points occur equally on both sides of the average. Statistical calculations, such as calculating the mean, median, and standard deviation, are used to confirm whether a data set follows a normal distribution. However, it's important to note that "normal" refers to what is typical for a specific process, and some processes naturally produce skewed distributions due to inherent limits.
- Skewed Distribution: Skewed distributions are asymmetrical due to a natural limit on one side. The peak of the distribution is shifted toward the limit, creating a tail that extends away from it. Skewed distributions can be either right-skewed or left-skewed, depending on the direction of the tail.

Figure 11.11 depicts normal and skewed distribution in histogram.

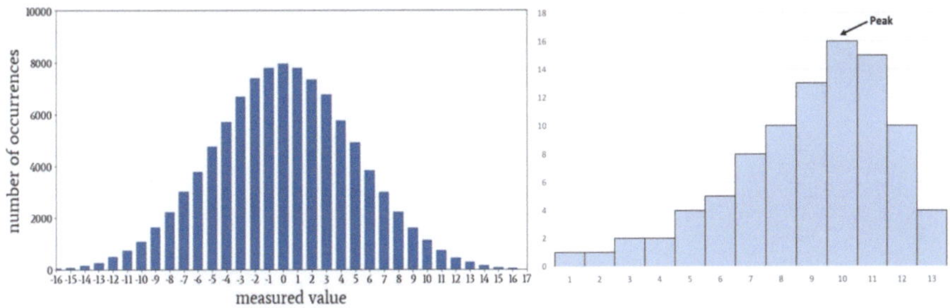

Fig. 11.11 Histogram shows normal and skewed distribution

11.4 Control Chart (Also Called Shewhart Chart, Run Chart)

A control chart, named after Walter A. Shewhart, is a statistical tool used to assess the consistency and predictability of a process in producing products or services with consistent measurable properties. By plotting data points over time, control charts visually depict the changes in a process and highlight the presence of common cause and special cause variations.

The primary purpose of a control chart is to monitor the performance of a process over time by analyzing the variation in data points. It helps in determining the stability and predictability of the process by identifying any outliers or deviations from the expected pattern.

Control charts distinguish between two types of variations: common cause and special cause. Common cause variations are the inherent, quantifiable variations that are part of the normal functioning of a system. They represent the typical patterns observed historically. On the other hand, special cause variations are unusual, non-quantifiable variations that are not part of the usual system behavior. They are considered signals of abnormalities or specific causes affecting the process.

When creating a control chart, data points are plotted in the order they occur. The chart includes a central line that represents the average, an upper line that indicates the upper control limit (UCL), and a lower line that denotes the lower control limit (LCL). These lines are based on historical data and act as reference boundaries. By comparing current data to these limits, one can determine if the process variation is within the expected range (in control) or if it shows signs of unpredictability or abnormal behavior (out of control). If the process is out of control, corrective measures can be taken to identify and address the root causes of quality issues.

Control charts provide a visual representation of process performance over time, allowing for the identification of common cause and special cause variations. They help in assessing the stability and predictability of a process, enabling quality control efforts to focus on improving consistency and addressing any abnormal variations. A table provides a static representation of numerical data, making it useful for quick comparisons and summaries, but it does not reveal patterns or trends over time. In contrast, a control chart is a dynamic tool used in statistical process control

Hospital	Baseline Median	Jan	Feb	Mar	Apr	May	Jun	Jul	Aug	Sep	Oct	Nov	Dec
Hospital 1	20	21	22	18	12	15	14	12	22	17	18	14	16
Hospital 2	12	15	17	14	18	17	15	13	11	9	8	15	12
Hospital 3	7	7	8	8	9	10	11	12	13	14	15	16	18
Hospital 4	30	32	35	30	32	29	19	22	26	28	29	33	35

Fig. 11.12 Stock out percentage of drugs over a period

(SPC) to monitor variability, trends, and stability in a process. For example, tracking stock-outs of medicines in a pharmacy in tabular format for example (Fig. 11.12) may list the number of stock-outs each month, but it does not indicate whether shortages are becoming more frequent or if variations are within acceptable limits. A control chart, however, plots stock-out rates over time, distinguishing between normal fluctuations (common cause variations) and unexpected disruptions (special cause variations) in supply. This visualization helps pharmacy managers identify patterns, anticipate potential shortages, and implement corrective actions before stock-outs become critical.

11.4.1 Control Chart Example

11.4.1.1 When To Use a Control Chart

1. Control Ongoing Processes: Control charts are useful for monitoring and controlling ongoing processes in real-time. They help identify and correct problems as they occur, such as internal quality control in a laboratory (e.g., Levey-Jennings Chart) or temperature monitoring in cold storage.
2. To Predict Expected Outcomes: Control charts can be used to predict the expected range of outcomes from a process. By analyzing historical data and patterns, control charts provide insights into the future performance of the process.
3. To Determine Process Stability: Control charts help assess whether a process is stable and in statistical control. They provide visual indicators of process stability over time, allowing for the identification of any abnormal variations or trends.
4. To Analyze Process Variation: Control charts aid in analyzing the patterns of process variation, distinguishing between variations caused by special causes (non-routine events) or common causes (inherent to the process). This analysis helps in understanding the sources of variation and making informed decisions for process improvement.
5. To Guide Quality Improvement Projects: Control charts assist in determining the focus of a quality improvement project. They help decide whether the project

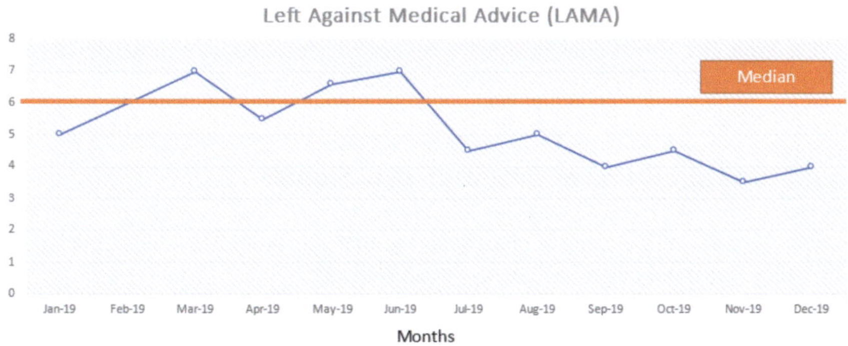

Fig. 11.13 Control chart of month-wise patients leaving against medical advice

should target specific problems within the process or aim for fundamental changes in the process itself.

Figure 11.13 illustrates a control chart of patients left against medical advice (LAMA) over a period of time.

11.4.1.2 Benefits of Control Charts

1. Understanding Process Variation: Control charts provide insights into the variation present in a process, helping to identify the extent and nature of variability.
2. Pattern Analysis: Control charts enable the analysis of patterns in process data, facilitating the identification of trends, cycles, or shifts in performance.
3. Monitoring Process Performance: By monitoring data points over time, control charts allow for the ongoing assessment of process performance, ensuring that the process remains within control limits.

11.4.1.3 Basic Procedure for Using Control Charts

1. Identify Critical Operations and Characteristics: Determine the key operations in the process that require inspection and identify the critical characteristics of the product or service.
2. Determine Variable or Attribute: Classify the critical characteristic as either a variable (measurable quantity) or an attribute (categorical characteristic).
3. Choose Appropriate Control Chart: Select the appropriate control chart based on the type of data (variable or attribute) and the desired analysis.
4. Determine Data Collection Period: Determine the time period for collecting and plotting data on the control chart.
5. Collect Data and Analyze: Gather the relevant data, create the control chart, and analyze the data points to evaluate process performance.
6. Establish Control Limits: Set the control limits on the control chart to indicate the acceptable range of variation.
7. These steps are essential for monitoring and maintaining the quality of processes, ensuring that any deviations are promptly identified and addressed.

8. Monitor and Improve: Use the control chart to monitor the process over time, detecting any out-of-control signals. Investigate the causes of these signals and take corrective and preventive actions (CAPA) as necessary.

11.4.2 Interpretation of Out-of-Control Signals

Various signals on the control chart indicate that the process is out of control or exhibiting abnormal behavior. These signals include:

1. Single Point Outside Control Limits: A data point falls outside the upper or lower control limits.
2. Successive Points on One Side: Two out of three successive points are on the same side of the centerline and beyond 2σ (standard deviations).
3. This refers to a situation where four out of five successive data points on a control chart are on the same side of the centerline and beyond one standard deviation (1σ). This pattern indicates a potential shift in the process and warrants investigation to determine the underlying causes and take corrective actions.
4. Runs of Points: A run of eight or more consecutive points on the same side of the centerline.
5. Consistent or Persistent Patterns: Any consistent or persistent patterns in the data that deviate from expected behavior suggest unusual circumstances in the process.

By interpreting these out-of-control signals, practitioners can investigate the causes, implement corrective actions, and ensure the process remains in control.

11.4.2.1 Interpretation of Out-of-Control Signals

1. Single Point Outside Control Limits: An out-of-control signal occurs when a single data point falls outside the established upper control limits (UCL).
2. When two out of three successive data points are on the same side of the centerline and farther than two standard deviations (2σ) from it, this signifies an out-of-control signal. This pattern suggests that the process may be experiencing unusual variation, and it is important to investigate and address the potential causes to bring the process back into control.
3. Four Out of Five Successive Points: When four out of five successive data points are on the same side of the centerline and farther than 1σ from it, it suggests an out-of-control signal.
4. Runs of Points: An out-of-control signal can be identified by a run of consecutive data points. For example, if there are eight or more consecutive points on the same side of the centerline, or if there are specific patterns like 10 out of 11, 12 out of 14, or 16 out of 20 points exhibiting a specific trend.

5. Obvious Consistent or Persistent Patterns: Any consistent or persistent patterns in the data that deviate from expected behavior suggest something unusual about the data and the process. These patterns may indicate an out-of-control signal.

11.4.2.2 Additional Considerations for Control Charts

6. Continue to Plot Data and Check for Signals: It is important to keep plotting new data points as they are generated and continuously monitor for new out-of-control signals.
7. When initiating a new control chart, it is possible that the process may initially be out of control. This could be due to natural variability or underlying issues that need to be addressed. Establishing control limits helps in setting the acceptable range of variation, which allows for monitoring and taking corrective actions to achieve and maintain process stability. By continuously analyzing data and making necessary adjustments, the process can be brought into control, ensuring consistent quality and performance. In such cases, the control limits are calculated using the first 20 points as conditional limits. However, once at least 20 sequential points are obtained from a period when the process is operating in control, it is essential to recalculate the control limits.
8. Six or More Consecutive Points: If there are six or more consecutive points above or below the median line, it suggests an out-of-control signal.
9. Five or More Consecutively Increasing or Decreasing Points: An out-of-control signal is indicated when there are five or more consecutively increasing or decreasing points on the control chart.

11.4.3 Control Chart: Out-of-Control Signals

Figures 11.14 and 11.15 depict the Control Chart Rules on out-of-control signals. By recognizing these out-of-control signals and promptly investigating their causes, practitioners can take appropriate corrective actions to bring the process back into control and ensure its stability and quality.

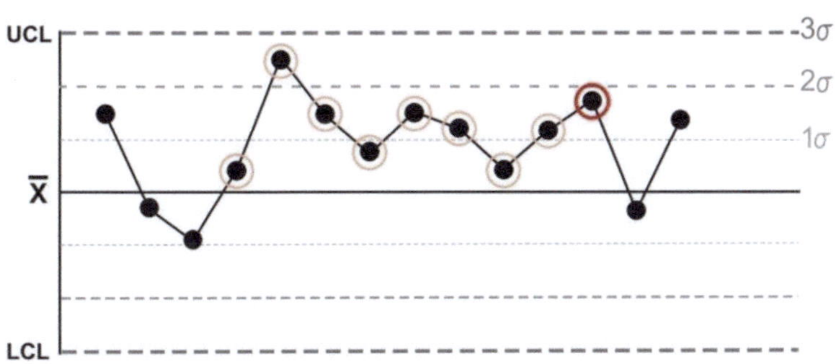

Fig. 11.14 Control Chart rule 2 with nine or more points in a row on the same side of the mean

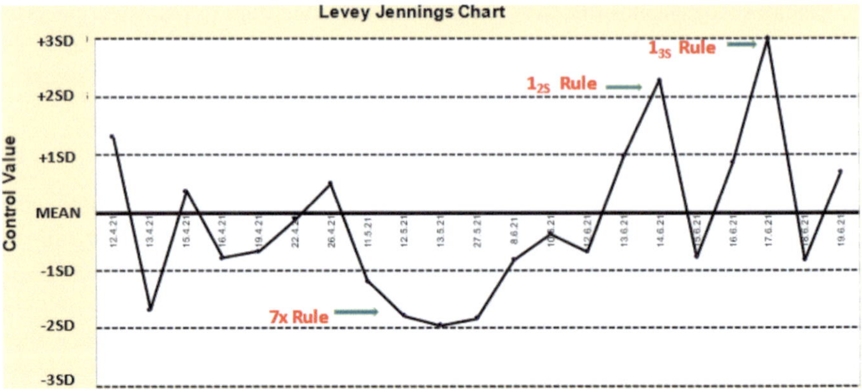

Fig. 11.15 Out-of-range signals on Levey Jennings Chart

11.5 Scatter Diagram (Also Called Scatter Plot, X-Y Graph)

A scatter diagram is a graphical representation that shows the relationship between two variables. It is a useful tool for identifying and understanding cause-and-effect relationships. By visually and statistically analyzing the data, scatter diagrams can assess the strength of the relationship between the variables.

For instance, after identifying multiple causes and effects using a fishbone diagram, a scatter diagram can be employed to determine the relationship between a specific cause and effect. The diagram plots the dependent variable on the vertical axis (Y-axis) and the independent variable on the horizontal axis (X-axis), with each data point represented as a dot at their intersection. Connecting these dots can reveal any existing relationship or equation, typically in the format $Y = F(X) + C$, where C is an arbitrary constant.

When variables are correlated, the data points will align along a line or curve on a scatter plot. The closer the points are to this line or curve, the stronger the correlation between the variables. This alignment indicates a predictable relationship, where changes in one variable are associated with changes in the other. Whether positive or negative, a strong correlation is reflected by tightly clustered data points around the trend line. Interpreting the scatter plot involves examining the data in the following ways:

1. Strength: The slope of the equation determines the strength of the relationship, categorized as strong, moderate, or weak.
2. Trend: The trend can be positive or negative. A positive correlation indicates that as values for both variables increase or decrease, while a negative correlation shows that as values for one variable increase, the other variable decreases. When there is no correlation between the variables, the data points tend to spread out in a pattern that resembles a circular shape on a scatter plot. This random

distribution indicates that there is no predictable relationship between the variables, and changes in one variable do not correspond to changes in the other.
3. Shape: The shape of the scatter plot can be linear, non-linear, or show no apparent pattern.

11.5.1 Uses of Scatter Diagrams

- Validating cause-and-effect relationships.
- Demonstrating the impact of one variable when another is changed.
- Understanding the causes of poor performance.
- Assessing the influence of the independent variable to understand causal relationships and making data-driven decisions. This involves examining how changes in the independent variable impact the dependent variable. By analyzing this relationship, we can determine the extent to which the independent variable affects the dependent variable.

11.5.2 When to Use a Scatter Diagram

Scatter diagrams are useful in the following scenarios:

1. Paired Numerical Data: When you have data points consisting of pairs of numerical values, a scatter diagram can effectively display the relationship between these variables.
2. Dependent Variable with Multiple Values: If your dependent variable can have multiple values corresponding to each value of the independent variable, a scatter diagram can help visualize this relationship.
3. Identifying Root Causes: When you are attempting to identify potential root causes of problems, a scatter diagram can be employed to examine the relationship between variables and identify potential connections.
4. Objective Evaluation of Cause and Effect: After brainstorming causes and effects using a fishbone diagram, a scatter diagram can be used to objectively determine whether a specific cause and effect are related.
5. Exploring Cause-Effect Relationships: A scatter diagram can be useful when determining if two effects that appear to be related actually occur together with the same cause. However, just because two things seem to happen together doesn't mean one causes the other. Correlation doesn't imply causation due to hidden variables, reverse causality, sample bias, measurement errors, etc. For example, research might find a correlation between the use of a particular medication and symptom improvement in patients or vice versa adverse effect. However, this correlation does not prove that the medication directly causes the improvement or worsening respectively. Other factors, such as the severity of the condition, adherence to treatment, and co-existing health issues, genetics may also influence health outcomes.

6. Autocorrelation Testing: Before constructing a control chart, it is important to test for autocorrelation. Scatter diagrams can be utilized in this process to evaluate the relationship between variables and assess the presence of autocorrelation.

By critically evaluating correlations and exploring alternative explanations, organizations can avoid costly mistakes and optimize their operations. It's important to remember that correlation serves as a starting point in the analysis process, but it's not the definitive end of the story. It's essential to recognize its limitations and explore alternative explanations before drawing conclusions. To effectively address issues in the factory setting, consider the following steps:

- Delve deeper into the data: Conduct a thorough analysis of data collected across different seasons and production lines to uncover any patterns or trends that may provide additional insights.
- Identify common factors: Investigate whether specific ingredients or production processes coincide with instances of breakdowns or issues, as this could point to underlying causes that need to be addressed.
- Consider external influences: Consider external factors such as weather patterns that may impact production and adjust schedules accordingly to mitigate potential disruptions.

Scatter diagrams are employed when analyzing paired numerical data, exploring relationships between variables, identifying root causes, evaluating cause and effect objectively, investigating relationships between effects and causes, and testing for autocorrelation in data.

11.5.3 Scatter Diagram Procedure

1. Gather pairs of data where a potential relationship is suspected.
2. Create a graph with the independent variable plotted on the horizontal axis and the dependent variable on the vertical axis. Represent each data pair with a dot or symbol at the intersection of the x-axis and y-axis values. If multiple dots coincide, place them adjacent to each other so that all points are visible. Ensure that the data pairs remain properly matched.
3. Examine the distribution of points to find out if a relationship is apparent. Assess the distribution of data points to identify if a relationship is evident. If the points form a distinct line or curve, suggesting a clear correlation, you may proceed with regression or correlation analysis. If no clear correlation is observed, proceed to steps 4 through 7 for further analysis.
4. Divide the points on the graph into four quadrants. When dividing the points on the graph into four quadrants:

- Horizontal Division: Count X/2X/2 points from top to bottom and draw a horizontal line through that point.
- Vertical Division: Count X/2X/2 points from left to right and draw a vertical line through that point.
- For Odd Number of Points: If there are an odd number of points, draw the lines through the middle point.

5. Count the number of points in each quadrant, excluding those on the lines.
   Sum of Points in Diagonally Opposite Quadrants: Calculate the sum of data points in each pair of diagonally opposite quadrants.
   Determine the Smaller Sum: Identify which of these sums is smaller.
   Total Number of Points: Calculate the total number of points across all quadrants.
   - A = points in the upper left + points in the lower right
   - B = points in the upper right + points in the lower left
   - Q = the smaller value between A and B
   - N = A + B

To determine the limit for NN in a trend test, consult a trend test table. If Q is less than the limit, it indicates a relationship between the two variables. If the value of Q is greater than or equal to the limit found in the trend test table, it indicates that the observed pattern in the data could have occurred by random chance suggesting that there is no significant trend or a relationship between the variables, and any apparent pattern may not be meaningful.

By following these steps, one can construct a scatter diagram and conduct a trend test to determine the presence of a relationship between the variables. Figure 11.16 depicts the Scatter Plot (Y, dependent variable; X, independent variable), their relationship and its interpretation

11.5.4 Disadvantages

1. Skewed Analysis: If the data covers a very narrow range, the results of the analysis can be distorted or biased, making it difficult to draw accurate conclusions.
2. Stratification Impact: Patterns in the scatter diagram can appear or disappear depending on how the data are stratified or grouped, which can affect the interpretation of the relationship between variables.
3. Correlation vs. Causation: While a scatter diagram shows a correlation between two variables, it does not provide evidence of causation. A relationship observed in the diagram does not necessarily imply that one variable causes the changes in the other variable.

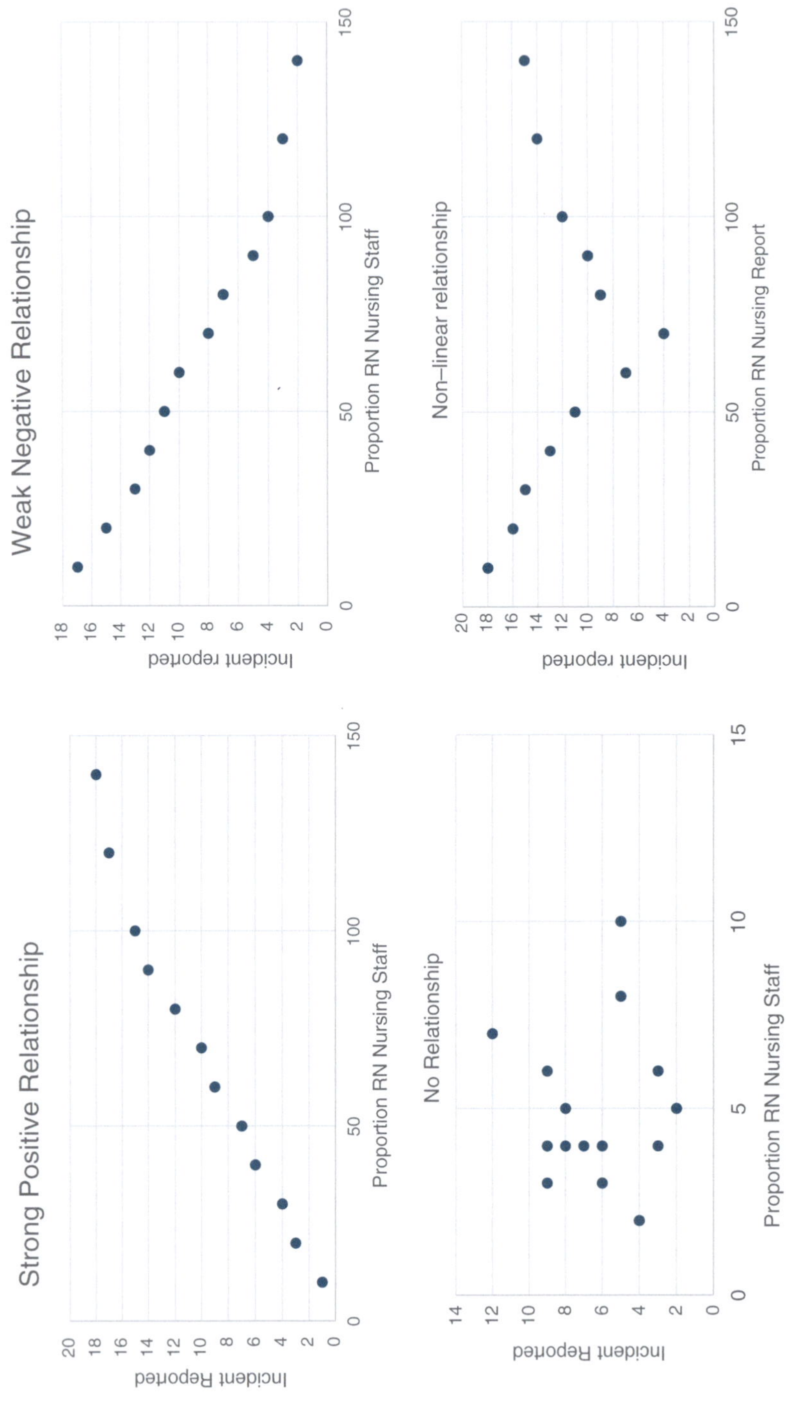

Fig. 11.16 Scatter plot (Y, dependent variable; X, independent variable), their relationship and its interpretation

11.5.5 Examples of Situations Where a Scatter Diagram Can Be Useful

1. Medical Errors and Hospital Stay: To investigate the suspected relationship between the number of medical errors reported (Variable A) and the length of hospital stay (Variable B), a scatter diagram can be plotted to visualize the data points and examine any correlation between the two variables.
2. Hand Hygiene and Healthcare-Associated Infections (HAIs): In an effort to improve healthcare workers' hand hygiene and reduce HAIs, a scatter diagram can be employed to demonstrate the relationship between hand hygiene compliance (Variable A) and the occurrence of HAIs (Variable B), highlighting the need for improved hand hygiene practices (Fig. 11.17).
3. Consultation Time and Patient Satisfaction: To assess the relationship between the length of time of a consultation with a doctor in an outpatient department (Variable A) and the patients' satisfaction level (Variable B), a scatter diagram can be utilized. The proximity of the data points on the diagram can indicate the strength of the one-to-one relationship between the variables.

In these examples, a scatter diagram serves as a valuable tool for visualizing the relationship between variables and exploring correlations. However, it is important to keep in mind the limitations of the analysis, such as the potential for skewed results, the impact of data stratification, and the distinction between correlation and causation.

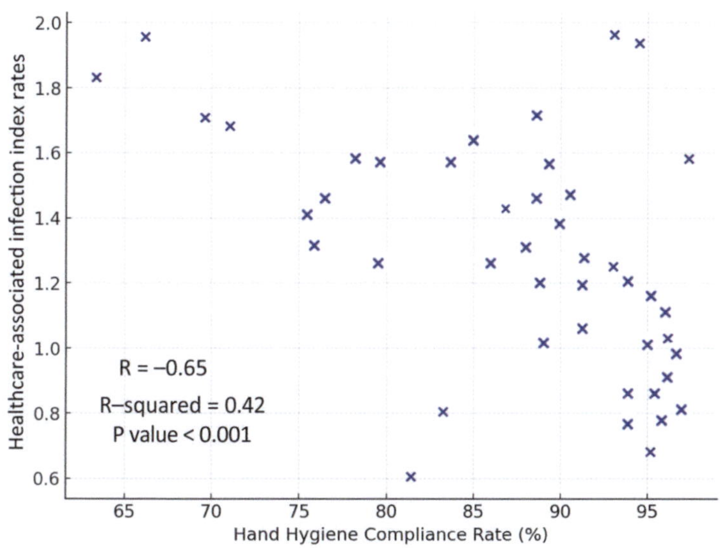

Fig. 11.17 Scatterplot of monthly healthcare-associated infection index rates and hand hygiene compliance rates

11.5.6 Scatter Diagram Considerations

- Cautious Interpretation: It's important not to assume causation based solely on a scatter diagram showing a relationship between two variables. Keep in mind that both variables may be influenced by a third variable, and further analysis is required to establish causal relationships.
- Linearity and Relationship Strength: When plotting the data on a scatter diagram, the closer the diagram resembles a straight line, the stronger the relationship between the variables. The linearity of the pattern can provide insights into the strength of the relationship.
- Statistical Analysis: If the scatter diagram does not exhibit a clear line or pattern, statistical measures such as N (number of points) and Q (sum of points in quadrants) can be used to determine the likelihood of a relationship. If the statistical analysis suggests no relationship, it indicates that the observed pattern could have occurred due to random chance.
- Consider Stratification: If the scatter diagram shows no apparent relationship between the variables, it is essential to consider whether the data might be stratified or grouped in a way that obscures the underlying relationship. Exploring data based on different stratification criteria may reveal hidden patterns.
- Range of Independent Variable: If the scatter diagram fails to reveal a relationship, it is worth considering whether the independent variable (X-axis) has been varied widely enough. Sometimes, a relationship may not be apparent if the data do not cover a broad range of values for the independent variable.

By keeping these considerations in mind when analyzing scatter diagrams, one can avoid making unwarranted assumptions, assess the strength of the relationship, employ statistical analysis for validation, account for potential stratification effects, and ensure an adequate range of values for the independent variable.

Different types of charts shown in Fig. 11.18 are used for measuring quality and patient safety, which help in improving healthcare outcomes and reducing risks.

What it is	When to use it	What to Remember	What it looks Like
Scatter Plot	Scatter plots' primary uses are to observe and show relationships between two numeric variables.	When using more than 10 data points, place on horizontal axis	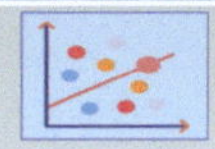
Column Chart	Column charts are frequently used when a single value for each category needs to be displayed.	Use when you want to show minimum & maximum values	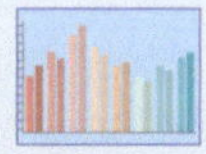
Bar Chart	A bar chart is used when you want to show a distribution of data points or perform a comparison of metric values across different subgroups of your data.	Place your values on the x-axis, categories on the y-axis; use when the number of the categories is between 5 & 8	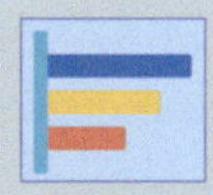
Line Chart	Line charts are useful for displaying data that changes over time or shows relationships between two numerical values.	great when the number of categories to plot is 8+ and works well in the combination with bar charts	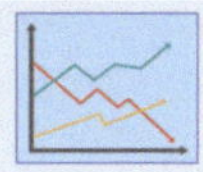
Combination Chart	The combination chart, also known as a combo chart, is a type of visualization used to compare data in a few varying categories over a period. It uses several lines/bars, each representing an individual category.	This follows the same rules as a pie graph, but with more categories	
What it is	**When to use it**	**What to Remember**	**What it looks Like**
Pie Chart	a pie chart can only be used if the sum of the individual parts add up to a meaningful whole and is built for visualizing how each part contributes to that whole.	The data set must represent parts of a while, & data should be of comparable sizes, so nothing is dwarfed	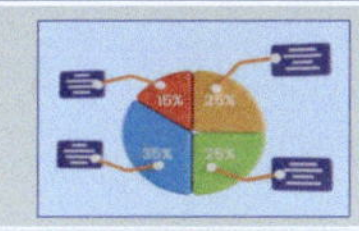
Stacked column	Stacked column charts are useful for representing data that has two categorical variables, where one variable is used to divide the columns into segments and the other is plotted as columns.	Use this to show the trend of composition	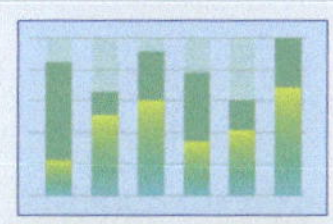
Histogram	A rule of thumb is to use a histogram when the data set consists of 100 values or more.	Use this when you want to show the distribution of data across groups	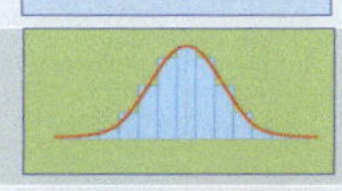
Venn Diagram	A Venn diagram provides a clear way to express the relationships between entities, and particularly where two or more things have overlapping attributes.	Consider this for attribution modeling. No overlapping = No attribution problems	

Fig. 11.18 Different types of charts and their uses

11.6 Pareto Chart (80:20 Rule)

A Pareto chart, named after Vilfredo Pareto, is a bar graph that represents data from the largest to the smallest frequencies. It is based on the concept of the 80-20 rule, which states that 80% of problems or failures are caused by a vital few (20% of major factors), while the remaining 20% of problems or failures are caused by trivial many (80% of minor factors). The Pareto chart helps prioritize improvement efforts by highlighting the most important factors contributing to a problem or failure. It can display any number of issues or variables along with their frequency of occurrences. The chart consists of bar graphs representing individual factors in descending order of their impact and a line graph showing the cumulative total. This visual representation enables easy identification of significant situations.

11.6.1 Advantages of Pareto Charts

- Distinguishes between vital few and trivial many factors.
- Identifies the relative importance of causes.
- Helps teams focus on causes with the greatest impact when solving problems.

11.6.2 When to Use a Pareto Chart

- When analyzing data on the frequency of problems or causes in a process.
- When there are numerous problems or causes and the focus is on the most significant ones.
- When analyzing broad causes by examining their specific components.
- When communicating data to others.

Emphasize Clarity: Make sure your categories are distinct and do not overlap.

Prioritize Impact: Focus on the categories that have the most significant impact on the issue, rather than solely considering their frequency.

Avoid Strict Adherence to the 80/20 Rule: While the specifics may differ, the essence of the principle remains important.

11.6.3 Disadvantages of Pareto Charts

- Cannot perform root cause analysis (RCA) to determine the underlying or major causes of a defect.
- Does not represent the severity of defects or problems, as it only shows qualitative data.
- Focuses on past data where damage has already occurred.
- Applicable only to certain cases and not universally.
- Other potential disadvantages include errors in scoring, incorrect application, etc.

11.6.4 Procedure for creating a Pareto chart (Fig. 11.19)

1. Determine the categories to group the items.
2. Determine the appropriate measurement, such as frequency, quantity, cost, or time.
3. Choose the time period the Pareto chart will cover (e.g., work cycle, day, week, month, year).
4. Collect or assemble the data, recording the category each time.
5. Calculate subtotals for each category based on the chosen measurement.
6. Create a table listing the causes and their frequencies as percentages.
7. Arrange the rows in descending order of the causes of importance (starting with the most important cause).
8. Add a cumulative percentage column to the table to show the running total of the percentages of the causes, helping to identify the most impactful causes visually.
9. Then, plot the causes on the x-axis and the cumulative percentage on the y-axis
10. Connect the plotted points to form a curve to a visual representation of the data, making it easier to analyze.
11. Additionally, plot a bar graph on the same chart with causes on the x-axis and percentage frequency on the y-axis.

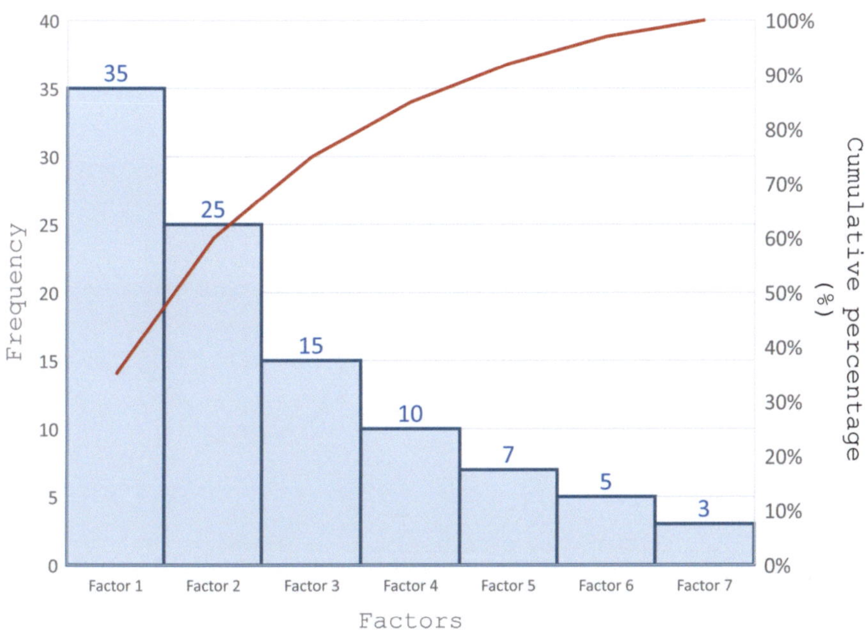

Fig. 11.19 Pareto chart example

11.7 Root Cause Analysis (RCA)

Root Cause Analysis (RCA) is a systematic process used in healthcare to identify the underlying causes of adverse events or quality issues. It begins with defining the problem or event clearly and gathering relevant data and information by addressing who, what, where, how, why, and when it occurred (Fig. 11.20). Who was involved, including healthcare providers, staff, or patients affected? What happened, detailing the specific event or failure? Where did the issue occur, pinpointing the location or process breakdown? How did the problem arise, analyzing contributing factors such as communication errors, system failures, or human errors? Why did it happen, identifying root causes like lack of training, poor protocols, or workflow inefficiencies? When did the event take place, determining patterns, timing, or trends that may have influenced the outcome? By systematically answering these questions, RCA helps in developing corrective actions, improving processes, and preventing future errors, ensuring better safety and quality in healthcare. The next step involves identifying potential causal factors through techniques such as brainstorming, fishbone diagrams (Ishikawa), or the "5 Whys" method. Once potential causes are identified, they are systematically analyzed to determine which factors contributed most significantly to the problem. The final stage of RCA focuses on developing and implementing corrective actions to address the root causes identified, aiming to prevent recurrence of similar issues in the future. Throughout the process, collaboration among healthcare professionals, thorough documentation, and continuous monitoring of implemented actions are essential to ensure effective resolution and sustained improvements in patient safety and quality of care.

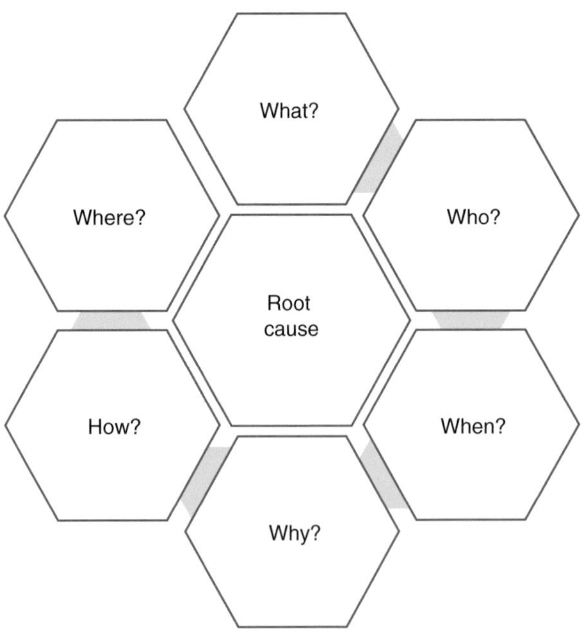

Fig. 11.20 Root Cause Analysis process in healthcare

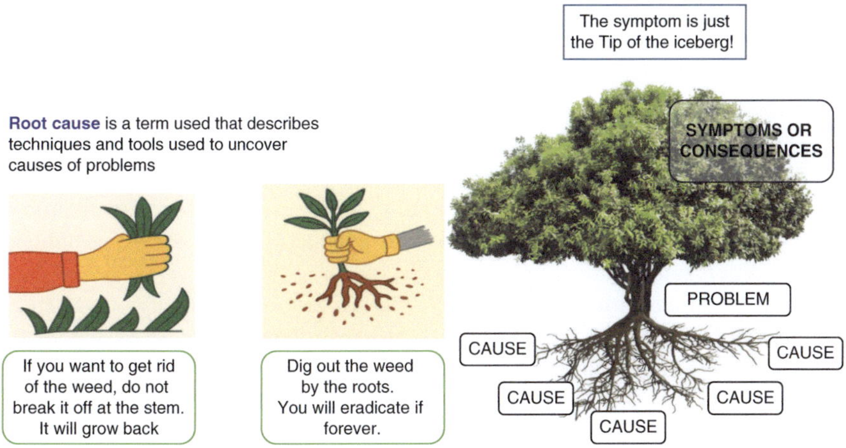

Fig. 11.21 Root Cause Analysis (RCA) beyond surface-level symptoms

Root Cause Analysis is a continuous improvement process that helps healthcare organizations enhance patient safety and quality of care by systematically addressing and mitigating the root causes of adverse events beyond surface-level symptoms (Figs. 11.20 and 11.21).

Root Cause Analysis is conducted in various scenarios within healthcare to address issues, prevent recurrence, and enhance patient safety and quality of care. Some specific situations when RCA should be conducted include:

1. **Adverse Events**: When a significant adverse event occurs, such as a patient death, severe injury, or a critical medical error, an RCA is conducted to understand the underlying causes and prevent future occurrences.
2. **Near Misses**: Even if an adverse event was narrowly avoided, an RCA can help identify potential vulnerabilities in the system that could lead to future incidents if not addressed.
3. **Unexpected Outcomes**: When patients experience unexpected outcomes or complications despite following standard procedures and protocols, an RCA can help identify any gaps or weaknesses in the processes.
4. **Repeated Issues**: If similar types of incidents or errors are recurring, an RCA can help uncover persistent systemic issues and implement effective corrective actions.
5. **Sentinel Events**: Joint Commission-accredited healthcare organizations are required to conduct an RCA for any sentinel event, which is a patient safety event that results in death, permanent harm, or severe temporary harm.
6. **Regulatory Requirements**: Sometimes, RCAs are mandated by regulatory bodies, accreditation organizations, or external quality oversight entities following specific types of incidents.

7. **Complaints and Claims**: Patient or family complaints and legal claims related to healthcare delivery may also trigger an RCA to understand the root causes and mitigate future risks.
8. **Process Failures**: When there is a significant breakdown in clinical or operational processes that could compromise patient safety or quality of care, an RCA can help identify and rectify the issues.
9. **Accreditation and Certification**: During the preparation for accreditation or certification processes, organizations may conduct RCAs to ensure compliance with safety and quality standards.
10. **Quality Improvement Initiatives**: As part of continuous quality improvement efforts, RCAs can be conducted proactively to identify and address potential risks before they result in adverse events.

Conducting an RCA in these situations helps healthcare organizations identify underlying issues, implement corrective actions, and foster a culture of continuous improvement and patient safety. Figure 11.22 depicts the steps involved in the RCA.

11.7.1 Aggregate Root Cause Analysis (Aggregate RCA)

Aggregate Root Cause Analysis (Aggregate RCA) is a process used to analyze multiple incidents or near-misses that are similar in nature. Instead of investigating each incident individually, the aggregate RCA looks for common patterns or systemic issues across a group of events. This approach helps in identifying broader organizational or system-wide problems and implementing comprehensive solutions.

11.7.1.1 Advantages of Aggregate RCA

- **Efficiency**: Analyzing multiple incidents together saves time and resources compared to conducting separate RCAs for each incident.
- **Comprehensive Solutions**: Identifying common root causes leads to broader, system-wide improvements rather than isolated fixes.
- **Preventative Approach**: By addressing systemic issues, aggregate RCA helps in preventing future incidents across the organization.

11.7.1.2 Examples of Use

- **Medication Errors**: Analyzing multiple medication errors to identify common factors such as labeling issues, storage problems, or communication breakdowns.
- **Patient Falls**: Examining several fall incidents to determine common environmental or procedural factors contributing to falls.
- **Infection Control**: Investigating multiple cases of hospital-acquired infections to find common lapses in infection control practices.

Aggregated review is necessary for potentially serious events such as patient falls, missing patients, para-suicidal events, and adverse drug events (ADEs). Severity assessment utilizes a matrix incorporating key factors like the extent of

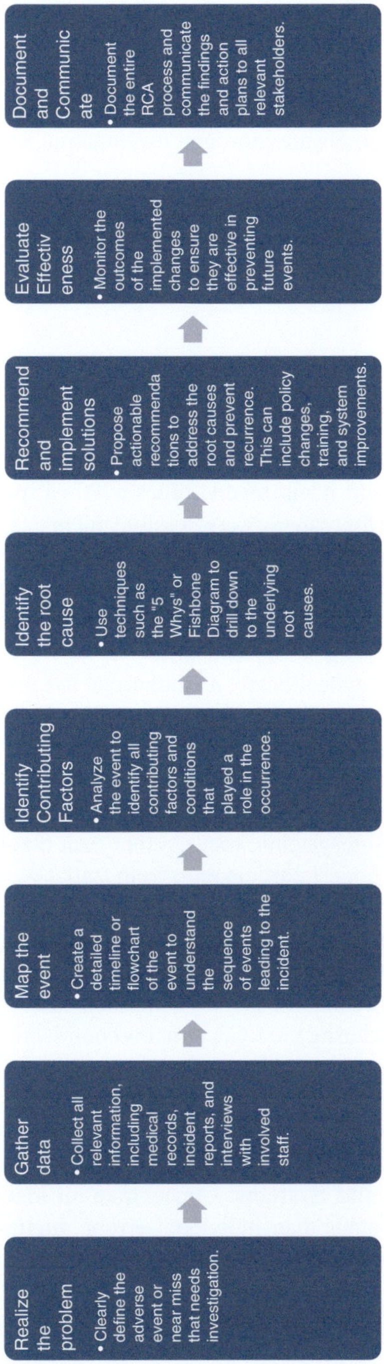

Fig. 11.22 Steps involved in the Root Cause Analysis (RCA) process

injury, length of stay, and the level of care needed for remediation. Events scoring 3 on this scale mandate an RCA to uncover underlying causes and prevent recurrence. For potential AEs, severity is gauged based on the most plausible worst-case scenario within the systems involved. In the case of potential medication-related incidents scoring 3, they can undergo individual analysis or be reviewed quarterly as part of aggregated RCAs to ensure thorough examination and effective corrective actions.

Probability categories are applied to both actual AEs and close calls to assess the likelihood of occurrence and help healthcare organizations prioritize preventive measures and allocate resources effectively, based on the probability and potential impact of various incidents:

- **Frequent**: Events in this category are likely to occur immediately or within a short period, typically several times a year, indicating a high probability of occurrence.
- **Occasional**: These events are expected to happen, though not as frequently as frequent events. They may occur several times within a span of 1-2 years.
- **Uncommon**: Events in this category are possible but not expected to occur often. They may happen intermittently, sometimes occurring every 2-5 years.
- **Remote**: Events in this category are unlikely to occur, with occurrences spread out over longer periods, typically ranging from 5 to 30 years apart.

When conducting RCA in healthcare, it's important to keep several key points in mind to ensure the process is thorough, effective, and leads to meaningful improvements in patient safety. Some special points to consider:

1. **Non-Punitive Approach**: Ensure the RCA process is non-punitive. The focus should be on identifying systemic issues rather than blaming individuals. This encourages open and honest participation from staff.
2. **Multidisciplinary Team**: Assemble a diverse team with members from various disciplines and departments. This ensures a comprehensive perspective and a thorough analysis of the event.
3. **Confidentiality**: Ensure that the privacy of participants is protected and that all information collected during the RCA remains confidential. This helps in building trust and encourages candid input.
4. **Thorough Data Collection**: Collect comprehensive data, including medical records, incident reports, interviews, and any other relevant documentation. The accuracy and completeness of the data are crucial for a successful RCA.
5. **Focus on Processes and Systems**: Concentrate on identifying flaws in processes and systems rather than individual errors. **Include what really happened, NOT what was supposed to happen. At each step of the event flow ask "Why?"**
6. Most adverse events result from multiple factors within the healthcare system.

7. **Use Structured Tools**: Utilize structured tools and methodologies like Fishbone Diagrams, Flowcharts, and the "5 Whys" technique to systematically identify root causes.
8. **Prioritize Root Causes**: Recognize that not all identified root causes will have the same level of impact. Prioritize root causes based on their significance and the likelihood of recurrence.
9. **Actionable Recommendations**: Ensure that the recommendations and action plans are specific, actionable, and feasible. Recommendations that are vague or too broad are less likely to be implemented effectively. Clear, specific, and actionable suggestions are essential for ensuring that proposed improvements are understood and put into practice efficiently.
10. **Implement and Monitor Changes**: Put the recommended changes into practice and continuously monitor their effectiveness. Adjust the action plan as needed based on feedback and results.
11. **Feedback Loop**: Establish a feedback loop to inform staff and stakeholders about the outcomes of the RCA and the effectiveness of the implemented changes. This helps in fostering a culture of continuous improvement.
12. **Education and Training**: Educate staff about the RCA process and its importance in improving patient safety. Training should be provided to ensure everyone understands their role in the process.
13. **Documentation**: Maintain detailed documentation of the entire RCA process, including findings, decisions, and implemented actions. This documentation can be a valuable resource for future reference and learning providing insights and guidance for continuous improvement and informed decision-making.

11.7.2 When NOT to Do an RCA?

There are specific situations where conducting an RCA may not be appropriate. Firstly, when dealing with minor incidents that have no significant impact on patient safety, quality of care, or operational processes, an RCA might be unnecessary, and simpler corrective actions could suffice. Additionally, if critical resources such as time, personnel, or budget are unavailable, it might be better to focus on more pressing issues where an RCA can be performed adequately. If an RCA is already underway for similar issues, initiating a separate RCA for each incident may lead to duplication of efforts; in such cases, aggregate incidents and conduct a single, comprehensive RCA. It's also important to have a clearly defined scope; if the scope of the problem is not clear, preliminary information should be gathered first. If incidents occur with overwhelming frequency, it may be impractical to conduct an RCA for each one; instead, consider aggregate RCA or trend analysis. Situations requiring immediate corrective action should prioritize those actions over an RCA, which can be conducted later if necessary. An RCA should also be halted if there is no commitment from leadership or key stakeholders to implement the findings and recommendations. Isolated human errors that do not indicate systemic issues might not require an RCA, but this should be confirmed to ensure broader problems are not overlooked. Additionally, situations where regulatory compliance or immediate external requirements take precedence might warrant a delay in conducting an RCA., or if an incident is caused by rare and unpredictable events, other risk management approaches may be more appropriate.

Lastly, if there are significant concerns about confidentiality or legal implications that cannot be adequately managed. If any of the following four situations arise during an RCA, the process is immediately halted, and the CEO or Facility Director is advised to take an independent administrative approach. The findings from the RCA remain confidential and protected, and they are shared with others only as prescribed by law.

1. **Intentionally unsafe acts**: the RCA must stop to allow for appropriate administrative intervention.
2. **Criminal acts**: the discovery of any criminal acts mandates that the RCA be paused so that legal authorities can handle the situation appropriately.
3. **Situations involving alcohol/substance abuse by employees:** situations involving alcohol or substance abuse by employees require halting the RCA to address the issue through relevant administrative or support mechanisms.
4. **Alleged or sustained patient abuse:** any alleged or sustained patient abuse calls for the immediate cessation of the RCA to ensure the safety and well-being of patients while administrative actions are taken.

These measures ensure that the integrity of the RCA process is maintained, and that legal and ethical responsibilities are appropriately managed.

11.7.3 The Root Cause Analysis (RCA) Team/Committee

The RCA committee typically comprises experts from various disciplines to ensure a comprehensive investigation as below. The multidisciplinary composition ensures that the RCA committee can thoroughly analyze incidents from various angles, identify root causes comprehensively, and develop effective strategies to enhance patient safety and quality of care.

1. **Clinical Experts**: Physicians, nurses, and other healthcare providers from the frontline services bring firsthand knowledge of patient care and treatment procedures. Their insights into clinical processes and direct patient interactions are crucial for understanding events leading to incidents.
2. **Quality Improvement Specialists**: Professionals skilled in quality improvement methodologies provide expertise in RCA processes, ensuring systematic investigation and implementation of corrective actions.
3. **Patient Safety Officers**: These individuals focus on identifying systemic issues affecting patient safety and play a pivotal role in guiding the RCA process toward prevention of future incidents.

4. **Frontline staff, Residents, and Trainees**: Involving frontline staff, residents, and trainees allows for fresh perspectives and encourages learning from incidents. They contribute by highlighting issues related to training, supervision, and procedural understanding.
5. **Administrative and Support Staff**: Including personnel from administration, IT, pharmacy, and other support services ensures a holistic investigation encompassing operational and logistical aspects.
6. **External Consultants**: Depending on the complexity of the incident, external consultants such as legal advisors or subject matter experts may be brought in to provide specialized knowledge or unbiased assessment.

The primary tasks of an RCA team revolve around identifying systemic features and processes that contribute to safety incidents. This involves asking the right questions to uncover underlying causes and contributing factors. The team's goal is to shore up systems and processes that enhance patient safety while identifying and eliminating or controlling those that pose risks. By focusing on systemic improvements, the RCA team aims to prevent recurrence of incidents and continuously improve the overall safety and quality of care provided within the healthcare organization.

The RCA team can efficiently and comprehensively determine the flow of events by revisiting the "initial understanding" and "Final Understanding" event flow diagram (Example depicted in Figs. 11.23 and 11.24). This diagram serves as a foundational tool, helping the team map out the sequence of events leading to an incident. By clearly stating their findings and detailing them sufficiently, the team ensures that system-level vulnerabilities, including root causes and contributing factors, are easily identified. This clarity is crucial for prioritizing corrective actions and implementing effective measures to prevent similar incidents in the future, thereby enhancing patient safety and overall organizational resilience.

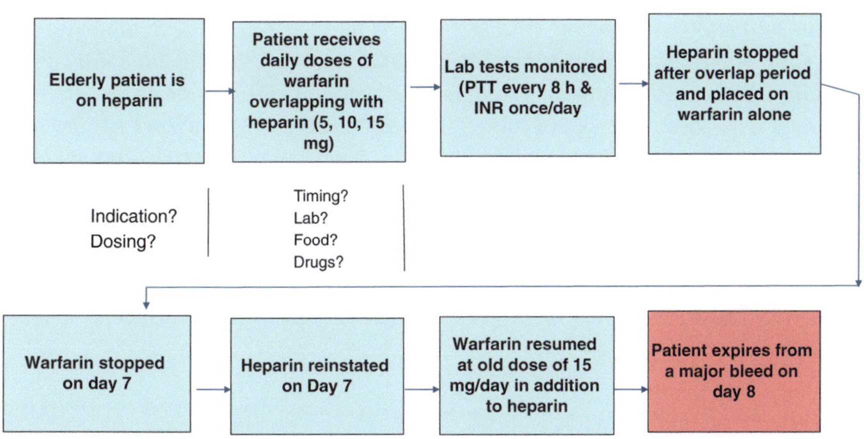

Fig. 11.23 Event flow diagram of initial understanding after Root Cause Analysis (RCA)

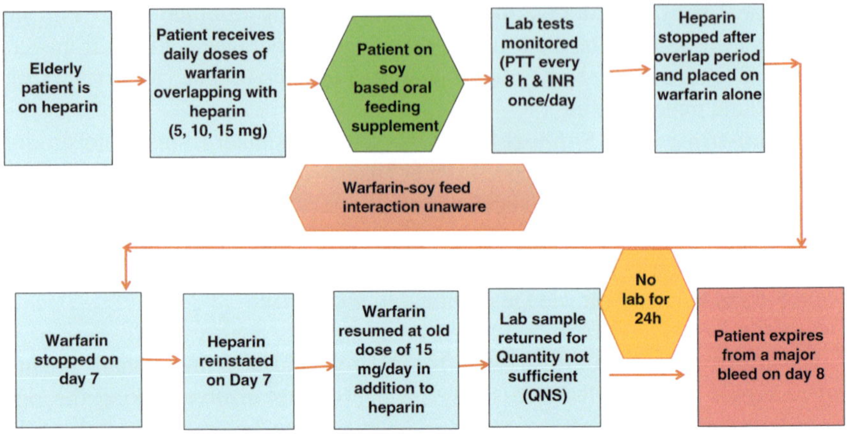

Fig. 11.24 Event flow diagram of final understanding after Root Cause Analysis (RCA)

When the problem is clearly understood, improvement is possible – **Taiichi Ohno**

11.7.4 Actions from RCA

Actions from RCA should focus on system-level vulnerabilities and human-factors engineering to enhance overall safety and performance. By identifying and addressing system flaws, RCA helps to prevent recurrence of incidents (Fig. 11.25). For example, separating look-alike and sound-alike medications reduces the risk of medication errors. Incorporating redundancy, such as double-check systems and fail-safe mechanisms, ensures that errors are caught before they affect patient safety.

Human-factors engineering further supports these efforts by designing systems that align with human capabilities and limitations, such as implementing user-friendly interfaces for electronic health records to minimize input errors and improve accuracy. The current design of tubing, catheters, and syringes allows for the possibility of accidentally causing patient harm by connecting the wrong components, which can result in delivering medication or fluids through an incorrect route. This highlights the need for improved design standards to enhance patient safety and prevent such errors. Clear labeling, color-coding, and unique connectors are among the potential solutions to reduce these risks and ensure that medical devices are used correctly. These combined approaches create a safer, more efficient healthcare system.

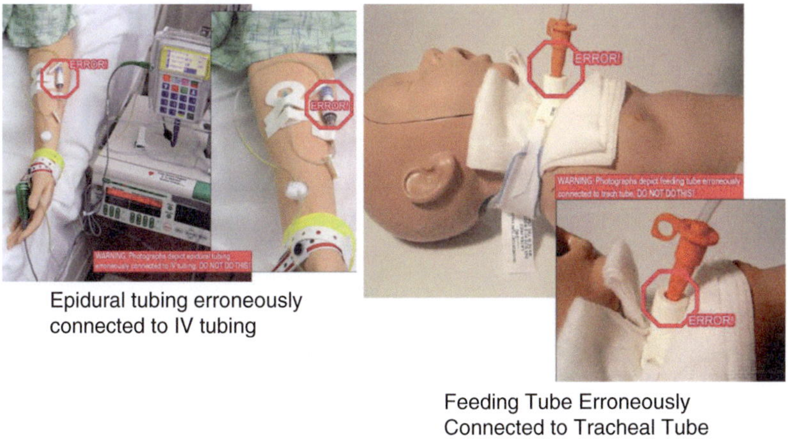

Fig. 11.25 System-level vulnerabilities and human-factors engineering after Root Cause Analysis (RCA) to avoid Medical Device Misconnections. (Adopted from Examples of Medical Device Misconnections|FDA)

11.7.5 The Latent Failure Model of Complex System Failure

Errors in healthcare can be attributed to both human and system causes. Human causes often stem from individual actions, such as fatigue, inattention, lack of knowledge, or miscommunication. These errors are usually the result of cognitive limitations, insufficient training, or failure to adhere to protocols. Systems causes, on the other hand, involve flaws within the organizational structure and processes. These can include inadequate workflow design, poor communication channels, insufficient staffing, lack of proper equipment, and suboptimal policies. While human errors are often immediate and visible, system errors are typically more insidious, resulting from underlying systemic weaknesses. Effective error prevention requires addressing both human and systems factors through comprehensive training, robust communication, and continuous improvement of processes and infrastructure.

The Latent Failure Model of Complex System Failure, developed by James Reason, posits that major accidents in complex systems are often the result of a combination of latent and active failures. Latent failures are hidden problems within the system's design, processes, or organizational structure, such as poor design, inadequate training, or insufficient maintenance. These issues can lie dormant for long periods and create conditions ripe for failure. When combined with active failures, like human errors or procedural mistakes, these latent failures can trigger serious incidents. Reason's model emphasizes the need to identify and address latent failures to enhance system safety and prevent accidents (Fig. 11.26).

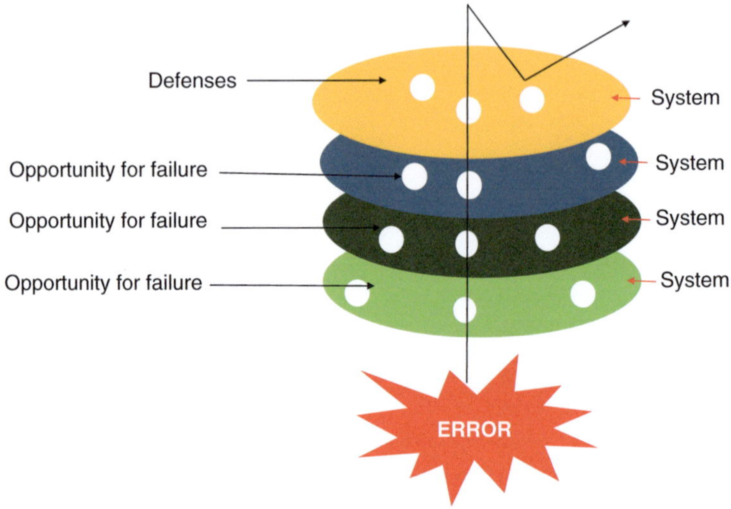

Fig. 11.26 Latent Failure Model of Complex System (Swiss Cheese model) of problems within the system's design, processes, or organizational structure

11.7.6 Basic Root Cause Analysis Tools

Basic RCA tools are essential for systematically identifying the underlying causes of problems and developing effective solutions. Some commonly used RCA tools are:

1. The "5 Whys": Suitable for straightforward issues without complex causality. Example when a patient received the wrong medication.
2. Fishbone Diagram (Ishikawa or Cause-and-Effect Diagram): Useful for complex problems with multiple contributing factors. Example: Identifying causes of a high infection rate in a hospital, with branches for equipment sterility, staff training, hygiene practices, etc.
3. Failure Mode and Effects Analysis (FMEA): Used to anticipate problems before they occur and design processes to mitigate risks. Example: Assessing potential failure points in a surgical procedure to prevent complications.
4. Pareto Chart: Helps focus on the most significant issues that will yield the greatest benefit if resolved. Example: Identifying the most common causes of patient complaints in a clinic.
5. Root Cause Tree Diagram: Suitable for detailed analysis of complex issues. Example: Analyzing a medication error by tracing the process from prescription to administration.
6. Scatter Diagram: Useful for identifying potential cause-and-effect relationships. For instance, examining the relationship between the incidence of errors and the staff working hours can reveal important insights.

7. Check Sheets: Used to gather data systematically to identify patterns and trends. Example: Recording the types and frequencies of errors during a specific shift.
8. Control Charts: Helps in identifying variations and trends that indicate potential problems. Example: Monitoring infection rates in a hospital to identify deviations from the norm.

Using these basic RCA tools helps healthcare organizations systematically identify root causes, develop effective interventions, and enhance overall patient safety and quality of care. By employing a flexible approach to RCA, healthcare organizations can continuously improve their systems, address underlying issues, and prevent the recurrence of errors. This proactive methodology ensures a safer and more reliable healthcare environment for patients (Fig. 11.27).

11.7.7 The 5 Whys Technique

The 5 Whys technique is a simple yet highly effective technique for problem-solving. It involves asking "why" five times in succession to reveal the underlying cause of an issue. The 5-Why methodology stands as a prime example of fundamental cause-and-effect analysis. Just as embarking on a journey of a thousand miles begins with the initial step, every inquiry, irrespective of its scope, commences with a single "Why" query. These inquiries then progress, iteratively traversing five stages, until a sufficient number of "Why" questions have been posed and resolved to adequately elucidate the incident at hand (Fig. 11.28).

Widely used across various industries, including healthcare, it aids in identifying and addressing core problems. It is particularly valuable in healthcare settings where patient safety and quality of care are paramount.

By "peeling the onion" with the 5 Whys, organizations can uncover and address fundamental issues, leading to more effective and sustainable solutions (Fig. 11.28). Each layer of the onion reveals more details about the problem, ultimately leading to the identification of the root cause and potential solutions. By using the 5 Whys to peel back the layers of the problem, the clinic can address the underlying issue of increased patient complexity, leading to more effective solutions that can improve patient wait times and overall service quality.

> By asking why five times and answering it each time, we can get to the real cause of the problem, which is often hidden behind more obvious symptoms - **Taiichi Ohno**

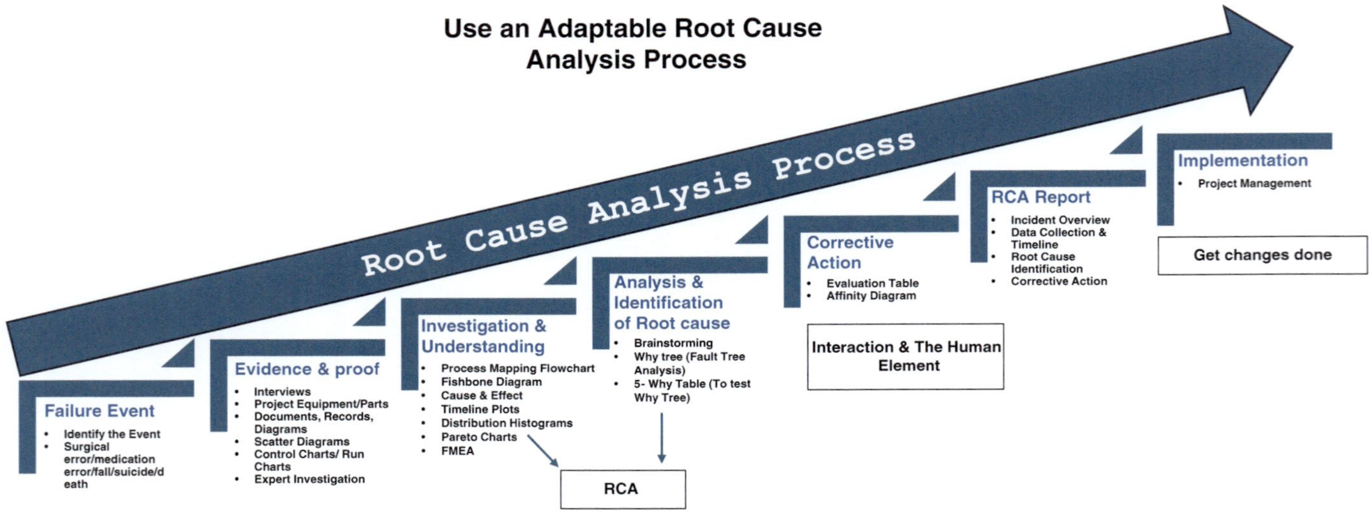

Fig. 11.27 Adaptable basic RCA tools and processes in healthcare

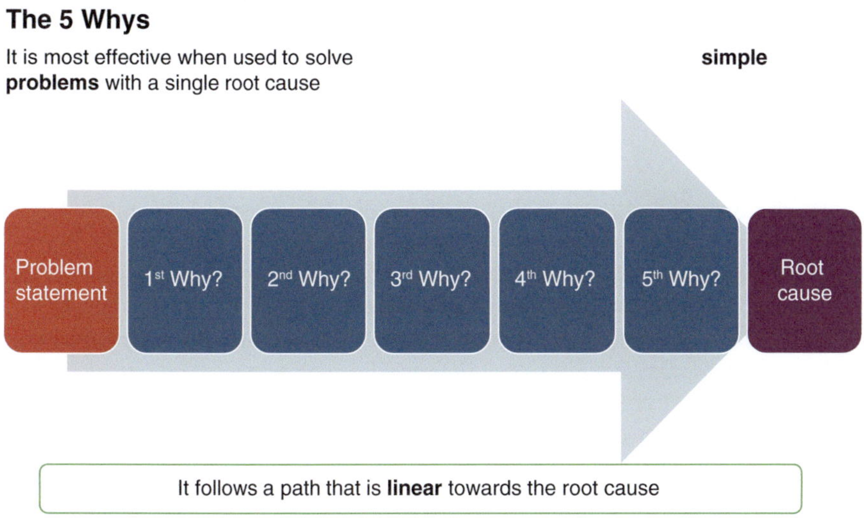

Fig. 11.28 Root Cause Analysis using the 5 Whys technique

11.7.8 Scenario: High Patient Wait Times in the Clinic

Problem Patients experience long waiting times for their appointments.

1. Why are patients experiencing long wait times?
- Because the clinic appointments are consistently delayed.

2. Why are the appointments delayed?
- Because the doctors are running behind schedule.

3. Why are the doctors running behind schedule?
- Because they are spending more time than allocated for each patient.

4. Why are doctors spending more time with each patient?
- Because the patients have complex cases that require additional discussion and examination.

5. Why are there so many complex cases that require more time?
- Because the clinic's patient load has increased significantly, and there has been an increase in the number of patients with chronic conditions needing more comprehensive care.

Root Cause Identified The root cause of long wait times is the increase in the number of patients with chronic conditions, which requires more time per patient. The clinic's current scheduling and staffing may not be adequately adjusted to handle this increased complexity.

Solutions

1. **Re-evaluate Scheduling**: Adjust appointment schedules to account for longer consultation times for patients with chronic conditions.
2. **Increase Staffing**: Consider hiring additional healthcare professionals or support staff to manage the increased patient load.
3. **Implement Triage System**: Introduce a triage system to prioritize patients based on the urgency and complexity of their needs.
4. **Patient Education**: Provide education and resources to help patients manage chronic conditions more effectively, potentially reducing the need for frequent and extended visits.

RCA using 5 Why's examples considering various scenarios in a healthcare setting:

Problem: Patient Falls in the Hospital

1. Why did the patient fall?
   Because they got out of bed unassisted.
2. Why did the patient get out of bed unassisted?
   Because they felt dizzy and disoriented.
3. Why did the patient feel dizzy and disoriented?
   Because they missed their medication for blood pressure control.
4. Why did the patient miss their medication?
   Because the medication schedule was not clearly communicated during shift change.
5. Why was the medication schedule not clearly communicated during shift change?
   Because there was no standardized protocol for handing over patient information between shifts.

Through this iterative process, the root cause of the patient fall is identified: the lack of a standardized protocol for shift change handover. With this understanding, healthcare providers can implement measures to improve communication and prevent similar incidents in the future.

Problem: Surgical site Infection (SSI) Following a Procedure

1. Why did the patient develop a surgical site infection?
   Because the wound was contaminated during the surgery.

2. Why was the wound contaminated?
   Because the surgical team did not adhere strictly to sterile procedures.

3. Why did the surgical team deviate from sterile procedures?
   Because there was a shortage of essential surgical supplies.

4. Why was there a shortage of surgical supplies?
   Because the hospital's inventory management system did not accurately track supply levels.

5. Why did the inventory management system fail to track supply levels accurately?
   Because it was outdated and prone to errors.

Root Cause: Outdated inventory management system leading to a shortage of surgical supplies and subsequent contamination of the surgical site.

The strength of the 5 Whys lies in its ability to prompt thorough investigation, leading to the discovery of the fundamental cause of a problem and enabling effective problem-solving.

11.7.8.1 Some Causes Are Linked with "AND" in Between

In a Cause Map diagram, some causes are linked with "AND" between them to show that more than one cause is needed to produce an effect. When an effect results from multiple causes, each cause is placed on the diagram and connected to the effect with an "AND" in between. While these causes are independent, they must both occur to produce the effect. The "AND" link is particularly useful when people provide different but valid explanations for a cause. This approach clarifies that cause-and-effect relationships are not always simple one-to-one connections; sometimes, multiple causes contribute to a single effect. In reality, every effect has causes.

Problem: Patient Gets a Wrong Medication

1. Why did the patient experience a medication error?
   Because the nurse administered the wrong dosage.
2. Why did the nurse administer the wrong dosage?
   Because the prescription was unclear.
3. Why was the prescription unclear?
   Because it was handwritten and difficult to read.
4. Why was the prescription handwritten?
   Because the hospital's electronic prescribing system was down.
5. Why did the nurse not check with the doctor before administering a medicine which was not clear
   Because the doctor was not available

Problem: The Patient Received the Wrong Medication

1. Why?
   The prescription label was incorrect.
2. Why?
   The pharmacist misread the doctor's handwriting.

3. Why?
   The handwriting was unclear.
4. Why?
   The doctor was rushing.
5. Why?
   The doctor was overloaded with patients.

Through this iterative inquiry, the root cause of the medication error is pinpointed: the electronic prescribing system failure and the doctor not being there.

The 5 Whys tool empowers healthcare teams to delve deeper into issues, identify underlying causes, and implement targeted solutions to enhance patient safety and quality of care.

11.8 Tree Diagram

A Tree Diagram is a graphical tool that breaks down broad categories or issues into finer levels of detail, helping to understand the hierarchy and relationships between different components or factors. It starts with a central idea and branches out into more specific elements, resembling the structure of a tree. This tool is particularly effective in emphasizing the chain of causes for any incidence or event, providing a clear visual representation of how different factors contribute to the overall issue. By using a Tree Diagram, one can systematically explore and analyze complex problems, making it easier to identify root causes and develop effective solutions. Avoid assumptions & logic traps. Encourage keeping drilling on down to the real root cause.

11.8.1 Advantages of Tree Diagram

1. Provides a clear, visual representation of a complex system or problem.
2. Helps in organizing thoughts and identifying all possible causes or solutions.
3. Useful in project planning and decision-making by showing the breakdown of tasks.

11.8.2 Disadvantages of Tree Diagram

1. It can become very complex and difficult to manage large issues.
2. It may require significant time and effort to construct comprehensively.

In terms of complexity handling, the Tree Diagram is better suited for complex problems with multiple layers and interrelated components, whereas the 5 Whys method is more appropriate for relatively straightforward issues with a single root cause. When it comes to visual representation, the Tree Diagram offers a clear, hierarchical visual structure that helps in understanding the relationships between different elements. In contrast, the 5 Whys method is more linear and textual, focusing on a sequential approach. Regarding the depth of analysis, the Tree Diagram provides a comprehensive view by mapping out all possible causes or solutions, offering a broad perspective. On the other hand, the 5 Whys method focuses on drilling down to the root cause through iterative questioning, which can be effective for in-depth analysis of a single issue. Finally, considering ease of use, the 5 Whys method is simpler and quicker to employ, making it suitable for immediate problem-solving. The Tree Diagram, while offering a more detailed analysis, requires more time and effort to construct and analyze, making it less convenient for quick resolutions. Both 5-Whys and Tree diagram are used alone or with any of the cause-and-effect diagrams to reach the root cause of a problem.

> It is surprising how easily we can uncover the causes of problems starting with a clean state and rethinking each premise that we used to accept unconsciously – **Shigeo Shingo**

11.9 Cause-and-Effect Diagram (Also Called "Ishikawa" or "Fishbone" Diagram)

A cause-and-effect diagram, introduced by Kaoru Ishikawa, is a graphical tool used to explore and visualize the potential causes or factors contributing to a quality-related problem or effect. Its purpose is to serve as the initial step in problem-solving by identifying all the root causes behind a problem. When creating a cause-and-effect diagram, all possible factors and sub-factors leading to the root cause are identified, resulting in a diagram resembling a fishbone with branches and sub-branches of causal factors (Fig. 11.29).

Brainstorming technique is utilized to populate a Cause-and-Effect Diagram seeking ALL possible causes for the issue of concern.

In the healthcare industry, causes are typically grouped into major categories such as Materials, Methods, Equipment, Measurements, Environment, and Man to identify the sources of variation.

11.9.1 When to Use a Cause-and-Effect Diagram

- When identifying potential causes for a problem.
- During problem-solving to uncover the root causes.
- To understand the relationships between different causes leading to a problem.

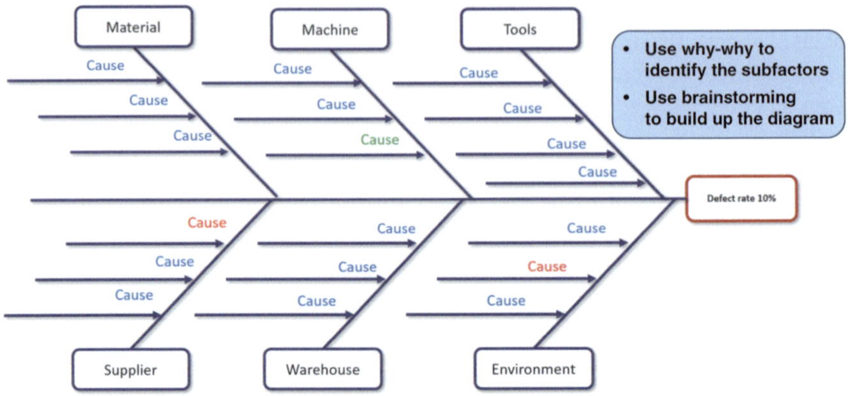

Fig. 11.29 How to use Cause and Effect (Fishbone) Diagram

11.9.2 Advantages of Cause-and-Effect Diagram

- It helps teams understand the complexity of causes and effects, emphasizing that multiple causes contribute to an effect.
- It facilitates a better understanding of factors, leading to faster solutions.
- It promotes group brainstorming and gathering diverse perspectives.
- It visually displays the relationships between causes, effects, and each other.
- It identifies areas for improvement.

11.9.3 Disadvantages of Cause-and-Effect Diagram

- It does not single out the root cause as all causes may appear equally important, leading to wasted effort on insignificant causes.
- It relies on opinions rather than evidence.

The cause-and-effect diagram works best when completed by a team, and success in identifying causes depends on the team's agreement on the central "why" question. It can be used to identify probable causes of gaps, problems, or patient dissatisfaction.

11.9.4 How to Use a Cause-and-Effect Diagram

1. Identify the problem area and define the problem or effect for which the causes need to be identified. Write it down at the head of the diagram and enclose it in a box. Create a horizontal arrow that points to it.
2. Brainstorm the broad areas related to the problem and write the categories of causes as branches from the main arrow. Label the main branches (e.g., methods,

materials, machinery, man, measurements, policies, procedures, environment). Then, for each cause, ask "Why does this happen?"

3. Identify plausible sub-causes of the main causes by asking "Why does this happen?" and attach them as sub-branches to the main branches. Continue asking "Why?" to generate deeper levels of causes. The layers of branches indicate causal relationships.
4. Brainstorm all possible causes of the problem. Write each idea as a branch extending from the relevant category. Causes can be written in multiple places if they relate to several categories.
5. Review the diagram to investigate major and minor causes in more depth. Pay attention to areas with fewer ideas.
6. Once the root cause is identified, set goals and create an action plan outlining strategies to overcome the problem.

Figures 11.30, 11.31 and 11.32 illustrate examples of various events and their potential contributors/causes.

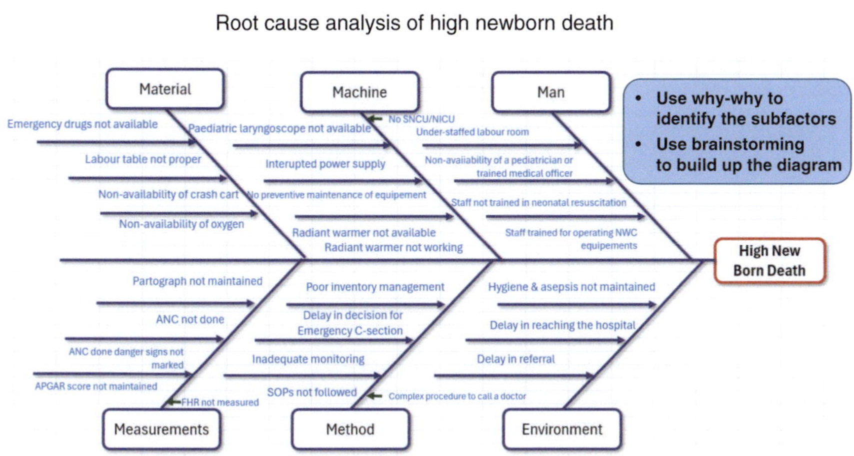

Fig. 11.30 Cause and Effect (Fishbone) Diagram for potential causes of high newborn death

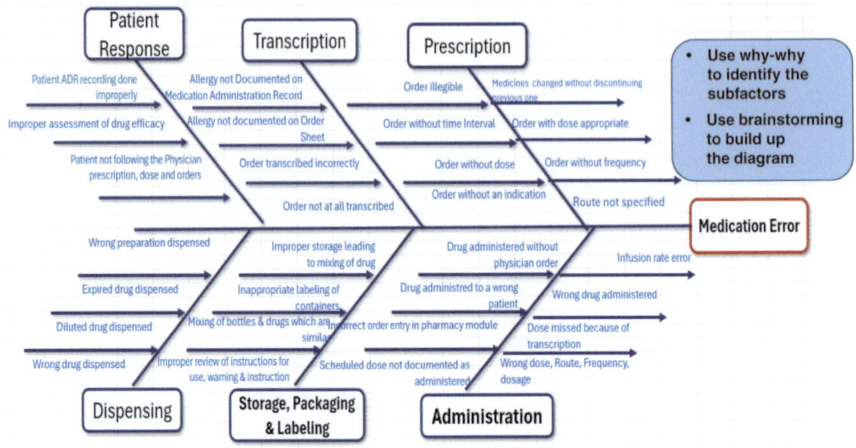

Fig. 11.31 Cause and Effect (Fishbone) Diagram for potential causes of medication error

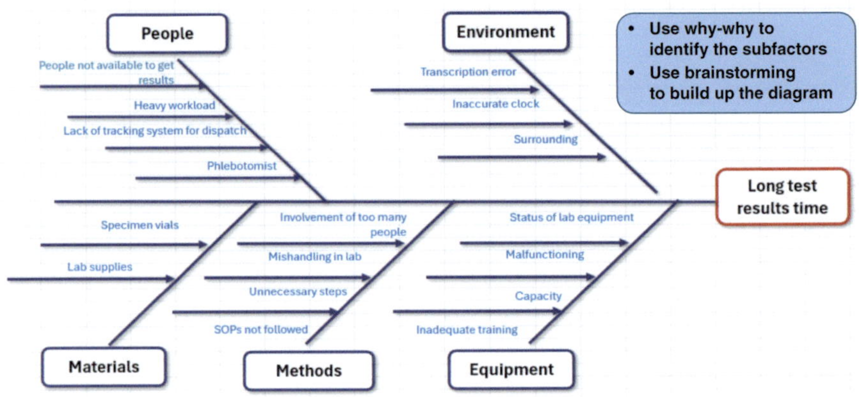

Fig. 11.32 Cause and Effect (Fishbone) Diagram for potential causes of long test result time

11.10 Risk Assessment

Hazards and uncertainties in healthcare are inherent due to the complexity of medical care, diverse patient conditions, and rapidly evolving medical knowledge. Scientifically, uncertainty can arise in the diagnosis, prognosis, and treatment of conditions. For example, while a myocardial infarction may be rightly diagnosed, uncertainties remain regarding the efficacy and safety of medical treatments or the prognosis, such as the patient's life expectancy post-surgery. The uncertainty is further compounded by questions about the patient's ability to maintain life changes, such as diet or exercise, after surgery. Hazards can arise from various sources, including human error, medical equipment malfunction, or adverse drug reactions. For example, medication errors such as incorrect dosage or wrong drug administration can lead to severe patient harm. Surgical complications, infections, and misdiagnosis are also potential hazards. Uncertainties in healthcare often involve unpredictable patient responses to treatments or the progression of diseases. For instance, while some patients may respond well to a specific therapy, others may experience unexpected side effects or limited effectiveness, as seen in cancer treatments. Additionally, evolving infectious diseases, like COVID-19, highlight the uncertainties healthcare providers face in managing new pathogens and emerging health crises. Both disease-centered and patient-centered dimensions of uncertainty must be carefully managed to provide holistic care and support optimal health outcomes. Figure 11.33 illustrates the various types of risks in the hospital. To manage these challenges, healthcare systems rely on evidence-based practices, risk assessments, and robust safety protocols to minimize hazards and navigate uncertainties.

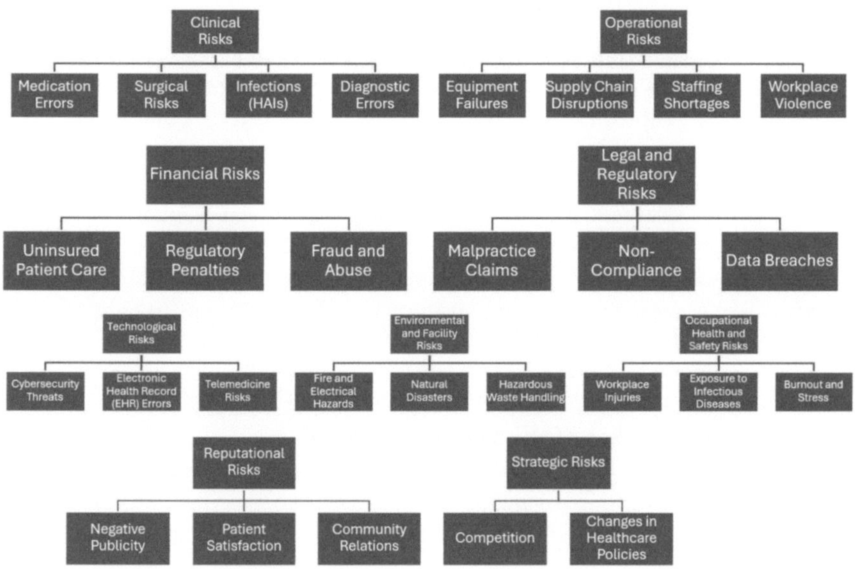

Fig. 11.33 Types of risks in a hospital

Having no problem is the biggest problem of All – **Taiichi Ohno**

11.10.1 Risk Identification Tools

11.10.1.1 London Protocol

Risk management tools are essential in identifying, analyzing, mitigating, and monitoring risks across various settings, including healthcare, finance, and business operations. These tools help in recognizing potential hazards, system vulnerabilities, and contributing factors that may lead to adverse events. These tools are used to recognize and catalog risks before they become issues.

- **SWOT Analysis**: This tool helps identify the strengths, weaknesses, opportunities, and threats that an organization faces.
- **Brainstorming**: Brainstorming encourages team discussions to identify potential risks in different areas.
- **Risk Registers**: A centralized tool for documenting identified risks, their potential impact, and mitigation strategies. It often serves as the starting point for risk management planning.
- **Hazard Identification (HAZID)**: Primarily used in industries like healthcare, this tool identifies potential hazards in the environment or processes.
- The **London Protocol:** The London Protocol, which is a structured method for investigating and analyzing adverse events in healthcare. Described below:

The **London Protocol** is a widely recognized structured framework used for investigating and analyzing adverse events in healthcare. Developed from James Reason's "Swiss Cheese Model" of accident causation, the protocol emphasizes understanding how complex systems and human factors contribute to failures rather than assigning blame to individuals. It is a vital tool for risk identification, aiming to improve patient safety through systemic learning and improvement.

11.10.1.2 Key Features of the London Protocol

- Systematic Investigation: The London Protocol provides a structured approach to analyzing incidents by breaking down events into specific contributory factors. This helps in thoroughly understanding the sequence of events that led to the adverse outcome.
- Holistic View of Incidents: Rather than focusing solely on the immediate error, the protocol examines multiple layers of contributing factors. These include organizational, environmental, team, and individual factors that may have created an unsafe situation.
- Contributory Factors: The protocol categorizes factors into groups such as:
- Patient Factors: Conditions or characteristics of the patient that may have influenced the event (e.g., complex medical history, language barriers).

- Task and Technology Factors: Issues related to how tasks were performed or the functioning of medical technologies.
- Team Factors: Communication and coordination between healthcare professionals.
- Work Environment: Aspects such as staffing levels, workload, and availability of resources.
- Organizational and Management Factors: Institutional policies, procedures, and culture that may have indirectly contributed to the incident.
- Root Cause Analysis (RCA): The London Protocol is closely associated with root cause analysis. It helps to identify both the immediate and root causes of adverse events. This process seeks to answer not just what happened, but why it happened and how it can be prevented in the future.

11.10.1.3 Process of the London Protocol

- Incident Description: A factual account of what happened is documented, including when and where the incident occurred, who was involved, and what the outcome was.
- Identification of Contributory Factors: A detailed investigation is conducted to identify the conditions that led to the incident. Factors may be identified through interviews, reviewing patient records, and examining workflows.
- Analysis and Categorization: Each factor is categorized into relevant domains (e.g., team, organizational, patient) to assess its role in the event.
- Recommendations: Based on the analysis, the protocol encourages making actionable recommendations to improve systems and prevent future incidents. This could include changes in policies, additional training, or adjustments to the work environment.

11.10.1.4 Examples of Using the London Protocol

- **Medication Error**: In the case of a patient receiving the wrong drug dosage, the London Protocol might reveal multiple contributing factors, such as ambiguous handwriting on a prescription, lack of pharmacy cross-checks, and inadequate nurse training. By investigating the system-wide issues, improvements like digital prescriptions or enhanced training could be recommended.
- **Surgical Site Infection**: After an unexpected post-operative infection, the protocol could uncover several factors, such as lapses in sterilization procedures, incomplete preoperative checks, and rushed surgical schedules due to staffing shortages. The recommendations might involve revising sterilization protocols and improving communication among the surgical team.
- **Missed Diagnosis**: For a missed diagnosis, the protocol could reveal gaps in clinical information flow, like missing lab results or poor handover between departments. Solutions could include establishing stronger communication protocols and implementing decision-support tools in electronic health records.

11.10.1.5 Benefits of the London Protocol

- Non-Punitive Approach: It fosters a culture of learning and safety by focusing on system improvement rather than blaming individuals for mistakes.
- Holistic Understanding: By considering a wide range of factors (from human to organizational), the protocol offers a comprehensive view of how incidents occur.
- Actionable Insights: The recommendations provided after using the London Protocol are geared toward preventing future incidents, fostering continuous improvement in healthcare settings.

11.10.2 Other Risk Assessment Tools

Other tools that help in analyzing and prioritizing risks based on likelihood and impact.

- **Failure Mode and Effects Analysis (FMEA)**: A proactive method used to identify and prioritize potential failure points within processes and their impact on outcomes, allowing organizations to put controls in place before issues arise.
- **Root Cause Analysis (RCA)**: A retrospective tool used to analyze the root cause of an adverse event. It helps in identifying underlying systems and process failures, especially in healthcare incidents.
- **Bow-Tie Analysis**: Visualizes risks, showing the path from causes to consequences, with preventive and mitigating controls on each side.
- **Risk Matrix/Heat Map**: Plots the likelihood of a risk occurring against the severity of its impact, helping prioritize which risks need immediate action.

11.10.3 Risk Mitigation and Control Tools

Once risks are identified and assessed, these tools help in designing and implementing controls to reduce their impact.

- **Standard Operating Procedures (SOPs)**: Define protocols and procedures to mitigate specific risks by standardizing actions to reduce variation in processes.
- **Checklists**: A simple tool to ensure that all safety, compliance, and quality measures are in place. Checklists are used extensively in surgical procedures and patient care to prevent errors.
- **Barriers and Safeguards**: Implemented to control risks, barriers such as alarms, physical guards, and protocols are used to prevent or minimize risk exposure.

11.10.4 Monitoring and Reporting Tools

These tools help in continuously tracking risks and ensuring that risk control measures are effective.

- **Key Risk Indicators (KRIs)**: Metrics that are used to track the status of known risks, alerting the organization when risks may materialize.
- **Incident Reporting Systems**: Systems that allow staff to report adverse events, near misses, or unsafe conditions, which can be used to detect trends or emerging risks.
- **Balanced Scorecard**: A strategic tool used for tracking an organization's performance across various dimensions, including risk management.

11.10.5 Risk Quantification and Financial Tools

These tools focus on quantifying risk and determining financial impacts.

- **Monte Carlo Simulation**: A probabilistic method used to simulate various outcomes of risk events, allowing organizations to estimate potential financial losses.
- **Cost-Benefit Analysis (CBA)**: Assesses the cost of implementing risk controls against the potential benefit in terms of risk reduction.
- **Scenario Analysis**: Involves imagining different future scenarios (best-case, worst-case, and most-likely outcomes) to assess how various risks could affect financial outcomes.

11.10.6 Compliance and Governance Tools

These tools help ensure that risks are managed in line with regulations and industry standards.

- **Regulatory Compliance Checklists**: Specific checklists designed to ensure that hospitals or organizations comply with laws, such as HIPAA (in the United States) or GDPR (in Europe) for data protection.
- **Audits**: Regular internal and external audits help in evaluating whether risk controls are effective and whether the organization is adhering to regulatory standards.
- **Track and Trace Systems**: In healthcare, these systems are used to monitor the supply chain and trace products to ensure their safety and compliance with standards (e.g., pharmaceutical track-and-trace).

11.10.7 Technology-Based Tools

With the advancement of digital technology, automated tools have been developed to improve risk management efficiency.

- **Electronic Health Records (EHRs)**: In healthcare, EHR systems can help track patient data, ensuring that clinicians have access to accurate information, which reduces risks of medical errors.
- **Cybersecurity Tools**: Risk management software that protects healthcare organizations from data breaches, such as firewalls, encryption systems, and intrusion detection systems.
- **Analytics**: Machine learning and AI tools analyze large datasets to predict where future risks may emerge, such as patient safety risks or financial risks.

11.10.8 Decision Support Tools

These tools assist decision-makers in understanding risks and deciding on appropriate mitigation strategies.

- **Delphi Technique**: A method where experts anonymously provide opinions on risk-related questions, which are then compiled to find a consensus.
- **Decision Trees**: Visual representations of decisions, showing various options and their potential risks and outcomes, helping to make structured choices.
- **Risk Decision Matrices**: A tool that allows decision-makers to prioritize actions based on the severity of risk, cost of mitigation, and likelihood of occurrence.

11.10.9 Risk Communication Tools

Communicating risks effectively is crucial for stakeholders at all levels.

- **Dashboards**: Real-time dashboards displaying key risk metrics and updates are useful for communicating the current risk profile to management and staff.
- **Incident Reports**: Structured reports that communicate the nature of adverse events, contributing factors, and recommendations for risk reduction.
- **Stakeholder Engagement Plans**: A tool to map and communicate with different stakeholders who may be affected by or involved in managing specific risks.

Risk management tools vary in complexity and scope but are essential in maintaining a proactive and comprehensive approach to identifying, assessing, mitigating, and monitoring risks in hospitals and other organizations. By using these tools effectively, healthcare organizations can ensure patient safety, regulatory compliance, and operational efficiency.

11.10.10 Risk Management

Risk management in healthcare involves identifying, assessing, and mitigating risks that could potentially impact patient safety, quality of care, financial stability, and reputation of healthcare organizations. It aims to proactively address potential hazards and uncertainties through systematic processes and strategies. Key aspects of healthcare risk management include patient safety initiatives, infection control measures, compliance with regulatory requirements, data security and privacy, effective communication among healthcare teams, and the implementation of protocols to handle adverse events or emergencies. By integrating risk management practices into daily operations, healthcare providers strive to enhance patient outcomes, minimize liabilities, and maintain a safe and trusted environment for patients and staff alike.

11.10.11 Failure Mode and Effects Analysis (FMEA)

Failure Mode and Effects Analysis (FMEA) is a structured method used in healthcare to proactively identify and address potential failures or risks within processes, systems, or services. It aims to improve patient safety, enhance quality of care, and reduce the likelihood of adverse events.

11.10.11.1 Key Steps in FMEA Include

1. **Identify Potential Failure Modes:**
   - Healthcare teams brainstorm and identify possible ways in which a process or system could fail.
2. **Determine Severity of Impact:**
   - Assess the potential impact or severity of each failure mode on patient safety or healthcare outcomes. For example, a failure mode in medication administration could be incorrect dosage leading to patient harm.
3. **Evaluate Likelihood of Occurrence:**
   - Estimate the likelihood of each failure mode occurring. This process involves considering factors such as frequency, probability, and historical data.
4. **Assess Detection Ability:**
   - Evaluate how likely it is for healthcare providers to detect the failure mode before it affects patients. This step emphasizes the importance of early identification and prevention.
5. **Calculate Risk Priority Number (RPN):**
   - Calculate the RPN by multiplying severity, occurrence, and detection scores. This calculation helps prioritize which failure modes to be addressed first. ensuring that the most critical issues are tackled promptly and effectively (Fig. 11.34).
6. **Implement Risk Mitigation Strategies:**
   - Develop and implement strategies aimed at mitigating or eliminating identified risks. This may involve process redesign, staff training, standardization of procedures, or introducing new technologies.

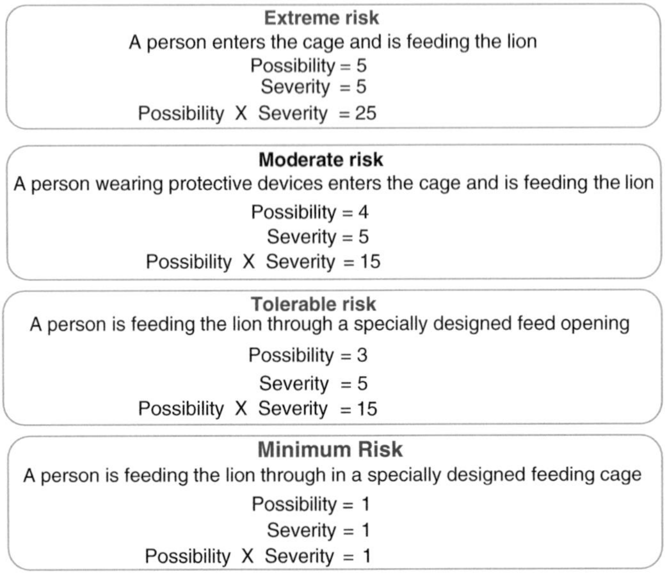

Fig. 11.34 Risk Management using Risk Priority Number (RPN) and prioritization of failure modes

11.10.11.2 Examples of FMEA in Healthcare

- **Medication Administration:** FMEA can be applied to the process of medication administration to identify potential failures such as incorrect dosing, medication mix-ups, or administration errors. By assessing severity (potential harm to patients), likelihood (frequency of errors), and detection (ability to catch errors before harm occurs), healthcare teams can implement safeguards like barcode scanning systems or double-check procedures to reduce risks.
- **Surgical Procedures:** FMEA can help analyze surgical processes to identify failure modes such as wrong-site surgery, equipment failures, or post-operative complications. Measures like pre-surgical checklists, standardized operating procedures, and team briefings can be implemented to enhance safety and reduce errors.
- **Infection Control:** FMEA can be used to assess infection control procedures to prevent healthcare-associated infections. Identifying failure modes like inadequate hand hygiene, improper sterilization techniques, or environmental contamination allows healthcare facilities to implement stringent protocols, staff training, and regular audits to mitigate infection risks.

11.10.12 Calculation of Risk Priority Number (RPN)

The formula for calculating RPN is as below:

$$\text{RPN} = \text{Severity} \times \text{Occurrence} \times \text{Detection}$$

Where:

- **Severity (S)**: This refers to the potential impact or seriousness of the failure mode on patients, processes, or outcomes. It is typically scored on a scale (e.g., 1 to 10), with higher scores indicating greater severity.
- **Occurrence (O)**: This assesses the likelihood or frequency with which the failure mode could occur. It is also scored on a scale (e.g., 1 to 10), with higher scores indicating higher occurrence.
- **Detection (D)**: This evaluates the ability to detect or identify the failure mode before it reaches the patient or causes harm. It is scored inversely (e.g., 1 to 10), where a higher score indicates better detectability.

11.10.12.1 Example of Calculating RPN

Let's walk through a hypothetical example of medication administration in a hospital setting:

1. **Failure Mode**: Incorrect dosage of medication administered to patients.
2. **Severity (S)**: Suppose the severity of this failure mode is rated as 8 on a scale of 1 to 10, indicating that incorrect dosage could potentially lead to moderate harm to the patient.
3. **Occurrence (O)**: The occurrence of this failure mode is assessed as 7, suggesting that it happens relatively frequently due to human error or system weaknesses.
4. **Detection (D)**: The ability to detect incorrect dosages before they reach the patient is rated as 6, indicating moderate capability in identifying errors through current checks and balances.

Now, calculate the RPN:

$$\text{RPN} = \text{Severity} \times \text{Occurrence} \times \text{Detection}$$

$$\text{RPN} = 8 \times 7 \times 6 = 336$$

Interpretation

In this example, the calculated RPN is 336. The higher the RPN, the higher the priority for addressing and mitigating the identified failure mode. Organizations typically focus on high-RPN items first to implement corrective.

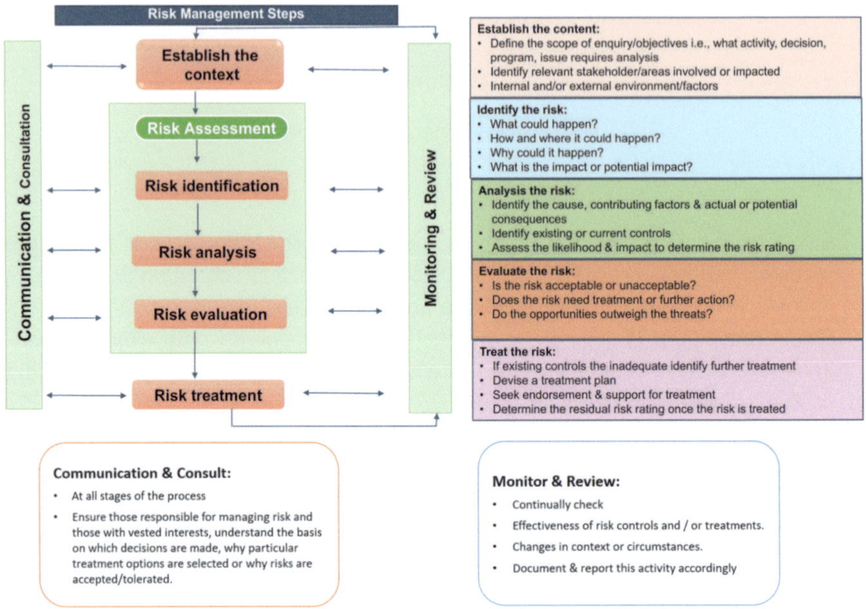

Fig. 11.35 Steps involved in risk management

Figure 11.35 summarizes the steps involved in risk management, and Fig. 11.36 illustrates key steps in FMEA.

11.10.13 Failure Mode and Effects Analysis (FMEA): Key Steps

By systematically applying FMEA, healthcare organizations can enhance patient safety, improve quality of care, and foster a culture of continuous improvement to minimize risks and errors in healthcare delivery.

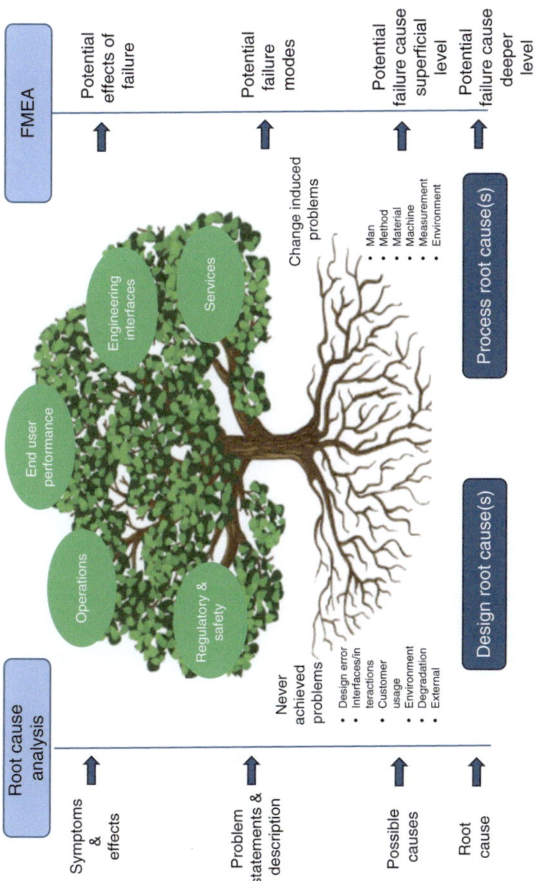

Fig. 11.36 Key steps in Failure Mode and Effects Analysis (FMEA)

11.11 Summary

These quality tools are valuable resources that can be utilized by top management, middle management, front-line supervisors, and departments to enhance process understanding and improvement within an organization. Each of these tools offers unique advantages, addressing different concerns and promoting efficiency.

These fundamental quality tools can be applied individually or in combination to effectively comprehend processes, analyze and investigate them, identify areas for improvement, and communicate quality enhancement efforts. However, it's important to note that not all tools are suitable for every problem-solving situation. Therefore, it is beneficial for all staff members to familiarize themselves with these basic quality tools and adapt them accordingly to various problem-solving opportunities.

Flowcharts and check sheets are particularly useful for identifying problems and clarifying processes, especially when steps may be unclear or missing. Cause-and-effect diagrams, histograms, scatter diagrams, and control charts, on the other hand, excel at problem analysis. Pareto diagrams prove effective in both problem identification and analysis. By effectively utilizing these tools, organizations can significantly enhance their problem-solving and quality improvement processes.

Flowcharts play a crucial role in facilitating process comprehension by outlining the steps involved. They help identify any potential gaps or ambiguities in the processes, enabling quality assurance activities to identify weaknesses and implement necessary changes for improved outcomes. Additionally, when combined with other data analysis tools, stratification allows for a better understanding of the data.

These quality tools serve as powerful aids in process improvement and quality enhancement efforts. Their appropriate application fosters a more efficient and effective problem-solving approach, leading to positive organizational outcomes. The seven quality tools are a set of essential tools used for quality improvement and problem-solving. Each tool serves a specific purpose in identifying and analyzing quality improvement opportunities.

11.12 Success Cause Analysis: Shifting Focus to Learn from Favorable Outcomes in Healthcare

Success Cause Analysis (SCA) applies RCA methodologies to understand the factors that contribute to favorable outcomes, rather than solely focusing on adverse events. Over the past three decades, the primary approach to improving healthcare safety has involved studying adverse outcomes to identify system vulnerabilities and implement corrective measures. While this approach has been valuable, it has inherent limitations. Focusing exclusively on unfavorable outcomes can stifle innovation, limit adaptability, and negatively affect worker morale and engagement.

In response, the Safety-II framework emerged, emphasizing that human adaptability and system resilience often ensure success in complex and unpredictable environments. Safety-II encourages learning from all events, including those that result in positive outcomes. However, operationalizing this framework has proven challenging. While healthcare teams routinely use structured processes like FMEA and RCA to address safety failures, few organizations have established systematic approaches to examine why and how favorable outcomes occur. This gap often prevents organizations from embedding successful strategies into their systems to promote safer outcomes in the future.

SCA utilizes RCA methodology to explore the factors contributing to a positive outcome and develop actionable plans for advancing systemic improvements. By examining the conditions that enabled teams to innovate, collaborate, and adapt effectively, SCA highlights how these success factors connect to broader, routine aspects of work. Trained RCA practitioners conduct SCAs to systematically review a successful outcome, identifying specific strategies that other units can adopt to enhance their performance and outcomes.

Even when success appears to result from heroic individual or team actions, SCA often reveals the underlying culture and structures that encouraged such behaviors. For instance, a strong safety culture, effective communication channels, and robust training programs may contribute to actions perceived as exceptional.

11.13 Benefits of Success Cause Analysis in Healthcare

Fostering Open Participation: Unlike traditional RCA, which is error-focused, SCAs create a non-threatening environment that encourages broader participation. Individuals feel empowered to share perspectives without fear, fostering a more inclusive and transparent review process.

Balancing Safety Efforts: SCAs complement traditional adverse event reviews, creating a more balanced approach to understanding safety. This participatory process enhances collegiality between risk management professionals and frontline staff.

Mitigating "Third Victim Syndrome": Quality and safety professionals who analyze only errors may experience stress and view their systems as inherently unsafe. SCAs provide a constructive alternative by highlighting positive outcomes, promoting psychological safety, and reinforcing a sense of purpose.

Engaging Patients and Families: SCA discussions are well-suited for involving patients and families in quality and safety efforts, making care more person-centered and aligned with their values and experiences.

Promoting Systemic Learning: By identifying and disseminating successful strategies, SCAs facilitate system-wide learning and improvement, helping organizations replicate and embed best practices.

11.14 Safety-I and Safety-II in Healthcare: Enhancing Patient Safety Through Complementary Approaches

Patient safety remains a critical concern in healthcare, where the consequences of errors can be severe. To address this, two primary frameworks have emerged: Safety-I and Safety-II. These paradigms offer distinct yet complementary perspectives on how to understand and improve safety within healthcare settings.

Safety-I is the traditional approach to safety management, focusing on preventing things from going wrong. It operates on a reactive basis, emphasizing the identification and elimination of errors, hazards, and system failures to minimize adverse outcomes. This approach relies heavily on methodologies such as RCA and incident reporting to investigate and address safety failures. For example, if a hospital experiences a series of surgical site infections (SSIs), a Safety-I approach would involve collecting data on these infections, conducting RCA to identify factors like improper sterilization or inadequate hand hygiene, and implementing corrective measures such as stricter sterilization procedures and enhanced training. The primary goal of Safety-I is to reduce errors and ensure compliance with established protocols, thereby directly targeting and mitigating specific failures.

In contrast, Safety-II represents a more proactive and holistic approach to safety. It acknowledges that healthcare systems are inherently complex and that variability in human performance can be both a challenge and an asset. Instead of focusing only on preventing errors, Safety-II aims to ensure that as many things as possible go right by understanding and enhancing everyday successful operations. This approach emphasizes resilience engineering, which involves building systems that can adapt to changing conditions and recover from unexpected disruptions. For instance, consider an emergency department (ED) that consistently manages high patient volumes with varying degrees of acuity while maintaining high patient satisfaction and low error rates. A Safety-II approach would involve identifying and analyzing the factors that enable this success, such as effective team communication, efficient task delegation, and adaptive problem-solving. By conducting Success Cause Analyses (SCA), the hospital can replicate these successful practices across other departments to enhance overall safety and efficiency.

The key difference between Safety-I and Safety-II lies in their focus and methodology (Table 11.2. While Safety-I is primarily concerned with identifying and fixing what goes wrong, Safety-II seeks to understand and amplify what goes right. Safety-I tends to view variability as a source of error, whereas Safety-II sees it as a source of adaptability and resilience. For example, in medication administration, a Safety-I approach might implement barcoding systems and double-check protocols to prevent errors. Meanwhile, a Safety-II approach would observe instances where medications are administered without errors, even under high-pressure conditions, and analyze how nurses manage their workflow and communicate effectively to maintain safety.

Integrating Safety-I and Safety-II can lead to a more comprehensive and robust safety management system in healthcare. By addressing both error prevention and the enhancement of successful practices, healthcare organizations can achieve a

Table 11.2 Contrasting Safety-I and Safety-II

Aspect	Safety-I	Safety-II
Focus	Preventing things from going wrong	Ensuring things go right
Approach	Reactive, error-focused	Proactive, performance-focused
Methodologies	Root Cause Analysis, incident reporting	Success Cause Analysis, resilience engineering
Perspective on Variability	Variability seen as a source of error	Variability seen as a source of adaptability
Goal	Reduce errors and adverse events	Enhance system resilience and adaptability

balanced approach that minimizes vulnerabilities while promoting resilience and continuous improvement. For example, in medication administration, combining Safety-I measures like error reporting and RCA with Safety-II strategies such as Success Cause Analysis and training in effective multitasking can significantly improve medication safety and foster a positive, resilient work environment.

However, each approach comes with its own set of benefits and challenges. Safety-I is effective for targeted error reduction and provides a clear framework for accountability and compliance. Nonetheless, it may overlook systemic strengths and create a culture of blame that discourages open reporting of errors. On the other hand, Safety-II offers a holistic improvement by focusing on successful operations and empowering the workforce, but it can be challenging to implement due to its complexity and the difficulty in measuring successes compared to errors.

11.15 Clinical Vignettes: Applying Success Cause Analysis in Practice

11.15.1 Clinical Vignette 1. Managing a Code Blue in a Cardiac Arrest

Scenario: A patient in a busy emergency department suffers a cardiac arrest. The Code Blue team responds swiftly, performing CPR, defibrillation, and administering medications. The patient is successfully resuscitated and transferred to the ICU without complications.

Success Factors Identified:

- Effective Teamwork: Clear roles were assigned during the code, with each team member executing their task flawlessly.
- Simulation Training: Team members had participated in regular high-fidelity simulation drills, increasing their confidence and coordination.
- Rapid Access to Resources: The crash cart was fully stocked and positioned optimally, ensuring no delays in accessing essential equipment.

Insights for Broader Application:

- Implement mandatory simulation-based training for all staff involved in emergency response.
- Conduct routine checks to ensure all crash carts are well-stocked and strategically located.

11.15.2 Clinical Vignette 2: Managing Postpartum Hemorrhage Successfully

Scenario: In a busy urban hospital, a patient undergoes a vaginal delivery complicated by significant blood loss, placing her at risk for postpartum hemorrhage (PPH). Despite the high-stress environment and the urgency of the situation, the obstetric team responds swiftly and effectively. They administer uterotonic medications promptly, perform uterine massage, and coordinate a timely transfusion of blood products. The patient's condition stabilizes without the need for surgical intervention, and she recovers fully without complications.

Success Cause Analysis: Following the successful management of this PPH case, the hospital's quality and safety team conducts an SCA to understand the factors that contributed to the positive outcome. Through structured interviews and observations, several key elements emerge:

1. Effective Team Communication: The obstetric team maintained clear and concise communication throughout the event. Roles were well-defined, and team members provided timely updates, ensuring that everyone was aware of the patient's status and the actions being taken.
2. Comprehensive Training and Simulation: The team had recently participated in high-fidelity simulation training for obstetric emergencies, including PPH scenarios. This training enhanced their preparedness, enabling them to execute protocols efficiently under pressure.
3. Availability of Essential Resources: The necessary medications, uterotonic agents, and blood products were readily accessible in the delivery room. The hospital's inventory management system ensured that critical supplies were always stocked and easily retrievable during emergencies.
4. Strong Leadership and Decision-Making: The lead obstetrician demonstrated decisive leadership, quickly assessing the situation and directing the team's actions. This clear leadership helped streamline efforts and minimize delays in intervention.
5. Supportive Culture and Psychological Safety: Team members felt empowered to speak up and suggest actions without fear of judgment or reprimand. This environment fostered collaborative problem-solving and ensured that all voices were heard during the critical moments of care.

Insights for Broader Application: The SCA revealed that the effective management of PPH was not solely due to individual expertise but was the result of a well-coordinated system that promoted teamwork, continuous training, and resource readiness. To replicate this success across the hospital, the following action plans are developed:

- Expand Simulation Training: Incorporate regular simulation exercises for various obstetric emergencies, ensuring that all relevant staff are proficient in managing PPH and other critical conditions.
- Enhance Communication Protocols: Implement standardized communication tools, such as SBAR (Situation-Background-Assessment-Recommendation), to maintain clarity and efficiency during emergencies.
- Optimize Resource Availability: Conduct periodic audits of essential supplies in all delivery units to ensure that medications and equipment needed for managing PPH are always available and easily accessible.
- Foster Leadership Development: Provide leadership training for senior staff to enhance decision-making skills and promote effective team management during high-stress situations.
- Cultivate a Supportive Culture: Encourage an open and non-punitive environment where team members feel comfortable sharing insights and suggestions, thereby strengthening collaborative efforts and continuous improvement.

Outcome: By implementing these strategies derived from the Success Cause Analysis, the hospital enhances its overall preparedness for managing PPH and other obstetric emergencies. The replication of effective communication, training, resource management, leadership, and supportive culture across various departments leads to improved patient outcomes, increased staff confidence, and a stronger safety culture within the institution.

11.15.3 Clinical Vignette 3. Preventing Medication Errors in Pediatric Dosing

Scenario: A hospital pharmacy ensures 100% accuracy in pediatric medication dosing despite a recent surge in patient volume.

Success Factors Identified:

- Double-Check Protocol: Pharmacists followed a strict double-check system for all pediatric prescriptions, with a second pharmacist verifying the calculations.
- Integrated EHR Alerts: The EHR system had built-in alerts for weight-based dosing anomalies, flagging potential errors before dispensing.
- Culture of Safety: Pharmacists felt empowered to speak up and escalate concerns without fear of reprisal.

Insights for Broader Application:

- Strengthen EHR systems with customized alerts for pediatric and high-risk medications.
- Foster a safety culture where employees feel supported to voice concerns.

11.15.4 Clinical Vignette 4: Successful Transition of Care for a High-Risk Patient

Scenario: A 75-year-old patient with multiple comorbidities is discharged after a complex hospital stay. The transition of care is seamless, and the patient experiences no readmission within 30 days.

11.15.4.1 Success Factors Identified

- Robust Discharge Planning: The care team developed a detailed discharge plan, including medication reconciliation, follow-up appointments, and patient education.
- Care Coordination: The primary care provider, specialist, and home health nurse were all informed and involved in the transition process.
- Patient Engagement: The patient and family were educated using teach-back techniques to ensure understanding of post-discharge instructions.

11.15.4.2 Insights for Broader Application

- Create standardized discharge planning protocols for high-risk patients.
- Leverage technology to enhance communication between care providers during transitions.

Table 11.3 describes the various components for SCA with examples.

Both Safety-I and Safety-II offer valuable insights for enhancing patient safety in healthcare. Safety-I remains essential for identifying and mitigating errors, ensuring protocols are followed, and preventing adverse outcomes. Meanwhile, Safety-II complements this by focusing on understanding and reinforcing the factors that lead to successful and resilient performance. By integrating both approaches, healthcare organizations can develop a more balanced and comprehensive safety management system that not only minimizes errors and vulnerabilities but also promotes a culture of continuous improvement, adaptability, and excellence in patient care. Embracing both paradigms allows healthcare providers to create safer, more reliable, and more patient-centered environments, ultimately leading to better health outcomes and enhanced patient satisfaction.

By studying and scaling the enablers of success—whether in emergency response, medication safety, or care transitions—healthcare organizations can build a proactive, resilient, and patient-centered safety culture. Through stories of

Table 11.3 Various components for success cause analysis (SCA)

Component	Description	Examples in healthcare
Event Identification	Recognizing and selecting instances of successful outcomes or positive events to analyze.	Identifying a case where a patient safely received high-risk medication without errors despite a busy shift.
Timing & Team Membership	Determining the appropriate time to conduct the analysis and selecting a diverse team with relevant expertise.	Scheduling the SCA shortly after a successful management of postpartum hemorrhage (PPH) with team members from obstetrics, nursing, pharmacy, and administration to provide comprehensive insights.
Data Collection	Gathering comprehensive information related to the event through various methods.	Collecting medical records, interviewing the healthcare staff involved, and reviewing relevant documentation and protocols related to the successful PPH case.
Fact Finding and Flow Diagramming	Systematically gathering facts and mapping the sequence of events to understand the workflow and interactions.	Creating a flow diagram of the steps taken during the PPH management, from initial detection of hemorrhage to administration of uterotonic medications and coordination of blood transfusion.
Contextual Understanding	Analyzing the situational factors and environment in which the successful outcome occurred.	Examining the specific conditions of the delivery room, patient load, available resources, and time of day during the successful PPH management.
Success Factors Identification	Identifying the key elements and actions that contributed to the favorable outcome.	Determining that effective teamwork, adherence to protocols, timely decision-making, and resource availability were critical to the successful management of PPH.
Development of Causal Statements	Formulating clear statements that explain how specific factors led to the successful outcome.	Effective team communication and recent simulation training enabled the rapid administration of uterotonic medications, stabilizing the patient quickly.
Systemic Analysis	Exploring how organizational systems, processes, and culture supported the successful outcome.	Analyzing how the hospital's electronic health record (EHR) system facilitated accurate medication administration and how the supportive culture encouraged proactive problem-solving.
Actionable Insights	Deriving practical recommendations and strategies based on the identified success factors and systemic analysis.	Developing standardized training programs that emphasize teamwork and protocol adherence based on the successful PPH management case.
Action Plan Development	Creating detailed plans to implement the actionable insights across the organization.	Designing a schedule for regular simulation training sessions focused on high-risk scenarios like PPH and updating communication protocols based on SCA findings.

(continued)

Table 11.3 (continued)

Component	Description	Examples in healthcare
Implementation Strategies	Executing the action plans through coordinated efforts and resource allocation.	Allocating time for staff to participate in training, updating protocols, and ensuring necessary equipment is available for managing PPH effectively.
Measurement and Feedback	Continuously tracking the effectiveness of implemented actions and making necessary adjustments.	Establishing key performance indicators (KPIs) to measure improvements in PPH-related outcomes and staff adherence to new communication protocols and providing regular feedback to teams.
Monitoring and Feedback	Continuously tracking the effectiveness of implemented actions and making necessary adjustments.	Setting up regular review meetings to assess the impact of new PPH management protocols and gathering feedback from staff to refine practices further.
Stakeholder Engagement	Involving all relevant parties, including frontline staff, administrators, and patients, in the SCA process.	Holding workshops with nurses, pharmacists, physicians, and patients to gather diverse perspectives on successful PPH management and to foster a collaborative improvement environment.
Documentation and Sharing	Recording findings and disseminating successful strategies and lessons learned throughout the organization.	Creating a centralized repository of SCA reports and conducting seminars to share best practices identified from successful PPH management cases.
Cultural Reinforcement	Promoting a culture that values learning from successes and encourages continuous improvement.	Recognizing and rewarding teams that consistently achieve positive outcomes, fostering an environment of psychological safety and ongoing learning.
Technology Integration	Utilizing digital tools and systems to support the analysis and replication of successful practices.	Implementing data analytics platforms to track and visualize success metrics across different departments and using EHR alerts to support protocol adherence.
Leadership Support	Ensuring that organizational leaders actively support and champion SCA initiatives.	Leaders participating in SCA meetings, providing resources for training, and advocating for a balanced safety approach that includes learning from successes.
Evaluation and Refinement	Periodically assessing the SCA process itself and refining methodologies to enhance effectiveness.	Conducting annual reviews of SCA processes to incorporate feedback and improve the framework based on organizational needs and evolving healthcare standards.

(continued)

Table 11.3 (continued)

Component	Description	Examples in healthcare
Collaborative Problem-Solving	Engaging multidisciplinary teams in identifying and addressing factors that contribute to success.	Bringing together nurses, pharmacists, and physicians to analyze and enhance PPH management strategies through collective expertise and shared experiences.
Continuous Improvement	Committing to ongoing evaluation and enhancement of practices based on SCA findings.	Regularly updating training programs and protocols based on insights from successful PPH management cases, ensuring that improvements are sustained and evolved over time.

excellence, clinicians not only learn to prevent harm but also to consistently deliver exceptional care. Capturing success stories requires intentional effort, as these may not be as readily documented as adverse events and organizations must invest in training and systems to support SCA without neglecting traditional safety efforts.

11.16 Data Analysis

In quality assurance, the use of data is essential for analyzing processes, identifying problems, testing solutions, and measuring performance. Data plays a crucial role in ensuring objectivity and providing valuable insights. Key reasons why data are important in quality assurance:

1. Objectivity: Data provides an objective basis for analysis and decision-making. By relying on factual information, organizations can avoid subjective biases and make more informed judgments.
2. Opportunity identification: Data analysis helps identify areas for improvement, allowing organizations to initiate quality assurance efforts in the right areas. By understanding current performance levels, organizations can prioritize their improvement initiatives.
3. Problem detection and assessment: Data enables the detection and assessment of problems within processes or systems. By analyzing relevant data, organizations can identify deviations, inefficiencies, or quality issues that need to be addressed.
4. Cause verification: Data analysis helps verify possible causes of problems. By examining correlations, trends, or patterns in the data, organizations can gain insights into the root causes of issues and develop targeted solutions.
5. Informed decision-making: Data plays a crucial role in quality assurance by providing a solid foundation for informed decision-making. By considering quantitative and qualitative data, organizations can identify patterns and trends in performance and quality metrics, evaluate potential risks and their impact on operations and outcomes, pinpoint areas for improvement and implement changes to enhance efficiency and effectiveness and evidence-based decisions to drive improvements.

6. Evaluation of interventions: Data allows organizations to evaluate the effectiveness of quality interventions. By comparing data before and after implementing changes, organizations can assess the impact and determine if improvements have been achieved.
7. Monitoring of processes over time: Data analysis helps monitor processes over time to assess if changes or improvements are sustained. By tracking key performance indicators and analyzing data trends, organizations can ensure that quality improvements are maintained.

The type of data collected and analyzed may vary depending on the specific quality assurance tasks. Quantitative data, such as service statistics, can provide numerical insights into performance metrics. Qualitative data, such as customer feedback or comments from healthcare workers, offer valuable insights into subjective experiences and perceptions.

By leveraging data effectively, organizations can drive continuous improvement, enhance decision-making, and ensure the delivery of high-quality services (Fig. 11.37).

11.16.1 Stratification

Stratification is the process of categorizing data, people, or products into distinct groups or layers based on specific criteria. It is a data analysis technique that works in conjunction with other tools. When data is aggregated from various sources or categories, it can be challenging to discern meaningful patterns. Stratification helps separate the data into meaningful subsets, allowing for clearer analysis. It is acknowledged as one of the seven fundamental quality tools. Stratification is useful in the following scenarios:

1. Before collecting data:
   - When data is gathered from various sources or conditions, such as different shifts, days of the week, departments, qualifications, suppliers, or population groups.
   - When data analysis necessitates the separation of different sources or conditions.

   Examples of sources that may require data stratification include:

- Equipment
- Days of the week
- Time of day
- Shifts
- Departments
- Services/products
- Materials
- Vendors

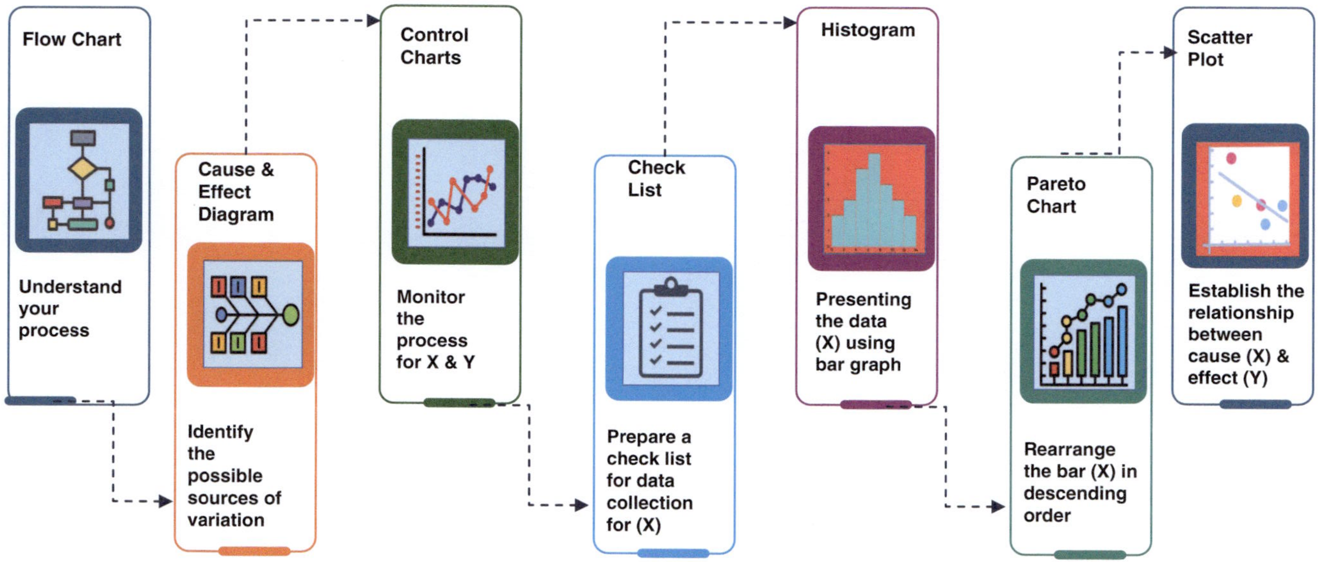

Fig. 11.37 Various quality tools and their utilization to drive continuous improvement, enhance decision-making, and ensure the delivery of high-quality services

The procedure for stratification is as follows:

1. Prior to data collection, identify and plan to collect information about the sources or conditions that may influence the results.
2. When creating a graph, such as a histogram, scatter diagram, or control chart, use different marks or colors to differentiate data from various sources. This differentiation within the data is referred to as "stratified."
3. Analyze these subsets of stratified data separately, focusing on each category or source individually to gain insights and draw conclusions.

By implementing the stratification technique, organizations can effectively analyze data subsets, identify patterns, and make informed decisions based on specific categories or sources.

Analysis of a problem involves a systematic approach to understanding and resolving the issue at hand. The following steps outline the process:

1. Study the problem: Thoroughly examine and gain a comprehensive understanding of the problem. Gather relevant information and data related to the issue.
2. Describe the problem in detail and define it: Clearly articulate the problem, its scope, and its impact. Define the problem statement to ensure a focused and specific approach.
3. Choose an appropriate management tool/technique: Select a suitable tool or technique from the available options that can effectively address the problem. Consider tools such as root cause analysis, SWOT analysis, fishbone diagrams, or process mapping.
4. Construct a mathematical model: If applicable, develop a mathematical model that represents the problem and its components. This model can help in analyzing the problem quantitatively and identifying potential solutions.
5. Analyze the problem through the model: Utilize the mathematical model or chosen technique to analyze the problem systematically. Identify patterns, trends, or relationships within the data to gain insights into the underlying causes of the problem.
6. Test the solution derived from the model: Validate the proposed solution or potential alternatives by testing them against the problem. Use simulations, prototypes, or other evaluation methods to assess the feasibility and effectiveness of the proposed solution.
7. Choose the right course of action: Evaluate the different alternatives or solutions generated through the analysis and select the most appropriate course of action. Consider factors such as feasibility, impact, cost-effectiveness, and alignment with organizational goals.
8. Execute the chosen solution by putting it into action.
9. Develop an implementation plan, assign responsibilities, and execute the necessary steps to implement the solution effectively.

10. Monitor as per requirements: Continuously monitor and evaluate the implemented solution to ensure its effectiveness and identify any necessary adjustments. Measure key performance indicators, gather feedback, and make improvements as needed.

11.17 Key Performance Indicators (KPIs)

KPIs are quantifiable measures that evaluate the efficiency, quality, and safety of healthcare services. These metrics help hospitals monitor performance, identify areas for improvement, and enhance patient care outcomes. Some critical KPIs include:

1. **Patient Safety Indicators**
   - **Hospital-Acquired Infections (HAIs):** Tracks the rate of infections like catheter-associated urinary tract infections (CAUTI) and surgical site infections (SSI).
   - **Medication Errors:** Measures the incidence of errors in prescribing, dispensing, or administering medications.
   - **Adverse Events:** Monitors occurrences of falls, pressure ulcers, or complications during care.
2. **Clinical Effectiveness**
   - **Mortality Rate:** Tracks in-hospital deaths, often adjusted for patient risk factors, providing insights into care quality.
   - **Readmission Rate:** Measures the percentage of patients readmitted within 30 days of discharge, indicating care quality and discharge planning effectiveness.
   - **Average Length of Stay (ALOS):** Tracks the average number of days patients spend in the hospital, influencing bed turnover and resource utilization.
3. **Operational Efficiency**
   - **Bed Occupancy Rate:** Measures the proportion of occupied beds, reflecting hospital capacity and resource utilization.
   - **Waiting Time for Services:** Monitors patient wait times in emergency departments, outpatient clinics, and surgeries, indicating operational efficiency.
   - **Surgical Turnaround Time:** Tracks the time between surgeries to optimize operating room use.
4. **Patient Satisfaction**
   - **Patient Feedback and Satisfaction Scores:** Evaluates patient perceptions of care quality, communication, and overall experience.
   - **Net Promoter Score (NPS):** Measures the likelihood of patients recommending the hospital to others, reflecting overall patient satisfaction.

5. **Staff Performance and Engagement**
   - **Staff-to-Patient Ratio:** Monitors the number of healthcare workers per patient to ensure adequate staffing for safe care.
   - **Staff Satisfaction and Turnover Rate:** Tracks employee engagement, burnout, and retention, directly affecting care quality.

Monitoring these KPIs helps hospitals maintain high standards, improve patient outcomes, and ensure operational efficiency. Figure 11.38 illustrates examples of various types of KPIs to evaluate the efficiency, quality, and safety of healthcare services.

By following these steps, organizations can systematically analyze problems, identify viable solutions, and implement effective measures to address the challenges they face. Table 11.4 depicts an appropriate quality tool depending on the objectives.

> If your KPI indicators have been meeting expectations for too long, it should be a common practice to purposely raise the bar on the standard – **Tracey & Ernie Richardson**

The PDCA (Plan-Do-Check-Act) cycle is a continuous improvement framework used to identify problems, implement solutions, monitor effectiveness, and refine processes. Proper documentation at each stage ensures transparency, accountability, and data-driven decision-making, helping organizations track progress and sustain quality improvements (Fig. 11.39). Proper documentation at each stage ensures transparency, accountability, and data-driven decision-making, helping organizations track progress and sustain quality improvements.

> Becoming satisfied with the way things are is the first step backward – **Jon Miller**

Table 11.4 Appropriate quality tool depending on objectives

Objective	Quality tool
To keep track of the facts	Use a check sheet or line graph.
To represent data visually	Use a histogram, line graph, or pareto diagram
To group ideas	Use a lotus flower diagram, fishbone diagram or affinity diagram
To figure out relationship	Use a fishbone or relations diagram
To see the steps in process	Use a flowchart
To determine a root cause of a problem	Use a fishbone diagram or 5 Whys
To evaluate strategic options	Use a SWOT (Strength, Weakness, Opportunity, Threat) analysis
To reach a conclusion that considers various elements.	Use a grid analysis or paired comparison.

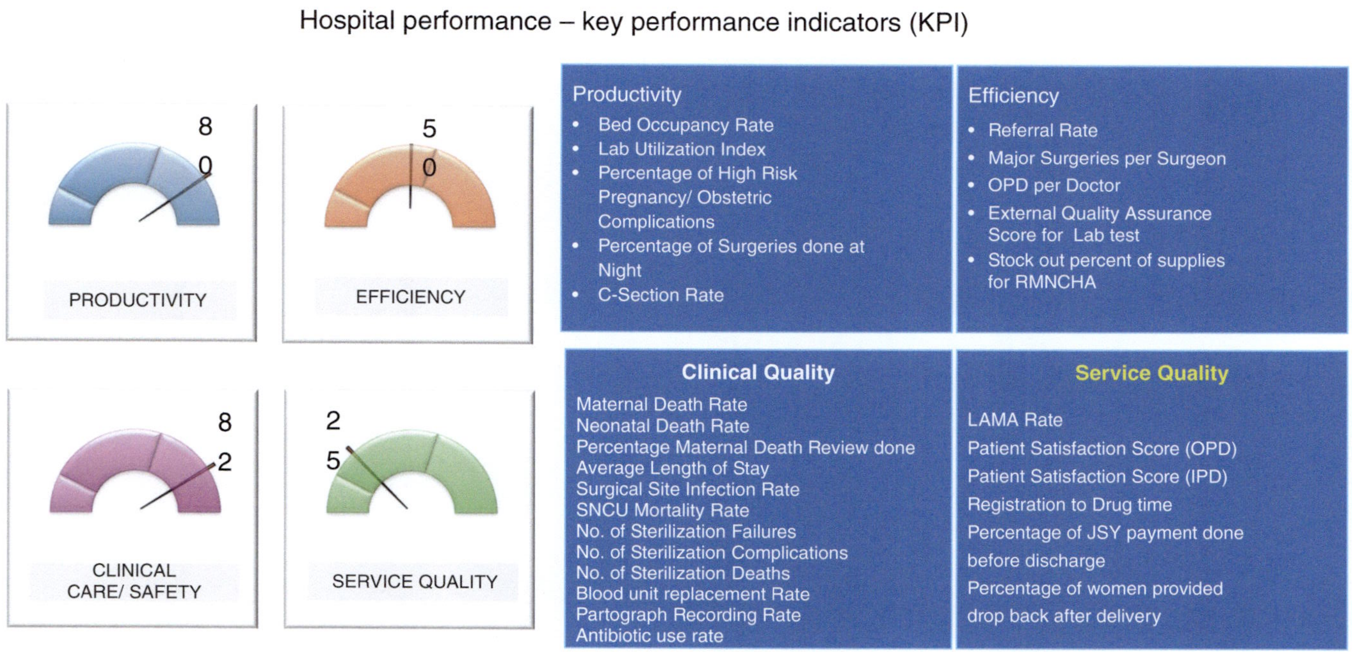

Fig. 11.38 Key performance indicators to evaluate the efficiency, quality, and safety of healthcare services

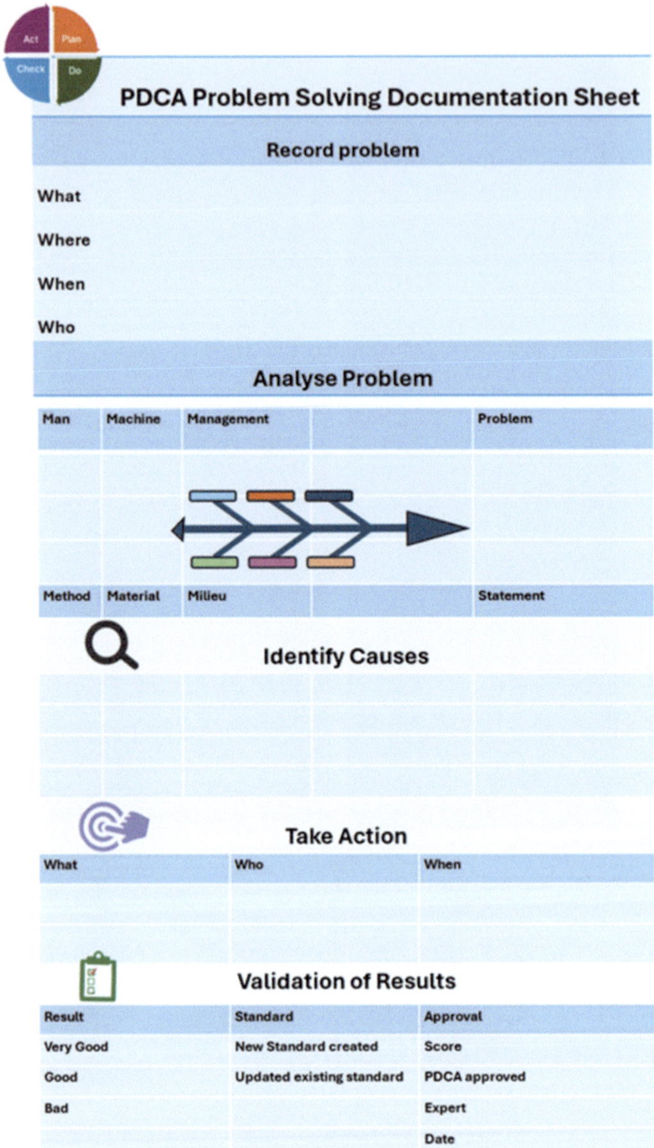

Fig. 11.39 Sample documentation sheet for progress of PDCA cycles

Quality Improvement Models 12

Contents

Quality improvement models provide structured approaches or methodologies to guide organizations in improving processes, products, or services. There are several quality improvement models that organizations can use to guide their improvement efforts. Some commonly used quality improvement models:

1. Plan-Do-Study-Act (PDSA) Cycle: The PDSA cycle is a simple and iterative model for continuous improvement. It involves four steps: plan (identify a problem and plan a change), do (implement the change on a small scale), study

S. Sharma, *Quality Demystified*, https://doi.org/10.1007/978-981-95-0339-1_12

(analyze the results and learn from them), and act (adjust and implement the change on a larger scale).

2. Six Sigma: Six Sigma is a data-driven methodology focused on reducing defects and enhancing process performance. It follows a structured approach known as DMAIC (Define, Measure, Analyze, Improve, Control) to identify and eliminate root causes of problems and achieve process excellence.
3. Lean: Lean is a systematic approach that focuses on eliminating waste and improving efficiency in processes. It aims to create value for the customer by optimizing resources, streamlining workflows, and reducing non-value-added activities.
4. Total Quality Management (TQM): TQM is a management philosophy that emphasizes the involvement of all employees in continuous improvement efforts. It aims to meet customer expectations by focusing on customer satisfaction, process improvement, and a culture of quality throughout the organization.
5. Model for Improvement: The Model for Improvement, developed by the Institute for Healthcare Improvement (IHI), is a framework that guides quality improvement initiatives. It involves three fundamental questions: What are we trying to accomplish? How can we ascertain that a change has led to an improvement? What specific changes can we implement to ensure positive results?
6. ISO 9001: ISO 9001 is an internationally recognized standard for quality management systems. It provides a systematic approach for organizations to ensure customer satisfaction, improve processes, and meet regulatory requirements. The standard emphasizes the importance of customer focus, leadership, and continual improvement.
7. Baldrige Excellence Framework: Developed by the U.S. National Institute of Standards and Technology (NIST), the Baldrige framework provides a comprehensive approach to organizational excellence. It includes criteria related to leadership, strategy, customers, measurement, analysis and knowledge management, workforce, operations, and results.

These models provide frameworks and tools for organizations to systematically identify, analyze, and implement improvements in their processes, products, and services. The selection of a model is contingent on the unique needs and context of the organization.

12.1 Plan-Do-Study-Act (PDSA) Cycle and SDCA (Standardize-Do-Check-Act) Cycle

The Plan-Do-Study-Act (PDSA) cycle and SDCA (Standardize-Do-Check-Act) cycle both are widely adopted and used as a follow-up model for implementing and monitoring improvements. Both follow a systematic series of steps to continuously enhance a product, service, or process. Rather than a closed circle, PDSA and SDCA cycle are best visualized as a spiral, emphasizing the ongoing nature of improvement.

PDSA Cycle:

1. Plan: Identify the problem or objective, set improvement goals, and develop a plan for change.
2. Do: Implement the planned change on a small scale, carefully documenting the process and any unexpected observations.
3. Study: Analyze the data and results collected during the "Do" phase to evaluate the impact of the change. Assess whether the desired improvements were achieved and identify any areas for further refinement.
4. Act: Based on the findings from the "Study" phase, make adjustments and modifications to the improvement plan. Refine the process and implement the revised changes on a larger scale, incorporating the lessons learned from the previous cycles.

On the other hand, the SDCA cycle emphasizes standardization and continuous improvement of existing processes. It starts with Standardize, establishing best practices and procedures to ensure consistency and reliability in operations. The Do phase involves executing these standardized processes in daily activities. In the Check phase, outcomes and processes are systematically monitored and evaluated against the established standards to identify deviations or opportunities for improvement. Finally, in the Act phase, adjustments are made to refine processes and maintain or improve performance over time.

While both cycles aim to drive improvement through iterative processes, the PDSA cycle is particularly suited for testing and implementing new ideas or changes, often in experimental or dynamic environments. In contrast, the SDCA cycle focuses on maintaining and refining existing practices through systematic monitoring and adjustment. Choosing between these cycles depends on the nature of the improvement goal—whether it involves testing new solutions (PDSA) or ensuring ongoing adherence to standards (SDCA)—while both contribute to enhancing quality and patient outcomes in healthcare settings.

By engaging in small, rapid cycles of change, the PDSA cycle enables teams to continually assess the effectiveness of improvement strategies. Repeating the cycle multiple times, particularly for complex changes, allows teams to thoroughly test the impact of the change and make necessary adjustments for further improvement (Fig. 12.1).

The PDSA cycle provides a structured approach to learning from experience, making informed decisions, and driving ongoing improvement. It is a valuable model for organizations seeking to foster a culture of continuous improvement and achieve sustained progress.

> Standards are not there mainly to maintain but as a basis for improvement. A standard is not the platform on which you stand, but a step on the ladder before the next one you take – **Jon Miller**
> Oftentimes, our trainers would see it as a problem if a standard remained stagnant for too long-they would say, "It's too easy!" – **Tracey & Ernie Richardson**

Fig. 12.1 Sequential improvement with multiple SDCA/PDSA cycles

Other quality improvement models that may be used when implementing change include the LEAN approach and Six Sigma. Both models place PDSA within a specific context to provide direction to the quality improvement process and make effective use of resources.

It is impossible to improve any process until it is standardized – **Masaki Imai**
One must standardize, and thus stabilize the process before continuous improvements can be made– **Jeffrey Liker**
Standardized work is the basis for empowering workers, sharing ideas for improvement, and driving innovation in the workplace – **Jeffrey Liker**
The standardized work must be specific enough to be a useful guide, yet general enough to allow for some flexibility – **Jeffrey Liker**

12.2 Lean in Healthcare

Muda, Mura, and Muri, Japanese terms, are three key concepts in the Toyota Production System (TPS), i.e., Lean manufacturing and management that are essential for identifying and eliminating inefficiencies within a system. Taiichi Ohno, "father" of the Toyota Production System, originally identified seven forms of *muda* or waste. One of the key steps in lean process and TPS is to identify which activities add value and which do not, then to progressively work to improve or eliminate them.

Muda (Waste) Muda refers to any activity that does not add value from the patient's perspective. There are two types of muda:

- **Muda Type I**: non-value-adding, but necessary for end-customers. These elements are typically more difficult to eliminate because, although they are classified as non-value-adding, they might still be essential.
- **Muda Type II**: non-value-adding and unnecessary for end-customers. These factors add to waste, generate hidden costs, and should be eliminated.

In healthcare, examples include unnecessary diagnostic tests, excessive inventory of medications or supplies, and waiting times for patients or staff. Eliminating Muda involves streamlining processes to focus on value-added activities.

Mura (Unevenness) Mura pertains to the variability in processes that can lead to inefficiencies. In a healthcare setting, this could manifest as inconsistent patient flow, fluctuating workloads for staff, or irregular scheduling of appointments. Addressing Mura involves standardizing procedures and levelling workloads to ensure a smooth, predictable workflow.

Muri (Overburden) Muri is about overloading resources beyond their capacity, which can cause stress, errors, and inefficiencies. In healthcare, this could mean

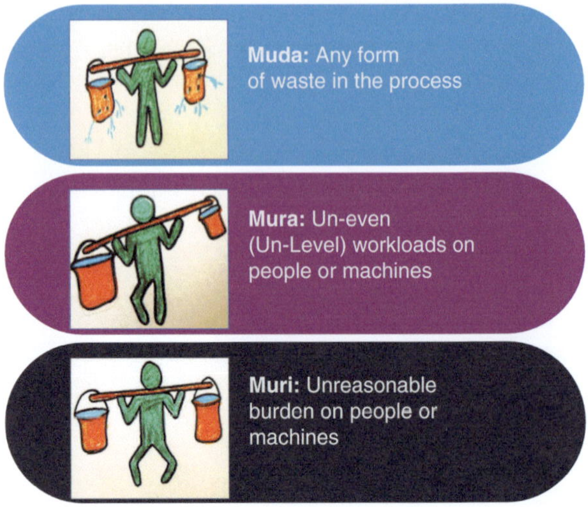

Fig. 12.2 Minimize Muda, Mura, Muri for efficiency

overburdening nurses with too many patients or expecting equipment to operate without proper maintenance. Reducing Muri involves balancing workloads and ensuring resources are used within their limits.

In Lean management, Muda, Mura, and Muri are interrelated and often addressed together to optimize processes. By systematically identifying and reducing these three types of inefficiencies, organizations can improve quality, reduce costs, and enhance overall operational efficiency (Fig. 12.2). This approach is widely applied in various sectors, including manufacturing, healthcare, and services, to create more streamlined and effective systems.

A systematic approach to eliminate the "3Ms" — Muda (waste), Mura (unevenness), and Muri (overburden/un-level)—is fundamental to lean management and process improvement. This involves a structured three-phase methodology that begins with Phase 1: Identify, where time studies are conducted to assess workload distribution (Muri), the entire process is mapped to detect inconsistencies (Mura), and Gemba walks are performed to observe waste in real settings (Muda). In Phase 2: Analyze, the focus shifts to identifying unbalanced workloads across people and machines, pinpointing areas for improvement within the process, and conducting root cause analysis to tackle each form of waste. Phase 3: Implement involves applying targeted solutions, standardizing work, and balancing workloads to reduce overburden, introducing level production and pull systems to minimize unevenness, and eliminating non-value-adding activities to cut down on waste. Supporting tools such as value stream mapping and Kaizen promote ongoing improvement, while engaging frontline staff fosters a culture of continuous problem-solving. Together, this integrated and systematic approach enhances operational efficiency, ensures consistent quality, and improves employee well-being.

> No matter how good our results look, there is still waste hidden in our processes, and we can never afford to stop looking for continuous improvement opportunities – **Tracey & Ernie Richardson**

12.3 The 8 Wastes in Healthcare

1. **Waiting—**Employee or equipment idle or down time for parts, components, raw materials, approvals, and paperwork (requisition to be approved, purchase order), or others to join a meeting, patients in waiting rooms or examination rooms, waiting for test results for patients in emergency room emergency department patients waiting for admission to the hospital.
2. **Transportation—**Unnecessary movement of materials or people that does not add value. For example, unnecessary movement of medicines supplies from storage to the floor then from the pharmacy to where it is needed; Patients are forced to walk from building to building for different aspects of their care. Though some of this transportation is considered "necessary" this waste to be minimized if it can't be eliminated. Transportation waste when patients are moved to different departments for tests or procedures that could have been done in one location.
3. **Over-processing (excess or unnecessary)** involves doing more work than necessary. It is different from the waste of overproduction. It includes treatment by specialists when primary care providers can manage; ordering complex/higher diagnostic tests such as (MRI) when a simpler method would suffice (X-ray or physical therapy); surgical intervention instead of an equally effective medical alternative.
4. **Motion—**Wasted walking or movement to adjust equipment as it is not conveniently located, or walking to fetch materials as hospital layout is not consistent with the workflow. For example, supplies might not be stored where patient care occurs. This can be addressed by rearranging the layout, tools, and personnel. Improper motion can lead to unnecessary fatigue and long-term injuries. Therefore, proper ergonomics should be applied when making adjustments to motion studies.
5. Only items required at the Point of Use (POU) such as necessary tools, materials, and paperwork should be arranged at the workstation. The less needed items be stored further away. Motion waste from healthcare providers searching for equipment or supplies in a disorganized workspace.
6. **Defects or poor "quality"—**Errors, rework, or repair activities are examples of defect waste. This type of waste includes the time and resources spent creating a defect, the efforts required to rework these defects, and the time spent inspecting them. For example, misdiagnosis; medication errors, Hospital-acquired infections, and other harm; wrong entries in records. Defects in medication administration leading to incorrect dosages or medication errors.
7. **Inventory—**Storing excess inventory/materials a critical metric that may expire, or become redundant over time, excess bedside equipment as it ties up working capital, creates extra handling, storage, and may never be recovered. For exam-

ple, carrying excess inventory to take care of the long lead time and difficulties known at start-up such as uncertainty of the supplies. Just-in-time inventory has been employed to take care of this type of waste. Inventory waste due to stockpiling excessive supplies or medications that may expire before use.

8. **Overproduction**—Servicing more, sooner, faster than required by the next step in the process. This type of waste is the most critical waste type to control since often it is done to cover an underlying problem and involves the other six wastes. Examples include unnecessary tests, ordering medications that the patients do not need, having doctors perform tasks other staff can handle effectively, and duplication of services. Overprocessing in administrative tasks, such as redundant paperwork or excessive documentation requirements. Overproduction of lab tests, ordering unnecessary tests that do not contribute to the diagnosis or treatment.
9. **Underutilized resources**—Underutilized resources in healthcare represent significant inefficiencies that can hinder optimal patient care and waste valuable assets. These resources include human capital, technology and equipment, data and information systems, community resources, and patient engagement. Healthcare professionals, such as nurses and doctors, may not be utilized to their full potential due to rigid job descriptions, bureaucratic processes, or insufficient training, which can lead to burnout and limit the quality of care provided. Advanced medical technologies and diagnostic tools are often underused because of a lack of training or integration into clinical workflows, resulting in wasted financial investments and missed opportunities for patient benefit. Furthermore, healthcare systems frequently gather vast amounts of data that remain underutilized; without effective data analysis and sharing, valuable insights into patient care, outcomes, and resource allocation may be lost. Many organizations also fail to leverage community resources, such as local support groups and public health initiatives, which can enhance patient education and promote preventive care. Additionally, patients are often not fully engaged in their own care plans, leading to the underutilization of available resources like telehealth services and self-management tools. Addressing these underutilized resources requires a strategic approach that fosters innovation, enhances training, and promotes collaboration among stakeholders. By maximizing the use of available assets, healthcare systems can improve efficiency, reduce waste, and ultimately enhance patient care.

Productivity in healthcare is crucial for delivering efficient and effective patient care while maximizing resource utilization and maintaining high-quality standards. Whether it's in the emergency department, operating room, outpatient clinic, or diagnostic imaging center, productivity plays a vital role in optimizing patient flow, reducing waiting times, and enhancing overall healthcare delivery.

> There is no waste in business more terrible than overproduction – **Taiichi Ohno**
> Lean manufacturing dramatically increases the importance of building things right the first time. For example, with very low levels of inventory, there is little buffer to fall back on in case there is a quality problem – **Jeffrey Liker**

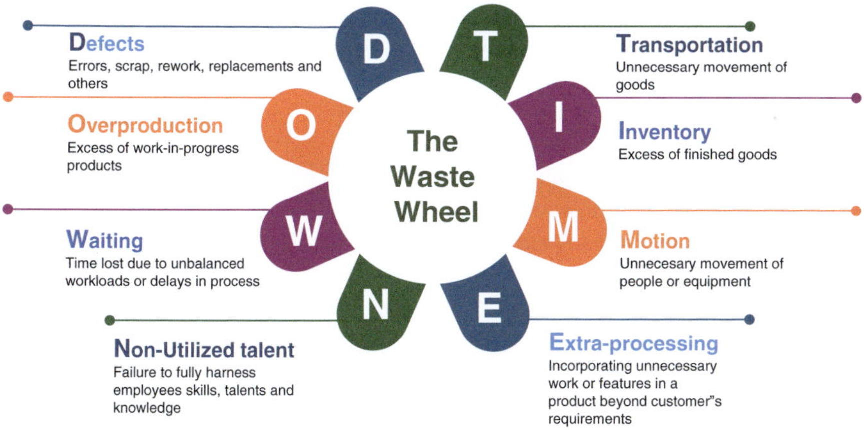

Fig. 12.3 The waste wheel with an acronym DOWNTIME

12.3.1 The Waste Wheel

By removing waste, organizations enhance their ability to deliver quality services while allowing staff to improve their work-life balance and job satisfaction (Fig. 12.3). Eliminating waste frees up time and resources from unproductive actions, increasing capacity to work more effectively and efficiently. A common acronym used to categorize wastes in healthcare is **DOWNTIME**, which stands for:

- Defects
- Overproduction
- Waiting
- Non-utilized talent
- Transportation
- Inventory
- Motion
- Extra processing

Examples of Some Improvement Activities for Lean

1. Streamlining Triage Process for Patients with Minor Injuries:
   - Identify and eliminate bottlenecks in the triage process to reduce waiting times. Patients with minor injuries often wait for the longest due to triage.
   - Implement standardized protocols for assessing and treating patients with minor injuries.
   - Optimize the flow of patients by grouping similar complaints and providing efficient assessment and treatment.
2. Creating an Acute Assessment Unit:
   - Establish a dedicated unit for patients with short stays.
   - Streamline the steps involved in assessment, diagnostics, treatment decision, and discharge.

- Incorporate specialist input to ensure comprehensive and efficient care.

3. Splitting the Emergency Assessment Unit:
   - Divide the Emergency Assessment Unit into two distinct areas.
   - Allocate a 24-bed rapid assessment area for initial evaluation and stabilization.
   - Designate a 16-bed short-stay inpatient facility for patients requiring further observation and treatment.
4. Additional Lean Tools Implementation:
   - Implement day care surgery to reduce hospital stays and improve efficiency.
   - Enable admission on the day of surgery to minimize waiting times and optimize resource utilization.
   - Establish a short-stay medical inpatient unit/ward for patients with less complex conditions.
   - Redesign inpatient flows to streamline processes and enhance patient flow.
   - Focus on reducing trauma incidents and improving turnaround times in operating theatres.

These examples illustrate how Lean principles and tools can be applied to specific areas of healthcare to eliminate waste, improve patient flow, enhance efficiency, and optimize resource utilization.
Efficiently Identifying Waste and Implementing Lean in Healthcare:

> It is a mistake to tie headcount reductions to the Lean process – **Mark DeLuzio**

1. Waste Eyes and Waste Walks:
   (a) Empower staff with minimal training to easily identify waste in their work areas (known as waste eyes).
   (b) Conduct waste walks as a quick and resource-efficient method to identify areas for improvement and establish a baseline of the current state.
   (c) Engage teams in problem-solving and finding solutions that benefit them, encouraging the implementation of PDSA cycles of improvement.
   (d) It is beneficial to have someone from outside the area accompany the walk to provide a fresh perspective.
2. The Role of Leadership in Lean Healthcare:
   - Leadership is a critical success factor for the effective adoption of lean principles in healthcare organizations.
   - Lean leadership is an integral part of the Lean Operating System, not an afterthought.
   - Understanding the implementation of lean healthcare involves recognizing the importance of Lean leadership in making a broader impact.
   - The concept of Lean leadership is exemplified through the Nine Leadership Behaviours framework, which guides leaders in developing and demonstrating Lean principles.
   - Taking a pragmatic approach, leaders can turn these principles into meaningful action, resulting in rapid and transformative changes.

- The personal changes in leaders' behaviors can have a contagious and astonishing effect, contributing to true Lean leadership.

By embracing waste identification through waste eyes and waste walks, and emphasizing the crucial role of Lean leadership, healthcare organizations can unlock the potential of Lean to significantly improve their operations, enhance efficiency, and drive positive change throughout the organization.

Quality means doing it right when no one is looking.– **Joseph M. Juran**
Leaders must have tolerance for allowing employees to make mistakes, always focusing on the process and not the individual. – **Mark DeLuzio**
Don't let perfect get in the way of better" is what Lean practitioners need to keep in mind. – **Mark DeLuzio**

There are two paths among organizations when it comes to shifting to a lean management system:

1. Organizations that implement lean principles adopt a top-down approach, guided by top managem`ent. This ensures leadership commitment and strategic alignment. Additionally, involve frontline employees, as their participation and insights for the effective execution of lean practices.
2. Organizations that are trying to make improvements here and there in the name of lean. Results in such case are mediocre or harmful in the long term.

In healthcare, some common pitfalls of implementing lean principles include:

Failure to Engage Frontline Staff: Without involving frontline healthcare workers in process improvement initiatives, the true benefits of lean cannot be realized. Lack of engagement can lead to resistance to change and limited sustainability of improvements.

Overemphasis on Cost Reduction: While reducing costs is important, solely focusing on financial gains without considering the impact on patient care quality and safety can result in compromised outcomes and staff dissatisfaction.

Ignoring Variation in Patient Needs: Healthcare is inherently variable, and applying standardized lean processes without considering the unique needs of individual patients can lead to ineffective or inappropriate care delivery.

Lack of Leadership Commitment: For a lean transformation to be successful, it demands unwavering leadership commitment at every organizational level. Without dedicated support and involvement from leaders, achieving the necessary cultural and operational changes for effective lean implementation becomes challenging. Without visible support from leadership, initiatives may falter, and progress may stagnate.

Short-Term Focus: Lean should be seen as a long-term cultural transformation, not merely a quick-fix solution. Rushing through improvement projects without adequate planning or sustainability measures can undermine the effectiveness of lean initiatives over time.

Addressing these pitfalls requires a holistic approach that prioritizes patient-centered care, engages frontline staff, fosters a culture of continuous improvement, and ensures strong leadership support throughout the organization.

Organizations that superficially adopt lean principles, also known as fake lean, often see mediocre or harmful results in the long term. In healthcare, fake lean refers to the misguided or superficial implementation of lean management principles without a genuine commitment to continuous improvement or patient-centered care. This approach involves using lean techniques for short-term gains or cost-cutting measures without addressing underlying issues or fostering a culture of continuous improvement and respect for people. For example, a hospital may adopt lean practices solely to reduce costs, neglecting the impact on patient outcomes or staff well-being. This can lead to streamlined processes and reduced expenses, but without addressing staff burnout or patient safety concerns. Ultimately, fake lean in healthcare prioritizes short-term financial gains over long-term improvements in quality of care and patient satisfaction.

While leaders must understand all forms of waste so they can actively take a role in eliminating waste, they must be attuned to the fact that they are often the direct cause of mura and muri – **Mark DeLuzio**

You can't make the transition to Lean unless the management team acts as one – **Mark DeLuzio**

The maniacal focus on cost alone has derailed many improvement initiatives **– Mark DeLuzio**

You must go beyond using the tools for improvements to applying Lean's principles in ways that they become indoctrinated into your business thinking and culture to impact the organization as a whole – **Mark DeLuzio**

12.4 Gemba Walk

The Gemba walk is a cornerstone of lean management, derived from the Japanese term "Gemba," which translates to "the real place" or "where the work is carried out." It involves leaders or managers going directly to the Gemba, such as the production floor, office, or service area, to observe operations, engage with frontline employees, and gain firsthand insight into processes and problems.

Key aspects of the Gemba walk include:

Observation: During the Gemba walk, leaders observe operations in real-time, paying attention to how work is performed, the flow of materials or information, and any issues or bottlenecks that arise. Observing processes firsthand allows leaders to understand the current state of operations and identify opportunities for improvement.

Engagement: Engaging with frontline employees is a crucial aspect of the Gemba walk. Leaders ask questions, listen to employees' perspectives, and encourage them to share their ideas and insights for improvement. By involving employees in the process, the Gemba walk fosters a culture of collaboration and empowers employees to contribute to continuous improvement efforts.

Problem-Solving: The Gemba walk is not just about observation; it also involves problem-solving. Leaders work with employees to address immediate issues or challenges that are identified during the walk. By addressing problems in real-time, the Gemba walk facilitates quick resolution and prevents issues from escalating.

Continuous Improvement: At its core, the Gemba walk is a tool for continuous improvement. By regularly visiting Gemba and engaging with employees, leaders demonstrate their commitment to continuous improvement and create opportunities for incremental changes that lead to greater efficiency, quality, and customer satisfaction.

Overall, the Gemba walk is a powerful practice that enables leaders to gain first-hand insight into operations, engage with employees, and drive continuous improvement efforts. By physically visiting the areas where patient care is delivered, leaders can observe workflow, identify inefficiencies, adherence to protocols, identify opportunities for standardization, and ensure that patient safety measures are being followed effectively and address potential issues in real-time. For example, during a Gemba walk in a hospital ward, leaders may observe nurses' medication administration practices, identify potential errors, and provide immediate feedback to enhance patient safety protocols. By embracing the Gemba walk, organizations can create a culture of accountability, collaboration, and excellence that leads to sustainable business success.

Additionally, the Gemba walk fosters a culture of collaboration and empowerment among healthcare staff. By engaging with frontline employees during the walk, leaders demonstrate their commitment to listening to staff concerns, recognizing their contributions, and involving them in decision-making processes. This collaborative approach encourages staff to share their insights, suggest improvements, and take ownership of quality and safety initiatives. Figure 12.4 reveals the steps to effective GEMBA walks.

Furthermore, the Gemba walk is particularly valuable during times of change or when implementing new processes or technologies in healthcare settings. Leaders can assess the impact of changes on workflow, identify any challenges or barriers to implementation, and provide support to staff as they adapt to new practices. For instance, when introducing a new electronic health record system, leaders can conduct Gemba walks to observe staff interaction with the system, address usability issues, and provide additional training or resources as needed. Do's and don'ts of Gemba walk are summarized in Table 12.1.

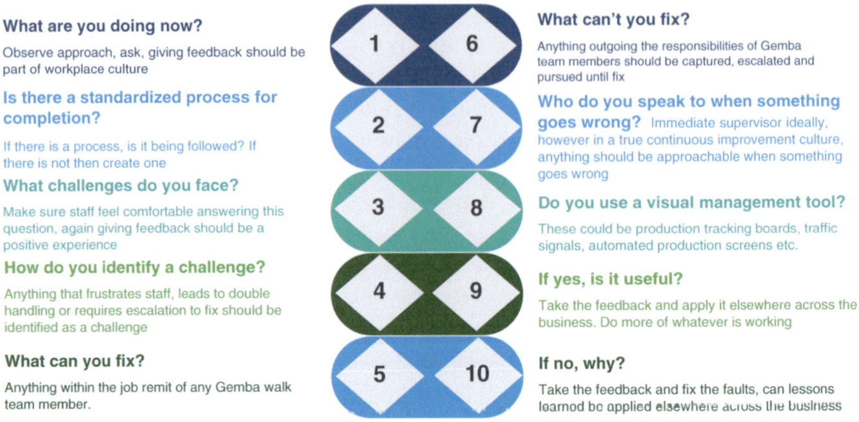

Fig. 12.4 10 Steps to effective GEMBA walks

> Direct observation is essential, and no combination of indirect methods, however clever, can possibly take its place – **Steven J. Spear**
> Gemba Kaizen is a human system that works only when it is people-centered – **Masaki Imai**
> Job titles are unimportant. In the end, the people who know the Gemba are the most respected – **Akio Toyoda**

The Gemba walk and Waste Walk are both essential tools in lean management, but they serve different purposes and focus on different aspects of improvement within an organization. The Gemba walk emphasizes understanding processes and engaging with employees, while the Waste Walk focuses on eliminating waste and improving efficiency (Table 12.2).

> Never allow yourself to move faster than the facts you confirm at the Gemba, and never let down your guard against the dangers of assumptions – **Tracey & Ernie Richardson**

12.5 Takt Time

Takt time helps in unmasking the hidden waste such as inefficiencies and delays. Takt time, in the context of healthcare, refers to the average time available for each patient encounter, procedure, or task to ensure that patient care activities are performed efficiently and in a timely manner to meet patient needs. It helps healthcare

Table 12.1. Do's and don'ts of Gemba walk

Do's of Gemba Walk	Don'ts of Gemba Walk
Engage with frontline employees: Interact with workers to gain insights and understand their perspectives.	Micromanage: Avoid dictating solutions or interfering with daily operations; instead, empower employees to find and implement solutions. Do not make it a fault-finding mission.
Observe processes: Pay attention to how work is performed and look for opportunities for improvement. Track waste habits	Overlook safety protocols: Ensure compliance with safety procedures and guidelines to prevent accidents or injuries during the walk.
Ask open-ended questions: Encourage employees to share their ideas and suggestions for enhancing efficiency and quality. Enquire about workflow obstructions.	Ignore employee feedback: Listen attentively to employees' concerns and suggestions and avoid dismissing their input. Do not judge people performance.
Focus on problem-solving: Address issues or challenges identified during the walk in real-time to prevent them from escalating.	Rush through the process: Take the time to thoroughly observe processes and engage with employees; rushing may result in overlooking important insights. Do not assume processes are working as documented. Pay attention to task transitions. Do not focus on critical areas only, while ignoring the whole process.
Lead by example: Demonstrate commitment to continuous improvement by regularly conducting Gemba walks and actively participating in improvement initiatives.	Lose sight of the bigger picture: While focusing on immediate issues, maintain awareness of broader organizational goals and objectives to ensure alignment with strategic priorities.
Mix-up the schedule: Organize Gemba walks at various times. Track walk outcomes to detect patterns. Observe how time variables impact the workflow.	Do not strictly adhere to a set schedule. Do not assume all processes work the same way. Do not neglect the impact of time on the operation.
Acknowledge all contributions. Share the outcomes with the team. Discuss potential action plans with key stakeholders	DO not ignore employee contributions. Make decisions without open discussions. Do not tailor action plans without involving the team

organizations align their resources and workflows with patient demand to optimize the delivery of care.

Some examples of Takt time are:

Emergency Department (ED) Takt Time: The average time allocated for each patient's evaluation and treatment in the emergency department. Hospitals may set a target Takt time to ensure timely care for patients presenting with various medical conditions, such as chest pain, trauma, or stroke.

Table 12.2 Differences between the Gemba walk and Waste walk

	Gemba Walk	Waste Walk (Waste Audit)
Purpose	It is a practice where managers or leaders go to the "Gemba," which is the place where work is done, to observe operations, engage with frontline employees, and gain firsthand insight into processes and problems.	It is a structured activity aimed at identifying and eliminating various forms of waste within processes.
Focus	To understand the current state of operations, identify opportunities for improvement, and support continuous improvement efforts. It emphasizes observing processes, asking questions, and listening to employees to gather insights and perspectives.	To identify the eight forms of waste in lean management: overproduction, waiting, unnecessary transportation, overprocessing, excess inventory, motion waste, defects, and underutilized talent.
Key benefits	Fosters a culture of collaboration, empowers employees to share ideas for improvement, and enables leaders to provide support and guidance based on direct observations.	It is a structured activity aimed at identifying inefficiencies and non-value-added activities within processes, leading to targeted efforts to eliminate waste, improve productivity, and streamline operations.

Operating Room (OR) Takt Time: The average time allocated for each surgical procedure in the operating room. Surgical departments may establish Takt time targets to optimize scheduling, minimize delays, and maximize OR utilization while ensuring patient safety and quality of care.

Appointment Takt Time: The average time allocated for each patient appointment in outpatient clinics or primary care practices. Healthcare providers may set Takt time goals to maintain efficient patient flow, minimize waiting times, and enhance patient satisfaction during clinic visits.

Medication Administration Takt Time: The average time allocated for nurses or pharmacy staff to prepare and administer medications to patients. Healthcare facilities may establish Takt time standards to ensure timely medication delivery, reduce medication errors, and improve patient safety.

Diagnostic Testing Takt Time: The average time allocated for each diagnostic test, such as laboratory tests, imaging studies, or diagnostic procedures. Diagnostic departments may set Takt time targets to minimize turnaround times for test results and facilitate timely diagnosis and treatment decisions.

Overall, Takt time in healthcare helps organizations optimize resource allocation, streamline workflows, and improve patient flow to deliver high-quality, patient-centered care efficiently and effectively.

Takt time and turnaround time (TAT) are both metrics used in process management, involve time management within processes, but they represent different aspects: Takt time focuses on pacing work to meet demand, while turnaround time measures the total time required to complete a process or service. Turnaround time refers to the total time elapsed between the submission of a request or initiation of a process and the completion or delivery of the final output. It is commonly used in healthcare to measure the time it takes for diagnostic tests, procedures, or services to be performed and results to be reported back to the ordering healthcare provider or the patient. Turnaround time is critical in healthcare settings to ensure timely diagnosis, treatment, and decision-making.

Identifying and addressing these wastes, time management can lead to improved efficiency, reduced costs, and enhanced patient satisfaction in healthcare settings.

The most dangerous kind of waste is the waste we do not recognize. – **Shigeo Shingo**

Lean thinking was introduced by Toyota in the 1960s as a systematic approach to identifying and eliminating waste or non-value-added activities in organizations. It focuses on continuous improvement with the aim of creating value. Waste is defined as any activity that consumes resources without adding value for the customer.

In the context of healthcare, Lean principles aim to deliver value without waste, ensuring high quality at a low cost. The goal is to eliminate waste in every aspect of operations, reducing inventory, cycle times, and costs. This enables the delivery of higher quality patient services in an efficient, effective, and responsive manner while maintaining the economic viability of the organization.

A Lean organization optimizes human effort, minimizes space requirements, reduces the need for capital investment, decreases the quantity of raw materials and supplies consumed, and streamlines production and delivery times.

The Lean Toolbox and the Waste Wheel are tools used in Lean thinking and process improvement, but they serve different purposes and have different focuses.

1. Lean Toolbox: The Lean Toolbox is a collection of various tools, techniques, and methodologies that organizations can use to identify and eliminate waste, improve processes, and drive continuous improvement. It includes a wide range of tools such as value stream mapping, 5S, kanban, standardized work, error-proofing (poka-yoke), and many others. The Lean Toolbox provides organizations with a comprehensive set of methods to analyze, optimize, and streamline their processes, ultimately aiming to create value for customers while minimizing waste and inefficiencies.
2. Waste Wheel: The Waste Wheel, on the other hand, is a specific tool within the Lean Toolbox that focuses on categorizing and visualizing the different types of waste, also known as the 7 Wastes or 7 Muda. The Waste Wheel helps teams and organizations identify and understand the various forms of waste that exist within their processes. It provides a structured framework to recognize and categorize waste into specific types such as overproduction, waiting, transportation, overprocessing, inventory, motion, and defects. By using the Waste Wheel, organizations can gain clarity on the types of waste present in their processes and target them for elimination or reduction.

When implementing Lean in healthcare, the focus is on improving quality from the patient's perspective. Processes that do not provide value to the patient are targeted for improvement. All team members are encouraged to identify areas of poor quality and waste in operations. A Lean Toolbox can be utilized to support waste elimination and streamline process flows. It is crucial to prioritize the needs of the patient, as the product or service provided can have a significant impact on their well-being.

12.6 Value Stream Mapping

To qualify as value-added, an activity must meet three essential criteria:

1. It should hold value from the patient's or customer's perspective.
2. It must transform the product or service in some meaningful way.
3. It needs to be executed correctly the first time.

Examples of value-added activities in healthcare include diagnostic time, active care time (interventions/procedures), passive care time (under observation without active intervention), joint working between health and social services, and patient journey/pathways.

Conversely, non-value-added activities should be minimized or eliminated. These activities may include check-in time, wait time, filling out multiple forms, and conducting inconclusive or duplicate tests.

12.7 Six Sigma

Six Sigma is a methodology and set of tools widely used in process improvement to minimize defects, variation, and waste, ultimately aiming for higher quality and efficiency. Developed by Motorola in the 1980s and later popularized by General Electric, Six Sigma is a methodology that focuses on achieving high levels of process performance. The term "Six Sigma" signifies a statistical concept aimed at reducing defects to just 3.4 per million opportunities, defining defects as units not belonging to the intended population. Since its inception, Six Sigma has been integrated into numerous TQM initiatives. Sigma (σ) represents the standard deviation, a measure of population variation, while the lowercase 's' estimates this variation based on a sample. A process meeting the Six Sigma standard is widely defined as producing only 3.4 defective parts per million opportunities (DPMO).

A process that is normally distributed will have 3.4 parts per million beyond a point that is 4.5 standard deviations above or below the mean (one-sided Capability Study). This implies that 3.4 DPMO corresponds to 4.5 sigma's, not six as the process name would imply.

The primary goal of Six Sigma is to identify and eliminate the causes of defects and errors in processes by employing a structured and data-driven approach. It involves a systematic framework known as DMAIC (Define, Measure, Analyze,

Improve, Control), which guides the improvement process. Brief overview of each phase is as under:

1. Define: Clearly define the problem, objectives, and scope of the project, as well as identify customer requirements and expectations.
2. Measure: Collect pertinent data and evaluate the current performance of the process to create a baseline. This step involves identifying key process metrics and developing measurement systems.
3. Analyze: Examine the collected data to pinpoint the root causes of defects and variations. This phase utilizes statistical analysis and other tools to identify patterns, trends, and potential sources of improvement.
4. Improve: Create and implement solutions to tackle the identified root causes. This phase focuses on designing and testing potential solutions and then implementing the most effective ones.
5. Control: Implement control mechanisms to maintain the improvements and prevent the process from returning to its previous state. This includes monitoring the process performance, implementing process controls, and providing ongoing training and support.

Six Sigma methodology utilizes various statistical tools and techniques, such as process mapping, cause-and-effect analysis, Pareto charts, statistical process control (SPC), design of experiments (DOE), and hypothesis testing. These tools help in analyzing process data, identifying improvement opportunities, and validating the effectiveness of implemented changes.

12.7.1 Key Concepts of Six Sigma

Key concepts of Six Sigma revolve around the following principles:

1. Critical to Quality (CTQ): Identifying the attributes that are most significant to the customer. This helps focus improvement efforts on meeting customer needs and expectations.
2. Defect: Defining a defect as any instance where the product or service fails to meet the customer demands. Six Sigma aims to minimize defects and improve overall quality.
3. Process Capability: Assessing what your process is capable of delivering in terms of meeting customer requirements. Understanding process capability helps set realistic goals and expectations.
4. Variation: Recognizing that variation in processes leads to variations in the end product or service experienced by the customer. Six Sigma aims to minimize process variation to ensure consistent and predictable outcomes.
5. Stable Operations: Establishing consistent and predictable processes to improve customer satisfaction. This involves implementing measures to reduce process variability and ensure stability over time.

6. Design for Six Sigma (DFSS): Applying Six Sigma principles during the design phase of a product or process to meet customer needs and process capability. DFSS strives to develop new designs that are more predictable, robust, and free of defects.

> Defects are not free. Somebody makes them, and gets paid for making them – **Deming, W. Edwards**

12.7.2 Six Sigma Implementation Methodology

In Six Sigma Implementation, Organizations Use Two Key Methodologies: DMAIC & DMADV

- DMAIC (Define, Measure, Analyze, Improve, and Control) is a methodology used to enhance existing business processes. It involves defining the problem, measuring process performance, analyzing data to identify root causes, implementing improvements, and establishing control measures to sustain the improvements.
- DMADV (Define, Measure, Analyze, Design, and Verify): DMADV is a methodology employed to develop new product or process designs that align with customer needs, resulting in more predictable and defect-free performance. It involves defining design goals, measuring and characterizing critical parameters, analyzing options, designing the process or product in detail, and verifying its performance before implementation.
  Figures 12.5 and 12.6 depict the DMAIC (Define, Measure, Analyze, Improve, and Control) and DMADV (Define, Measure, Analyze, Design, and Verify) methodology steps.

In the context of Six Sigma, quality improvement in healthcare focuses on identifying sources of variation, including defects, in clinical practice processes and working toward reducing this variation to enhance patient outcomes and satisfaction. Six Sigma emphasizes the importance of key roles in its successful implementation, in addition to leadership. These roles include:

1. Executive Leadership: The chief executive officer and other top management team members play a crucial role in setting the vision for Six Sigma implementation. They provide the necessary resources and freedom to explore innovative ideas for breakthrough improvements.
2. Champions: Champions are responsible for driving Six Sigma implementation across the organization in an integrated manner. They are usually drawn from upper management and act as mentors to Black Belts. They are also called "Quality Leaders."
3. Master Black Belts: These experts, identified by the champions, serve as in-house coaches on Six Sigma methodologies. They dedicate their full time to Six Sigma and support champions, guide Black Belts and Green Belts, and ensure the seamless deployment of Six Sigma across various functions and departments.

DMAIC	DMADV
Define the process improvement goals that are consistent with customer demands and enterprise strategy.	*Define* the goals of the design activity that are consistent with customer demands and enterprise strategy.
Measure the current process and collect relevant data for future comparison.	*Measure* and identify CTQs (critical to qualities), product capabilities, production process capability, and risk assessments.
Analyze to verify relationship and causality of factors. Determine what the relationship is, and attempt to ensure that all factors have been considered	*Analyze* to develop and design alternatives, create high-level design and evaluate design capability to select the best design.
Improve or optimize the process based upon the analysis using techniques like design of experiments.	*Design* details, optimize the design, and plan for design verification. This phase may require simulations.
Control to ensure that any variances are corrected before they result in defects. Set up pilot runs to establish process capability, transition to production and thereafter continuously measure the process and institute control mechanisms.	*Verify* the design, set up pilot runs, implement production process and handover to process owners.

Fig. 12.5 DMAIC (Define, Measure, Analyze, Improve, and Control) and DMADV (Define, Measure, Analyze, Design, and Verify) methodology

4. Experts: Experts work beyond organizational boundaries to improve services, processes, and products for suppliers, campuses, and customers. This level of expertise is particularly utilized in Aerospace and Defense sectors. They collaborate across multiple sites and business divisions, sharing lessons learned throughout the company. Raytheon is an example of a company that has implemented this role.
5. Black Belts: Black Belts work under the guidance of Master Black Belts and are responsible for applying the Six Sigma methodology to specific projects. They devote their full time to executing Six Sigma projects, whereas Champions and Master Black Belts concentrate on pinpointing projects and functions that are ideal for Six Sigma initiatives.
6. Green Belts: Green Belts are employees who take on Six Sigma implementation alongside their regular job responsibilities. They work under the guidance of Black Belts and provide support in achieving overall project results.

In Six Sigma methodology, professionals are certified into different belt levels, similar to martial arts, based on their expertise and role in process improvement such as Master Black belts, Black belt, Green belts, and Yellow belts (Fig. 12.7). These individuals are empowered to initiate, expand, and lead improvement projects within their respective areas of responsibility.

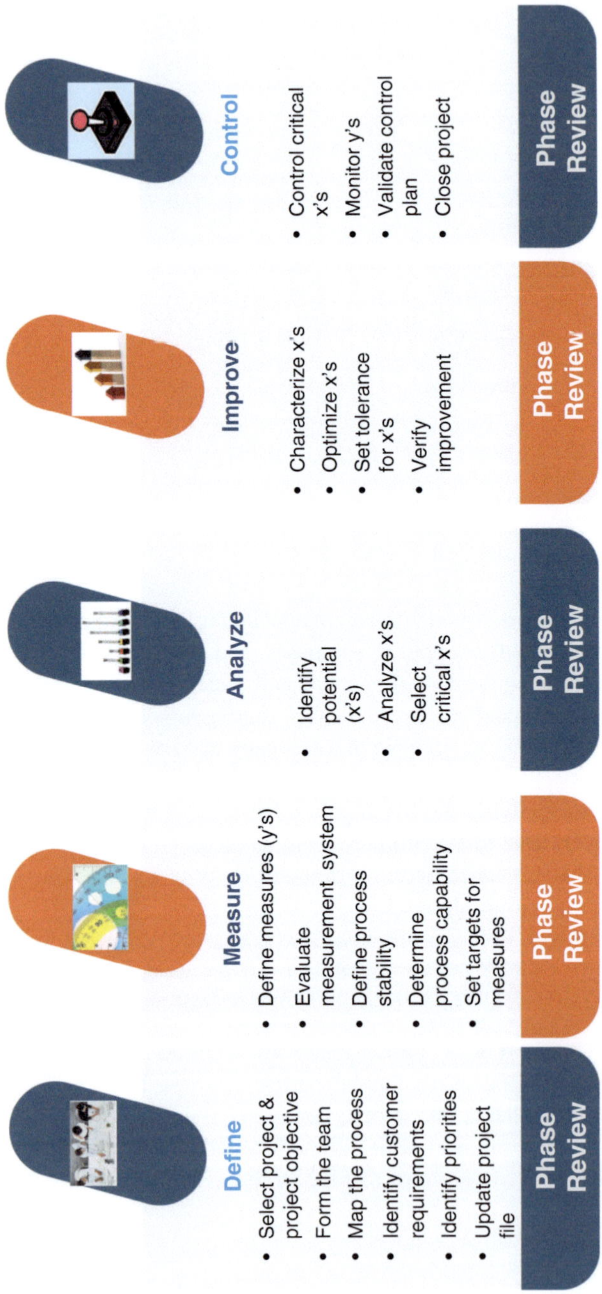

Fig. 12.6 Six Sigma DMAIC: Define, Measure, Analyze, Improve, Control methodology

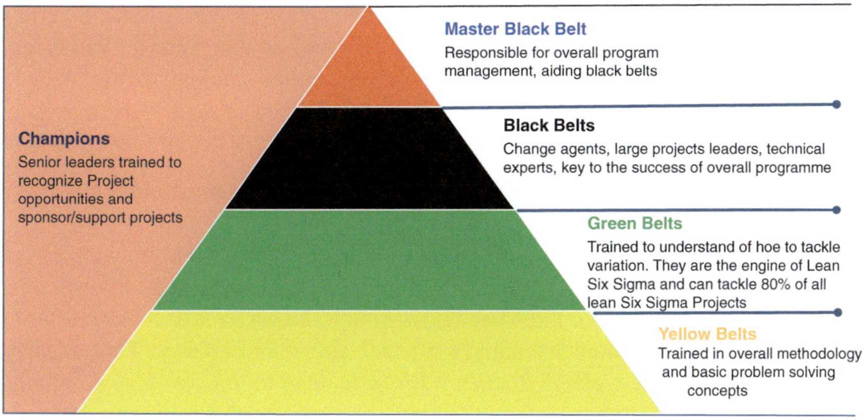

Fig. 12.7 Certification of professionals into different belt levels based on their expertise and role in process improvement

12.7.3 The Six Sigma Cycle of Change

"**Nothing endures but change.**" (Heraclitus (fl. 500 BC). A Six Sigma Champion aptly put it: "Change is not an overnight process." approach)

> The central issue is never strategy, structure, culture or systems. The core of the matter is always changing the behavior of people – **John Kotter**

Change is a gradual process, and in Six Sigma, it is driven by the results achieved through projects. When a Six Sigma project demonstrates positive outcomes, it helps individuals involved in or associated with the project recognize the value of Six Sigma as a disciplined problem-solving technique that relies on data-driven decision-making. Sustaining change effectively involves a cycle that revolves around three key points: Change Behavior, Experience Results, and Culture Change. Through repeated experiences, change can be reinforced and become lasting, whether it is in terms of behavior or culture.

For Six Sigma, the lasting impact of successful Black Belt and Green Belt projects, experienced repeatedly, plays a significant role in driving and institutionalizing change within an organization's culture. However, the challenging aspect lies in initiating the Six Sigma change implementation cycle. The Six Sigma change process operates as a closed cycle, and the question that arises is, "how do we enter the cycle?" Consider the three possible entry points:

1. Experience Results: Results require time, effort, and investment, so it is not feasible to start here.
2. Culture Change: Culture change is a gradual process that requires the experience of repeated results. It is also not a suitable starting point.
3. Change Behavior: Is it possible for us to simply decide one day to start doing things differently? Yes, we can, but it is not an easy task. It demands a deliberate

and concerted effort to decide to change our behavior. However, sustaining this conscious effort indefinitely is challenging. Unless our new behavior yields reasonably fast results, which subsequently drives a culture change, we are likely to revert to our old habits. This culture change will then lead us to permanently adopt new ways of doing things.

The starting point, therefore, is our own behavior, as it is something within our control.

Another aspect of the model is represented by the wheel, symbolizing the forces at work (Fig. 12.8). A large wheel at rest requires significant energy to overcome inertia and resistance. Once the wheel is in motion, however, it becomes easier to keep it rolling by periodically adding a small amount of energy. This analogy reflects the Six Sigma cycle of change: it is challenging to initiate change initially, but once the wheel of change is set in motion, it becomes easier to sustain. Put simply, Six Sigma becomes a self-fulfilling prophecy.

Six Sigma offers a structured and data-based approach to process improvement, empowering organizations to achieve higher levels of quality, efficiency, and customer satisfaction.

Example: Applying Six Sigma to the Sterilization Process of Surgical Instruments
Problem Identification:

A hospital identified that its surgical instruments sterilization process had a high defect rate, leading to increased costs and dissatisfaction among surgeons and operating room personnel. Issues included instruments not being fully sterilized, resulting in delays, rescheduling of surgeries, and potential infections.

12.7.4 Six Sigma Implementation

1. **Define:**
   - Goal: Reduce the defect rate in the sterilization process to improve quality and efficiency.

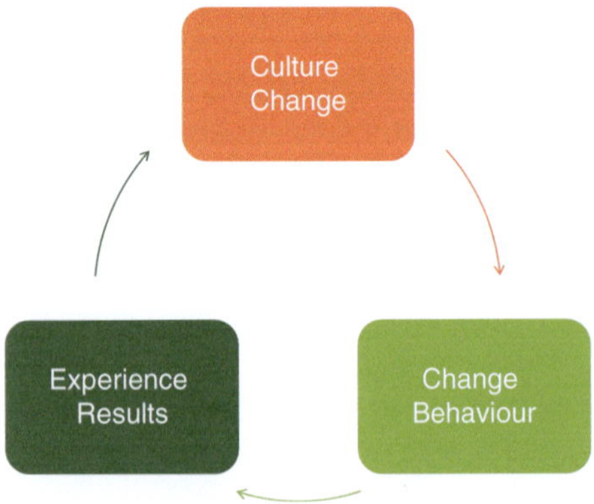

Fig. 12.8 The Six Sigma change process cycle

- Metrics: Defect rate, sterilization cycle time, and costs related to re-sterilization and surgical delays.

2. **Measure:**
   - Collect data on the current sterilization process, including the frequency and types of defects (e.g., incomplete sterilization, contamination).
   - Measure cycle times and costs associated with re-sterilization and delayed surgeries.
3. **Analyze:**
   - Identify root causes of defects using tools like cause-and-effect diagrams and Pareto analysis.
   - Common issues included improper loading of sterilization machines, inadequate maintenance, and insufficient training of staff.
4. **Improve:**
   - Implement corrective actions, such as standardizing loading procedures, regular maintenance schedules, and comprehensive training programs for staff.
   - Introduce automation to monitor sterilization parameters in real-time and alert staff of any deviations.
5. **Control:**
   - Establish control mechanisms, such as ongoing monitoring of sterilization cycles and regular audits, to ensure sustained improvements.
   - Implement feedback loops to gather input from surgeons and operating room personnel regarding the quality of sterilized instruments.

Results:

- **Reduced Defect Rate:** By addressing the root causes, the defect rate in the sterilization process significantly decreased.
- **Cost Savings:** Reduced need for re-sterilization and fewer surgical delays led to substantial cost savings.
- **Increased Satisfaction:** Surgeons and operating room personnel reported higher satisfaction due to the increased reliability of sterilized instruments, leading to smoother and more efficient surgical procedures.

  By applying Six Sigma to the sterilization process, the hospital improved the quality and efficiency of its operations, resulting in both cost savings and enhanced satisfaction among its staff.

12.7.5 Lean Six Sigma (L6S)

Lean Six Sigma (L6S) is a fusion between *Lean and Six Sigma*. It is a problem-solving framework rooted in a culture of continuous improvement, providing substantial benefits to an organization. Lean Six Sigma integrates Lean and Six Sigma principles to enhance performance by systematically eliminating waste (non-value adding activities) and minimizing process variation and maximizes productivity helping to do more with less. Lean 6 Sigma strives to eliminate the **errors** and **defects**, which invariably is a waste.

By focusing on root causes and implementing data-driven solutions, Lean Six Sigma ensures consistent quality and customer satisfaction. It reduces cost by identifying and eliminating unnecessary expenses, leading to significant cost savings and improved profitability. It improves customer experience to deliver products and services that meet or exceed customer expectations, building loyalty and trust. Equips employees with problem-solving skills and tools to drive positive change and take ownership of improvement initiatives. Lean Six Sigma accelerates decision-making as insights are data-driven, more informed, and thus reduce delays and inefficiencies. It proactively identifies and addresses potential risks and uncertainties, safeguarding against costly errors and setbacks. Lean Six Sigma initiatives when aligned with organizational objectives help achieve strategic goals and long-term success. Finally, it provides a competitive edge by continuously improving processes, staying agile, and adapting to changing market dynamics. Figures 12.9 and 12.10 illustrate Lean Six Sigma objectives to streamline processes, improve quality, and reduce waste for continuous improvement. Figure 12.10 lays emphasis on the differing aspects that provide answers to the questions; *what is Lean* and *what is 6 Sigma*:

> Six Sigma is a statistical problem-solving process that fits under the Lean umbrella. Rather than competing with Lean, it should sit alongside all of the other tools in the Lean toolbox – **Mark DeLuzio**

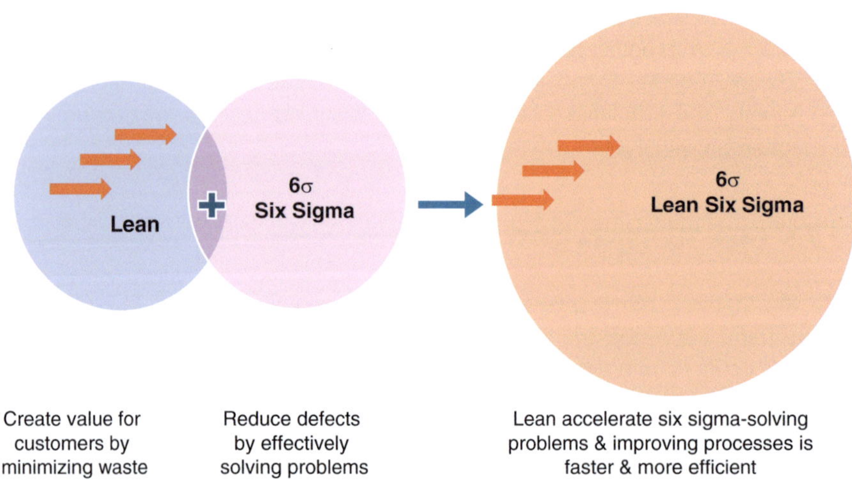

Fig. 12.9 Lean Six Sigma objectives to streamline processes, improve quality, and reduce waste

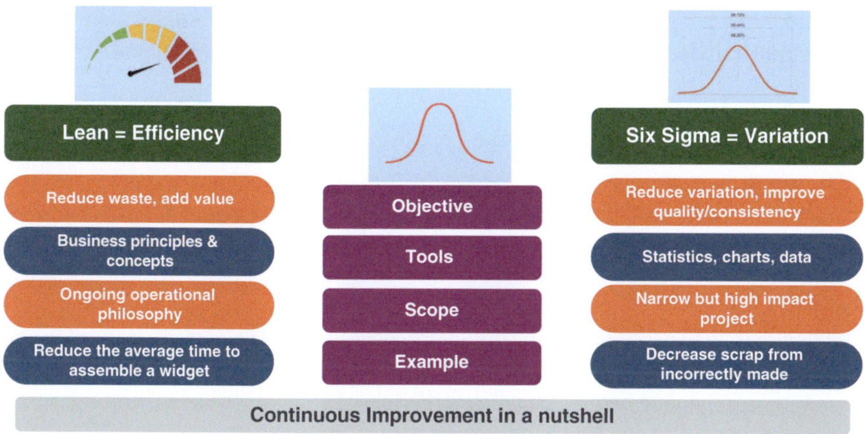

Fig. 12.10 Continuous Improvement with Lean Six Sigma

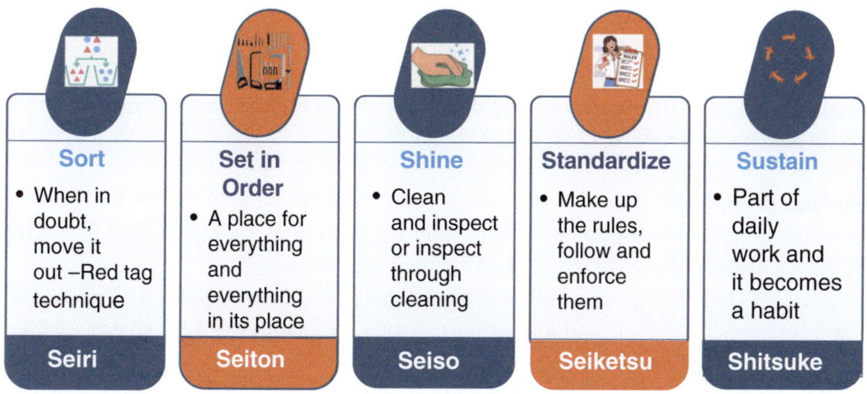

Fig. 12.11 5 "S" concept—A Visual Housekeeping SystemEnglish Equivalent of Japanese Terms

12.8 The 5S Methodology

The 5S methodology is a systematic approach to workplace organization and standardization that originated from Japanese manufacturing practices. It has been widely adopted in various industries, including healthcare, to improve efficiency, safety, and quality of care. The 5S methodology stands for Sort, Set in Order, Shine, Standardize, and Sustain (Fig. 12.11).

Below are the five steps explained with examples relevant to healthcare settings:

Sort (Seiri)
Objective: Eliminate unnecessary items and ensure that only essential items are kept in the workspace.

Example: In a hospital pharmacy, regularly review and remove expired medications and supplies to declutter the shelves and improve inventory management.

Set in Order (Seiton)
Objective: Organize the remaining items in a logical and accessible manner to minimize search time and waste.
Example: Arrange medical instruments and equipment in designated areas based on frequency of use to facilitate easy access and efficient workflow in an operating room.

Shine (Seiso)
Objective: Clean and maintain the workspace to create a safe and pleasant environment.
Example: Implement regular cleaning schedules for patient rooms, waiting areas, and equipment to ensure hygiene and prevent infections.

Standardize (Seiketsu)
Objective: Establish standardized procedures and protocols to maintain the improvements achieved through the first three steps.
Example: Develop and implement standardized operating procedures (SOPs) for medication administration to ensure consistency and reduce errors across different shifts and staff members.

Sustain (Shitsuke)
Objective: Foster a culture of continuous improvement by sustaining the gains achieved and regularly reviewing and refining the processes.
Example: Conduct regular audits and performance reviews to monitor adherence to the established standards and identify areas for further improvement in patient care and operational efficiency.

By implementing the 5S methodology in healthcare settings, organizations can create a more organized, efficient, and patient-centered environment, ultimately leading to improved quality of care and patient safety.

Tables 12.3 and 12.4 depict examples of 5S in OPD and pharmacy respectively.

> Visual control at the worksite should make it clear at a glance what the standard is and if anything is out of standard – **Jeffrey Liker**

12.9 Kanban In Healthcare

Kanban, which originated from Japanese manufacturing practices, is a visual scheduling system used to manage workflow efficiently. In healthcare, the Kanban tool can help streamline processes, reduce waste, and improve patient care. Breakdown of how Kanban can be applied in healthcare settings is as below:

Table 12.3 Application for 5S in outdoor department

5S Principle	Description	Example in Queue Management in OPD
1. Sort (Seiri)	Eliminate unnecessary items and streamline essential ones.	**Identify and remove bottlenecks:** Analyze the patient flow to eliminate unnecessary steps in the queuing process.
		Remove redundant check-in procedures: Simplify forms and paperwork required at registration.
2. Set in Order (Seiton)	Arrange necessary items for optimal efficiency and accessibility.	**Organize waiting areas:** Designate specific areas for different patient categories (e.g., new patients, follow-ups, urgent cases).
		- **Implement clear signage:** Use directional signs to guide patients through the OPD process.
3. Shine (Seiso)	Maintain cleanliness and order in the workspace.	**Keep waiting areas tidy:** Regularly clean seating areas and ensure magazines or informational materials are neatly arranged.
		Ensure electronic systems are functioning: Regularly check kiosks or displays used for queue management.
4. Standardize (Seiketsu)	Establish standardized procedures and protocols.	**Develop a standard queuing system:** Use ticket numbers or electronic displays to manage patient order.
		Create uniform check-in processes: Ensure all staff follow the same procedure for registering patients.
5. Sustain (Shitsuke)	Maintain and review standards through regular training and audits.	**Continuous staff training:** Educate staff on the importance of the queuing system and how to manage it effectively.
		Regularly review processes: Conduct periodic assessments to identify areas for improvement in queue management.

Table 12.4 Application of 5S in pharmacy and examples

Japanese English Actions	Examples
Seiri (Sort)	Sort, identify, and remove unwanted/unused items from the workplace; and reduce clutter (Removal/organization)
	Waste disposal, both infectious and non-infectious
	Timely condemnation of unwanted stores and a separate condemn items stores.
	Initiation of "Reduce, Reuse, Recycle Concept"; Items, which can be provided to the safe recycling process, for instance, inner wrapping paper of disposable surgical gloves, glass bottles of IV antibiotics/equipment and so forth.
	Benefits encompass more efficient space utilization, task simplification, hazard reduction, and a notable decrease in distracting clutter.

(continued)

Table 12.4 (continued)

Japanese English Actions	Examples
Seiton (Set in Order "Set" based on perfection of "Sort").	Arrange all necessary items systematically for smooth functioning (Orderliness).
	An efficient and organized storage system allows everyone to effortlessly find important items, leading to a more relaxed work environment. Examples include:
	Items are arranged according to alphabetical order or numerical order. All the items should be kept in a specific place following a system, so that anybody in need of these items can find them easily
	Name tag, board, and symbols development and installation. Determine the names of each room and set up a straightforward sign for easy identification by staff and visitors.
	Digital token, separate queue at pharmacy counter and separate waiting space at OPD and pharmacy to avoid confusion, congestion, and conflict.
	To classify patients and visitors coming to the hospital (as patients with urgent attendance, on the first visit, and the patients seeking re-examination).
	Tagging and labelling of all the items, instruments and devices
	Specific locations for the items, arranging workable instrument sets, storage of these sets, and color coding system for easy handling.
Seiso (Shine)	Maintain a high standard of cleanliness (Cleanness). Staff members are likely to experience increased comfort in a tidy and organized setting.
	Checklist of cleaning activities
	Daily 10 minutes morning "Shine" practice before starting routine work.
	Achieve excellence by renewing cleaning tools, designating tool storage, organizing space, and providing a small office along with improved uniforms for the cleaning staff.
	"Shine" at waste separation, collection, storage, transport and final treatment system
	Walls must be finished in smooth, impervious, washable material
	Prevent entry of insects, animals (especially rodents) or birds
Seiketsu (Standardize)	Set up the above 5S as norms in every section of the workplace (Standardize)
	Pre-packaging of frequently used drug.
	Organizing: Placing essential items in their designated spots with correct numbering, labeling, and color coding facilitates quick retrieval.
	Uniform color code system. e.g., "blue" to indicate sterile materials, while "red" is used for unsterile items.
	Regular supervisory visits
	Informal site visits to supervise the ongoing 5S activities
Shitsuke (Sustain)	Train and maintain discipline of the personnel engaged (Self-Discipline) - Short (usually one topic) but practical training
	Create positive competition
	Safety issues and 5S activities
	Effective hospital accident or incident (to hospital staff, visitors or patients) reporting system

Visualize Workflow

- Objective: Create a visual representation of tasks, helping teams to understand the flow and identify bottlenecks.
- Example: Use a Kanban board to track patient admissions, transfers, and discharges, allowing staff to quickly identify patient status and bed availability.

Limit Work in Progress (WIP)

- Objective: Restrict the number of tasks in progress at any given time to maintain focus and enhance efficiency.
- Example: Limit the number of ongoing surgeries in an operating room to ensure adequate staff and resources are available, preventing overburdening and potential errors.

Manage Flow

- Objective: Ensure a smooth, seamless, and continuous flow of tasks through the system, aiming to minimize delays and enhance throughput.
- Example: Implement a Kanban system for medication replenishment in hospital pharmacies, ensuring timely restocking and reducing stockouts. Sample Kanban card is shown in Fig. 12.12.

Make Policies Explicit

- Objective: Clearly define and communicate guidelines, expectations, and standards to ensure consistency and clarity.
- Example: Establish explicit protocols for patient triage in emergency departments, guiding staff on assessment criteria and treatment prioritization.

KANBAN CARD

Item Name: Paracetamol
Item number: XYZ
Supplier name: ABC
Reorder quantity: 2 boxes (100 packs)

Procedure

1. **Request order:** to place it in items 'to order bin' when you see Kanban
2. **Place order:** once order is placed put Kanban in 'Awaiting delivery bin'
3. **Receive delivery:** Once item is delivered, check 'Awaiting delivery' bin for the matching Kanban
4. **Return Kanban:** Stock item at a correct location using FEFO; place new item behind existing inventory

Fig. 12.12 Sample Kanban system for medication replenishment in hospital pharmacies

Feedback Loops for Continuous Improvement

- Objective: Encourage regular review and reflection on processes to identify areas for improvement and innovation.
- Example: Conduct weekly team meetings to discuss challenges encountered, share insights, and collaboratively identify solutions using Kanban metrics.

Benefits of Kanban in Healthcare

- Enhanced Visibility: Real-time tracking of tasks and statuses improves transparency and accountability.
- Improved Efficiency: Streamlined workflows and reduced bottlenecks lead to quicker response times and enhanced patient care.
- Increased Flexibility: Adaptability to changing demands and priorities ensures optimal resource utilization.
- Empowered Teams: Engagement in continuous improvement fosters a culture of collaboration, innovation, and excellence.

By integrating the Kanban tool into healthcare operations, organizations can achieve greater operational efficiency, improved patient outcomes, and enhanced staff satisfaction, aligning with the principles of Lean management.
The biggest value of the kanban system is to help visualize the flow, study it, and find inventory banks, parts shortages, and missed shipments. - **Jeffrey Liker**

12.10 Error-Proofing in Healthcare

Error-proofing, or "Poka-yoke" in Japanese, is a technique designed to prevent errors by creating processes and systems that either make mistakes impossible or immediately noticeable and correctable. In healthcare, where mistakes can have serious consequences, error-proofing plays a crucial role in enhancing patient safety and improving the quality of care. These systems can take various forms, such as physical devices, visual cues, or procedural steps, and are used across industries to improve quality, productivity, and efficiency. The goal of poka-yoke is to make processes more reliable and consistent by minimizing the possibility of human error.

Designing Safe Processes The objective is to create systems and workflows that minimize the risk of errors through design. Example: Implementing standardized protocols and checklists for surgical procedures to ensure all necessary steps are followed and errors are prevented.

Standardization of Tools and Equipment The objective is to ensure consistent and reliable use of tools and equipment to reduce variability and potential errors. Example: Using color-coded labels and standardized medication packaging to prevent medication errors and enhance identification. Physical Shapes, i.e., designing parts that only fit together in the correct way, like USB connectors, Enteral feeding tube connector.

Automating Manual Tasks The objective is to reduce reliance on manual processes by automating repetitive tasks to minimize human error. Example: Implementing barcode scanning systems for patient identification and medication administration to ensure accuracy and reduce medication errors.

Visual Cues and Alerts The mechanisms that alert operators to potential errors, such as alarms or visual signals indicating a problem, potential errors, or deviations from standard procedures. Example: Using color-coded wristbands for patient identification, alerting staff to specific conditions or allergies to prevent medical errors. Sensors and alarms detecting errors such as a sensor stopping a machine if a part is missing or improperly placed.

Education and Training Equips healthcare professionals with the knowledge and skills to recognize and prevent errors. Example: Conducting regular training sessions on error recognition, reporting mechanisms, and preventive measures to empower staff in error prevention.

Benefits of Error-Proofing in Healthcare

- Enhanced Patient Safety: Reduction in errors leads to improved patient outcomes and reduced harm. For example, implementing incompatible designs for enteral feeding tube connectors to prevent misconnection with IV lines, ensuring error-proofing and patient safety.
- Improved Quality of Care: Standardized and consistent processes result in higher quality care delivery.
- Increased Efficiency: Streamlined workflows and reduced rework enhance operational efficiency.
- Cost Savings: Prevention of errors reduces costs associated with adverse events and rework.

12.10.1 Challenges in Implementing Error-Proofing

- Resistance to Change: Overcoming resistance to new processes and systems can be a hurdle in implementation.
- Resource Constraints: Limited resources and funding may impact the ability to invest in error-proofing initiatives.
- Complexity of Healthcare: The multifaceted nature of healthcare can make error-proofing challenging due to the diverse range of tasks and interactions involved.

Error-proofing in healthcare is a proactive approach to preventing errors before they occur, emphasizing the importance of system design, standardization, automation, and education. By integrating error-proofing strategies into healthcare practices, organizations can significantly enhance patient safety, improve quality of care, and optimize operational efficiency, ultimately fostering a culture of continuous improvement and excellence.

Stop production so that production never has to stop. It is OK to stop production to ferret out the root causes of defects – **Toyota Proverb**
Automation applied to an inefficient operation will magnify the inefficiency – **Bill Gates**

12.10.2 Implementing Poka-Yoke

Implementing Poka-Yoke involves a structured approach to systematically identify and eliminate errors in processes. Figure 12.13 depicts a six-step framework for implementing Poka-Yoke which includes:

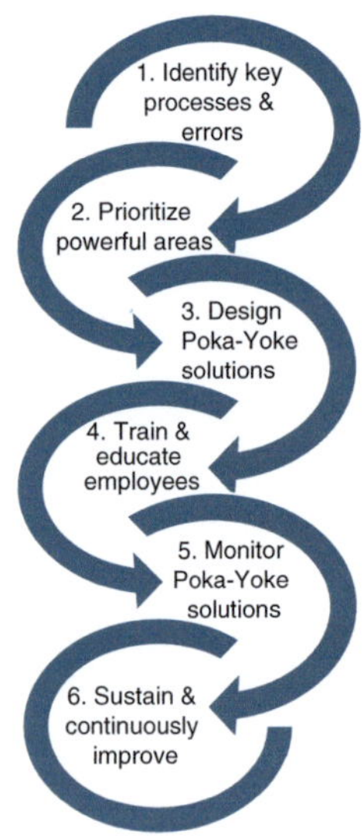

Fig. 12.13 A 6-step framework for implementing Poka-Yoke to prevent errors and improve processes

1. **Identify the Problem:** Clearly define the issue or error that needs to be prevented.

Actions:
- Collect data on defects or errors.
- Analyze the process to identify where and how errors occur.
- Engage with employees to gather insights into common mistakes and potential causes.

2. **Understand the Process:** Gain a thorough understanding of the entire process where the error occurs.

Actions:
- Map out the process step-by-step using flowcharts or process diagrams.
- Identify critical control points and areas where errors are likely to happen.
- Observe the process in action and document any variations or inconsistencies.

3. **Determine Root Causes:** Identify the underlying causes of the errors to address them effectively.

Actions:
- Use root cause analysis techniques such as the 5 Whys or Fishbone (Ishikawa) diagrams.
- Focus on finding the actual cause of the error, not just the symptoms.
- Involve cross-functional teams to get different perspectives on potential root causes.

4. **Develop Poka-Yoke Solutions**: Create mistake-proofing solutions to prevent or detect errors.

Actions:
- **Generate Ideas:** Brainstorm potential Poka-Yoke devices or mechanisms.
- **Assess Solutions:** Evaluate the feasibility and effectiveness of each proposed solution.
- **Design Implementation:** Develop solutions that are straightforward, cost-effective, and easy to implement.

5. **Implement the Solutions:** Integrate the Poka-Yoke solutions into the process.

Actions:
- Pilot the solutions in a controlled environment to test their effectiveness.
- Train employees in the new procedures and the importance of the Poka-Yoke solutions.
- Make any necessary adjustments based on feedback and observations during the pilot phase.

6. **Monitor and Improve:** Ensure the Poka-Yoke solutions are effective and continuously improve them.

Actions:
- Establish monitoring systems to track the performance of the Poka-Yoke solutions.
- Regularly review the process and collect data on any errors that occur.
- Brainstorm, analyze, design intervention, evaluate and make iterative improvements to the Poka-Yoke solutions.

Implementing Poka-Yoke using this six-step framework helps organizations systematically eliminate errors, enhance quality, and improve efficiency. By identifying the problem, understanding the process, determining root causes, developing solutions, implementing them, and continuously monitoring and improving, companies can create a robust mistake-proofing system that enhances overall performance.

> You must go beyond using the tools for improvements to apply Lean's principles in ways that they become indoctrinated into your business thinking and culture to impact the organization as a whole – **Mark DeLuzio**

Selecting the Quality Tool 13

Contents

Choosing the most suitable quality improvement approach depends on various factors, including the organization's goals, resources, culture, and specific challenges. When selecting a quality tool, it is important to consider long-term effectiveness rather than opting for quick fixes. While standalone data quality tools may offer immediate solutions, they are unlikely to address underlying problems in the long run. Specialized data quality tools often demand a high level of expertise for their successful deployment. They can be complex and require comprehensive training before they can be effectively utilized. While these tools can be powerful, relying on them for short-term data quality priorities may result in missing deadlines and falling short of expectations.

It is crucial to match the tool with the skill set and experience of the user. Just as a rookie pilot would struggle to operate a jumbo jet due to the complexity of the flight instruments, inexperienced users may face challenges when using highly sophisticated quality tools. It is important to consider the capabilities and

S. Sharma, *Quality Demystified*, https://doi.org/10.1007/978-981-95-0339-1_13

proficiency of the individuals involved to ensure the successful implementation and utilization of the chosen quality tool. Taking the time to select the appropriate tool and provide the necessary training and support will contribute to achieving long-term success in improving data quality.

13.1 The Critical Incident Technique

The Critical Incident Technique is an advanced tool used for root cause analysis, specifically focused on identifying causes in complex situations. Unlike basic cause analysis tools like flowcharts, brainstorming, or fishbone diagrams, which aim to find causes, the critical incident technique involves conducting interviews with individuals involved in the process where an event occurred. This technique is particularly useful for examining adverse events and understanding the breakdowns within a system. It delves into the reasons behind the incident, the circumstances surrounding it, and why the system failed at that point. Additionally, reporting near misses is crucial as it helps identify at-risk situations or factors, preventing future incidents and injuries. A near miss refers to an unplanned event that had the potential to cause harm but did not result in injury, illness, or damage. The feedback and dissemination of findings from critical incident analysis play a vital role in creating awareness about system errors. This awareness enables improvements in system design, specifically targeting the reduction or elimination of medication errors. It is important for healthcare organizations and professionals to actively participate in voluntary reporting systems as a significant aspect of their commitment to patient safety. By sharing and learning from critical incidents, the healthcare industry can continually enhance its practices and ensure safer outcomes for patients.

13.2 Prioritization

Priority setting in healthcare systems poses a challenge at all levels, be it global, national, or local, as it often lacks a systematic approach, leading to suboptimal utilization of resources. When making decisions on resource allocation and rationing in healthcare, multiple criteria such as costs, effectiveness, equity, and feasibility are taken into consideration. Table 13.1 depicts the commonly used criteria for prioritization. The complexity of these decisions necessitates the consideration of various factors.

In order to maximize the impact of interventions and improve overall population health, different improvement ideas or interventions are evaluated based on their cost-effectiveness and return on investment. These considerations may include enhancing coverage, addressing health disparities among disadvantaged or vulnerable groups, and responding to life-threatening situations. However, there are instances where multiple solutions exist for a given problem, making it challenging to determine the most suitable option to pursue. Additionally, there may be cases

Table 13.1 Commonly used prioritization criteria

Criteria to identify priority problem	Criteria to identify intervention for problem
Cost and/or return on investment	Expertise to implement solution
Availability of resources (staff, time, equipment, funds) to solve problem	Effectiveness of solution
Magnitude of problem (e.g., number of affected individuals)	Feasibility of intervention
Availability of solution	Implementation readiness
Impact of a problem	Calendar time to implement
Urgency of solving problem (COVID/H1N1 or patient waiting time)	Ease of implementation/maintenance
	Implementation leadership return on investment
	Potential negative consequences
	Impact on systems or health
	Systems/technology complexity
	Legal considerations

where various improvement ideas have been implemented without achieving the desired outcomes.

When confronted with such difficult decisions and the need to balance competing health issues, employing a defined prioritization technique can provide a structured approach for objectively ranking issues and making informed decisions. This allows for the optimization of results with limited resources, considering all aspects and facets of the competing health priorities. By utilizing a prioritization technique, healthcare systems can ensure a more systematic and transparent process for resource allocation, leading to the achievement of maximum benefits and improved overall healthcare outcomes.

13.3 The PICK (Possible, Implement, Complex, Kill) Chart Is a Popular Visual Tool Used in Lean Six Sigma Methodology

It serves the purpose of organizing, prioritizing, and selecting the most appropriate improvement ideas or solutions for a given problem. The chart considers the potential impact or positive outcomes of each idea as well as the resources needed for implementation.

Using the PICK chart, improvement ideas or solutions are categorized into four main quadrants: Possible, Implement, Complex, and Kill. The Possible quadrant represents ideas that are feasible and have a high potential for success. These ideas are considered viable options for implementation. The Implement quadrant includes ideas that are both feasible and have been selected for implementation.

The Complex quadrant contains ideas that may pose challenges or require significant resources, making their implementation more difficult. These ideas need careful evaluation and planning before proceeding. The Kill quadrant represents ideas that are deemed unfeasible or impractical based on various considerations.

These ideas are not pursued further, allowing resources to be focused on more promising alternatives.

By using the PICK chart, teams can visually assess and prioritize improvement ideas, ensuring that the most suitable options are selected based on their potential impact and resource requirements. It provides a structured approach to decision-making and enables efficient allocation of resources for maximum effectiveness in problem-solving and process improvement initiatives.

The PICK chart utilizes two primary criteria commonly relevant to healthcare organizations, namely "importance/urgency" and "cost/impact," or other similar criteria such as "need/feasibility." This chart is employed to evaluate and compare competing activities, projects, or programs based on how well they meet these criteria. The example below demonstrates how "Need" and "Feasibility" are used as specific criteria.

The PICK chart consists of two axes, each ranging from low to high, representing different aspects:

- The vertical axis signifies the impact of an intervention.
- The horizontal axis represents the resources required to implement the idea, which can include manpower, machinery, materials, methods, or money.

The PICK chart is divided equally into four quadrants, each labelled with one of the PICK letters:

1. Possible: This quadrant encompasses ideas that have a low impact but are relatively easy to implement.
2. Implement: Ideas falling within this quadrant have a high impact and are also easy to implement.
3. Challenge: The quadrant labelled as "Challenge" contains ideas with a high impact but are difficult to implement due to resource constraints or other challenges.
4. Kill: The "Kill" quadrant represents ideas with a low impact and are also hard to implement. These ideas are generally not pursued further.

Figure 13.1 showcases a strategy grid that organizes tasks according to their importance and urgency, alongside their implementation and impact. By using the PICK chart, healthcare organizations can visually assess and prioritize various activities, projects, or programs based on their impact and resource requirements. This allows for a structured evaluation and decision-making process, enabling the allocation of resources in a manner that maximizes effectiveness and efficiency.

To label the quadrants, begin by setting up a grid with four quadrants and assigning one broad criterion to each axis. Create arrows on the axes to indicate "high" or "low" as depicted below. Based on the axes, label each quadrant as below:

- High Need/High Feasibility
- High Need/Low Impact

- Low Need/High Feasibility
- Low Need/Low Feasibility.

Organize and prioritize competing activities, projects, or programs by assigning them to the appropriate quadrant according to the labels. The following example uses "Need" and "Feasibility" as the criteria, and the items have been prioritized as follows:

1. High Need/High Feasibility: These items have a high demand and offer a high return on investment, making them the highest priority. Adequate resources must be dedicated to the upkeep and ongoing enhancement of these items.
2. Low Need/High Feasibility: These items are often of low need but difficult to eliminate. They may require redesigning to reduce investment while still maintaining impact.
3. High Need/Low Feasibility: These are long-term projects with great potential but will require significant investment, such as infrastructure development. Focusing on too many of these items simultaneously can overwhelm a healthcare organization.
4. Low Need/Low Feasibility: These items offer minimal return on investment and are of the lowest priority. They should be phased out to reallocate resources to higher priority items.

How to Use a PICK Chart

Step 0: Begin by assembling a team and selecting a facilitator to guide the process.
Step 1: Individually brainstorm and identify ideas for a specific problem.
Step 2: Identify unique ideas and remove any duplicates.
Step 3: Discuss each unique idea one by one and paste them on the chart at a position determined by the rating of that idea, as shown in Fig. 13.1.

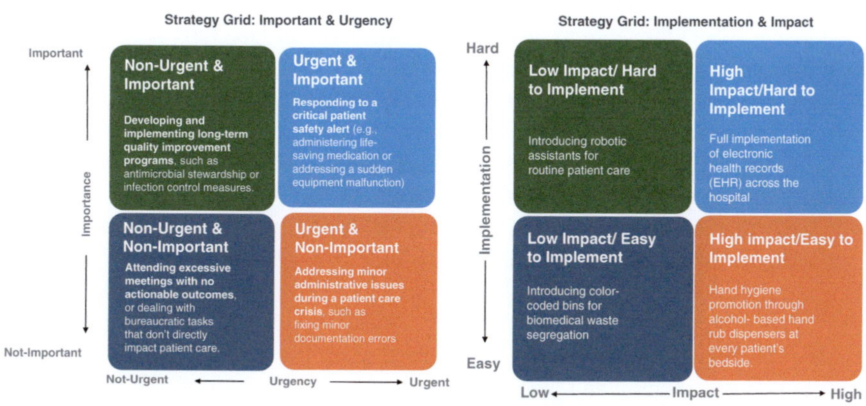

Fig. 13.1 Strategy grid for organizing tasks according to their importance and urgency, alongside their implementation and impact

PICK Chart Example

The facilitator can ask the following questions during this step to ensure clarity among team members and aid in rating the ideas:

1. The number of team members required to complete a task can vary depending on the complexity and scope of the task.
2. Has a similar activity been done before?
3. Is there a need for assistance from other departments or external experts?
4. What is the estimated time needed to complete the activity?
5. Can you identify any potential bottlenecks or obstacles for this activity?
6. Do team members have any prior experience with similar activities in another setting or hospital?

Asking these questions will help the team members evaluate and rate the ideas more effectively, leading to a better understanding of the potential feasibility and impact of each idea.

Rating of an Idea

To rate each idea, consider the following two criteria:

1. Impact: Rate the idea higher on the vertical axis if it brings greater benefits.
2. Resources Required: Rate the idea lower on the horizontal axis if it consumes higher resources. Factors to consider while rating an idea include:
   - Manpower required for implementation.
   - Need for specific machines, materials, skills, methods, or budget allocation.

Advantage of the PICK chart: The PICK chart simplifies the rating process by focusing on two dimensions (impact and resources), making it easier to rate ideas in both aspects.

Do's:

- Avoid placing an idea on the line or between quadrants, as it hinders decision-making. Allow the team to decide which quadrant an idea falls into, and sticky notes can be moved as ideas are discussed.
- Limit prolonged discussions on a single idea. If positioning an idea on the most relevant quadrant is challenging, discuss it after Step 5.
- Keep the PICK chart simple. Avoid subdividing quadrants or allowing participants to strategically place their sticky notes.
- Clarify that the "kill" quadrant means the idea is challenging and has low payoff, not that it should be completely disregarded.

Step 4:

- Once all ideas are placed on the PICK chart, reveal the quadrant names to prevent bias in the evaluation and placement of ideas.
- Note any disagreements regarding the idea positions, especially after disclosing the quadrant names. Document the reasons and discuss with experts.

Names of the Quadrants and Their Importance:

1. Q1: Possible (Just do it): Activities in this quadrant require minimal effort and have a low impact. These ideas/activities are typically lower in priority.
2. Q2: Implement (Top priority): Activities in this quadrant require minimal effort but have a high impact. Focus on implementing these activities first to maximize benefits.
3. Q3: Complex (Third priority): Activities in this quadrant have a high impact, but the required resources are substantial. They are generally considered third in priority.
4. Q4: Kill (Last priority): Despite the name, activities in this quadrant are not completely neglected. Instead, the team needs to evaluate these activities and find ways to shift them to other quadrants.

Step 5:

- Review and reconsider the positions of ideas on the PICK chart, finalizing their placements.

Step 6:

- Prioritize all ideas based on their positions in the chart, following the guidelines provided in the quadrant section.
- If multiple ideas fall within a quadrant, prioritize the idea positioned closer to the top left corner. Select the next nearest idea to the top left corner as the second priority.

The following problems have been identified in each health facility, prioritize this list according to different quadrants:

Q1: Possible (Just do it):
1. Keeping facility clean
2. Measuring patient satisfaction
3. Segregation of biomedical waste
4. Involving local community leaders for managing services
5. Ensuring hand hygiene in hospital
6. Calibration of equipment in laboratory

Q2: Implement (Top priority):
1. Reducing waiting time at clinic
2. Medical audits
3. Starting kitchen services in the hospital
4. Medication oversight—avoiding medication mistakes and excessive use of antibiotics.

Q3: Complex (Third priority):
1. Care coordination (cross-cutting)

Q4: Kill (Last priority):
1. Reducing unnecessary referrals to another health facility

Step 7: Create a project plan and start implementing the ideas based on their prioritization, one by one. Monitor the results and document the priority list and action plan for reference.

Table 13.2 provides examples of PICK chart based on need and feasibility

These examples help illustrate how healthcare administrators can categorize projects based on their impact and the resources needed to implement them, ensuring the right focus on both urgent and practical initiatives. Figure 13.2 depicts some improvement ideas.

Quality improvement initiatives are most effective when issues are tackled systematically with a consistent and analytical method; the approach should remain unchanged even if the problem varies. Simplifying the problem-solving steps helps workers understand and effectively use the process and tools.

> 94% of problems in business are systems driven and only 6% are people driven – **W. Edwards Deming**

Table 13.2 PICK chart based on need and feasibility

High need/high feasibility	
Example: Expanding vaccination campaigns during a disease outbreak. Need: High to control the spread of infectious diseases and protect vulnerable populations. Feasibility: Easily deployable with existing infrastructure, vaccines, and trained staff. Action: Should be implemented immediately to address the public health need. **Example: Implementing barcode medication administration (BCMA) to minimize medication errors.** Need: High, as medication errors are a significant cause of preventable harm in hospitals. Feasibility: Technology is readily available, and implementation can be straightforward with training. Action: High priority, implement to improve patient safety.	**Example: Training staff on infection control protocols in a hospital setting.** Infection control is important for minimizing the transmission of multidrug-resistant organisms (MDROs). Feasibility: Can be easily achieved with on-site training sessions and clear guidelines. Action: Prioritize this initiative for immediate execution. **Example: Establishing a proper triage system in emergency departments.** Need: High to prevent overcrowding and ensure critical patients are treated promptly. Feasibility: Easily implemented with trained staff and minimal cost. Action: Important and feasible, so it should be implemented quickly.

(continued)

Table 13.2 (continued)

High need/low feasibility	
Example: Establishing advanced trauma care units in rural hospitals. Need: High, as trauma care saves lives in emergency situations, especially in underserved areas. Feasibility: Difficult due to lack of specialized equipment, trained staff, and infrastructure. Action: Long-term planning is needed with phased implementation; it may require significant funding and support. **Example: Implementing artificial intelligence (AI) for diagnostic imaging.** Need: High, as AI can enhance diagnostic accuracy and speed. Feasibility: Challenging due to high costs, lack of trained personnel, and integration with existing systems. Action: Consider as a long-term goal, requiring investment in technology and training.	**Example: Introducing widespread genetic testing for personalized medicine.** Need: Could significantly improve patient outcomes by tailoring treatments. Feasibility: Difficult to implement due to high costs, lack of infrastructure, and need for specialized personnel. Action: Long-term goal may require external funding and gradual roll-out. **Example: Building a comprehensive electronic health record (EHR) system across all government hospitals.** Need: Would improve patient data management, coordination, and quality of care. Feasibility: Major infrastructure and training challenges require significant government investment. Action: Important, but long-term, with phased implementation needed.
Low need/high feasibility	
Example: Replacing hospital furniture with ergonomic designs. Need: Low in terms of immediate patient care impact. Feasibility: Easily doable with available budget and minimal disruption to operations. Action: Could be implemented if resources are available but not a priority for immediate patient care improvement. **Example: Creating a staff wellness program with fitness classes.** Need: Low, as this doesn't directly impact patient outcomes. Feasibility: Easy to organize, with existing facilities and programs that can be adapted. Action: Consider implementing if budget allows, but not urgent.	**Example: Introducing new patient feedback kiosks in outpatient departments.** Need: Can improve patient satisfaction but has minimal direct impact on clinical outcomes. Feasibility: Simple to install, inexpensive, and easy to use. Action: Can be done, but not urgent or essential. **Example: Switching to eco-friendly hospital uniforms.** Need: Low impact on patient safety and quality, more related to environmental sustainability. Feasibility: Easy to implement with new suppliers. Action: Consider if there are resources, but not a priority for patient care.

(continued)

Table 13.2 (continued)

Low need/low feasibility	
Example: Building a luxury wing for VIP patients in a public hospital. Need: Low, as it serves only a small portion of the patient population and doesn't improve care overall. Feasibility: Difficult due to high costs and the need for specialized services. Action: Likely to be abandoned or deprioritized due to lack of justification. **Example: Integrating virtual reality (VR) for routine patient consultations.** Need: Low, as it offers minimal improvement over traditional consultations for most cases. Feasibility: Complex to implement with high costs and minimal added value for typical patient needs. Action: Consider dropping the idea unless future technological improvements make it more relevant and accessible.	**Example: Implementing robotic systems for patient meal delivery.** Need: Low impact on patient care as meal delivery doesn't influence clinical outcomes significantly. Feasibility: Expensive and complex, requiring significant investment in infrastructure and maintenance. Action: Low priority and resource-intensive, unlikely to be pursued. **Example: Introducing virtual reality (VR) for patient entertainment during hospital stays.** Need: Does not directly improve healthcare outcomes. Feasibility: High cost, technology-heavy, and require training and maintenance. Action: Low priority, unnecessary for most healthcare settings.

Change concept	**Prioritize improvement**
• Eliminate waster • Improve workflow • Optimize inventory • Enhance the service provider–customer relationship • Change the work environment • Manage time • Manage variation • Design systems to prevent errors • Focus on the design of products and services	• Keeping Facility Clean • Reducing Waiting time at Clinic • Measuring Patient Satisfaction • Segregation of Biomedical Waste • Care coordination (cross-cutting) • Involving local community leaders for managing services • Starting kitchen services in the hospital • Reducing unnecessary referrals to another health facility • Medication management—preventing medication errors and overuse of antibiotics • Medical Audits • Calibration of equipment in laboratory

Fig. 13.2 Prioritization of improvement ideas—some examples

13.4 Systems Approach

In healthcare, blaming frontline staff for quality and safety issues is common but misguided. Lucian Leape emphasized that human error often stems from flawed systems, not individual actions. Yet, organizations often scapegoat individuals, ignoring systemic issues.

Medical errors often result from systemic failures rather than individual negligence. For instance, miscommunication between healthcare providers during patient handoff due to inadequate protocols or unclear procedures can lead to medication errors or delayed treatment. Similarly, deficiencies in electronic health record systems that fail to alert clinicians about potentially harmful drug interactions can result in adverse events.

These examples illustrate how errors in healthcare are often the consequence of systemic flaws rather than solely human error. Punitive cultures in healthcare environments can have detrimental effects on patient safety and staff morale. For example, when healthcare professionals are afraid to report errors or near-misses due to fear of reprisal or disciplinary action, vital opportunities for learning and improvement are lost. This fear-based culture can lead to underreporting of incidents, hindering efforts to identify and address systemic issues that contribute to errors. Additionally, punitive responses to errors can create a climate of blame and mistrust among healthcare team members, impeding collaboration and communication essential for delivering high-quality care. To break this cycle, organizations must foster genuine empathy and address systemic issues alongside individual accountability.

> The single greatest impediment to error prevention in the medical industry is that we punish people for making mistakes – **Lucian Leape**
> A bad system will beat a good person every time – **Deming, W. Edwards**

13.5 System Engineering Initiative for Patient Safety (SEIPS)

The System Engineering Initiative for Patient Safety (SEIPS) is a framework developed to improve patient safety by focusing on the interactions within healthcare systems. It adopts human factors and systems engineering approach to understand and enhance the complex processes involved in delivering safe care. SEIPS emphasizes the interconnections between key elements such as people (healthcare providers and patients), tasks, technology, physical environment, and organizational structures (Fig. 13.3).

The framework identifies how these elements impact healthcare processes and outcomes, encouraging a holistic view of patient safety beyond just focusing on individual errors. By analyzing work systems and the interactions within them, SEIPS provides a structured way to design safer healthcare processes, identify potential hazards, and implement sustainable improvements. This approach has been widely used in developing tools, interventions, and studies aimed at enhancing care quality and safety.

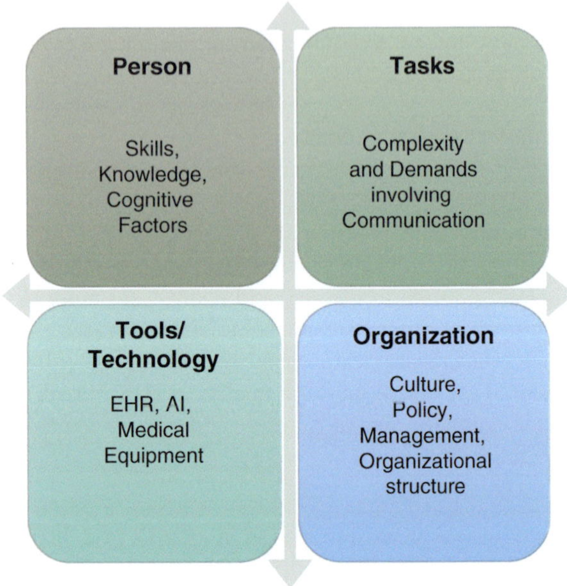

Fig. 13.3 System Engineering Initiative for Patient Safety (SEIPS) model for human factors

13.6 Risk Management in Critical Care and Systems Approach

Though risk management is vital for critical care and applied to individual patient management for patient safety and optimal outcomes, its understanding and application at the systems level are limited or non-existent. Critical care settings are prone to higher errors and risks due to the complexity and severity of patient conditions, rapid decision-making, and potential for complications. Risk management plays a crucial role in mitigating these risks by implementing measures such as severity of illness scores, predictive scoring systems, early warning systems, and involvement of patients and families in the care process. By identifying, assessing, and addressing risks systematically, risk management helps enhance patient safety and optimize outcomes in critical care settings. Understanding and implementing organizational-level risk management measures is essential for effective critical care. Implementing robust risk management in hospitals requires systemic changes and educational enhancements. Example of risk management in critical care is shown in Box.

Risk Management in Critical Care

Several measures which can help in assessing severity

- Severity of Illness Scores: Tools like APACHE assess mortality risk based on physiological parameters.
- Predictive Scoring Systems: SOFA and MODS track organ function and assess multiple organ failure risk.
- Early Warning Systems: MEWS identifies deteriorating patients for early intervention.
- Clinical Frailty Scale: Assesses patient's functional status and ability to cope with illness.
- Complication Risk Assessment: Evaluates risks of specific critical care-related complications.
- Patient and Family Involvement: Engaging patients in care can enhance safety and decision-making.

Systemic challenges:

- Breaks in communication and teamwork: Effective teamwork reduces errors and adverse events.
- Human Factors and Ergonomics: Designing workflows to reduce cognitive load and fatigue is crucial.
- Medication Safety: Measures to prevent medication errors are vital, as they're a major safety concern.

By embracing a holistic, systems-based approach, organizations can empower staff to speak up and collaborate on solutions. Cultivating mutual trust and continuous learning can lead to transformative improvements in patient-centered care.

13.7 Human Behavior and Cognitive Biases

Human behavior in healthcare settings exhibits inherent variability, impacting quality outcomes. Strict adherence to protocols often overlooks this variability, causing disparities between intended and actual results. Confirmation bias exacerbates these challenges, potentially masking areas needing improvement. These biases and behaviors can impact clinical decision-making, patient care, and overall healthcare outcomes, highlighting the importance of awareness and mitigation strategies within healthcare settings.

Addressing this requires integrating insights from human psychology into quality systems. Implementing changed processes, reminders, and fostering a culture of reflection through training programs are vital steps. Leadership is crucial in promoting flexibility and continuous learning, enhancing quality assessments and resilience. Embracing human variability can make healthcare systems more effective and humane, ensuring sustained quality improvement. These biases and behaviors can impact clinical decision-making, patient care, and overall healthcare outcomes, highlighting the importance of awareness and mitigation strategies within healthcare settings.

Do the right thing. It will gratify some people and astonish the rest – **Mark Twain**
The Central issue is never strategy, structure, culture, or systems. The core of the matter is always about changing the behavior of people – **John Kotter**
People are entitled to joy in work – **Deming, W. Edwards**

13.8 Culture vs. Perks

In healthcare, culture plays a far more significant role than perks in improving quality because it shapes the behaviors, values, and attitudes of healthcare professionals, directly influencing patient care. A strong culture that prioritizes patient safety fosters adherence to best practices, encourages open communication, and promotes error reporting without fear of retribution, which leads to better outcomes. Collaboration and teamwork, essential for quality care, thrive in environments where a culture of mutual respect and cooperation is prioritized. This level of trust and seamless coordination cannot be achieved through perks alone.

Culture drives sustainable quality improvements, as it promotes continuous learning and adaptation to new protocols, ensuring that high standards are maintained over time. While perks may offer short-term motivation, they do not foster the deep, long-term commitment to excellence that a positive culture instills. Healthcare workers in a strong, supportive culture are intrinsically motivated, finding purpose and pride in delivering quality care. This sense of mission leads to higher job satisfaction and retention, both critical for long-term improvement in healthcare. Ultimately, while perks may provide temporary incentives, it is the culture that sustains meaningful and lasting improvements in healthcare quality (Fig. 13.4).

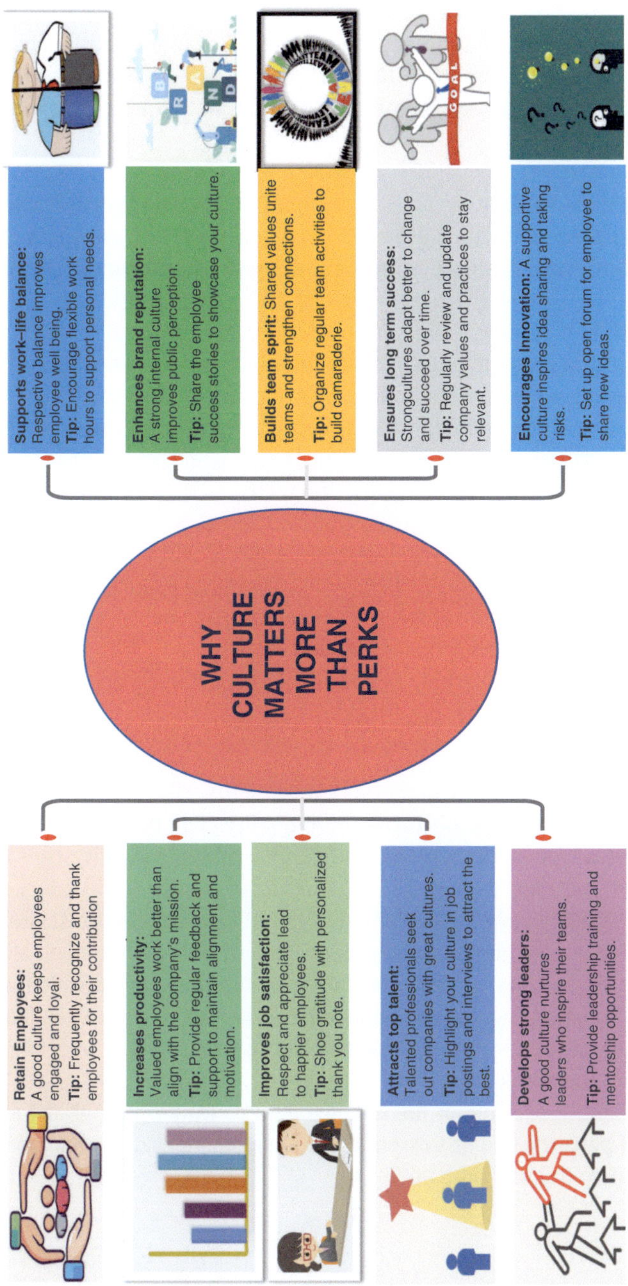

Fig. 13.4 The lasting impact of culture vs. perks in healthcare

13.9 Training and Education

Hospitals prioritize training to ensure adherence to policies, enhance outcomes, and drive quality systems. Training updates staff on knowledge, skills, attitudes, and competencies, enabling them to perform efficiently and effectively while fostering ongoing improvement. However, organizing effective training programs poses numerous challenges:

- Organizing logistics such as venue, equipment, and catering for multiple sessions.
- Finding qualified instructors, especially consultants, willing to spare time.
- Ensuring close supervision during training events to guarantee attendance and quality.
- Ensuring the content aligns with current treatment guidelines and best practices.
- Inconsistent evaluation methods lead to unreliable assessment of learning.
- Difficulty in organizing induction training for individual new hires.
- Documenting and maintaining training records for accreditation.
- Hidden costs associated with lost working hours and manpower.
- Lack of objective measurements for HR use in evaluating training effectiveness.

> Studies prove that the more employees are involved in kaizen events, the more satisfied they are with their jobs – **Mark DeLuzio**
> … the minute you quit learning is the minute your value starts going down – **Tracey & Ernie Richardson**

Lastly, more training doesn't mean more quality. The belief that more training automatically leads to better care is flawed. Prioritizing quantity over quality in training sessions neglects the underlying issues driving quality and safety outcomes. Attendance at training sessions doesn't always translate into improved behaviors or tangible outcomes, leading to a misplaced sense of accomplishment. Rather than addressing the root causes, organizations often use training as a distraction from the real work of transforming culture and empowering staff. True excellence lies in fostering a patient-centric, people-centric culture that engages and empowers all healthcare team members.

These challenges can hinder the effectiveness of training programs and impact organizational performance. It's essential for hospitals to evaluate and address these issues to optimize training outcomes.

> Quality is everybody's job, but because it is everybody's job, it can become nobody's job without the proper leadership and organization – **Armand Feigenbaum**

Manifestations of cognitive bias and human behavior in healthcare include

Confirmation Bias: Healthcare professionals may selectively seek or interpret information that confirms their pre-existing beliefs or diagnoses, potentially overlooking contradictory evidence.

Anchoring Bias: Clinicians may rely too heavily on initial impressions or information, anchoring their subsequent decisions and failing to consider new or evolving data.

Availability Heuristic: Healthcare providers may overestimate the likelihood of certain outcomes based on recent or vivid experiences, leading to biased decision-making.

Overconfidence Bias: Healthcare professionals may exhibit unwarranted confidence in their judgments or abilities, potentially leading to errors in diagnosis or treatment planning.

Stereotyping: Clinicians may make assumptions or judgments about patients based on their demographic characteristics, potentially leading to disparities in care.

Illusion of Control: Healthcare providers may overestimate their ability to control or influence patient outcomes, leading to unrealistic expectations and potential disappointment or frustration.

People always need to feel challenged, and to even be pushed a little so that they will step outside of their comfort zone – **Tracey & Ernie Richardson**

13.10 Teamwork

Teamwork is essential to patient safety, as it fosters collaboration, communication, and coordination among healthcare professionals to prevent errors and ensure optimal care. Effective teamwork allows for the sharing of knowledge, early identification of risks, and timely interventions, reducing the likelihood of adverse events. It promotes a culture of openness, where team members feel comfortable reporting mistakes or concerns, leading to continuous learning and improvement. By combining diverse skills and perspectives, teamwork ensures comprehensive patient care, enhances decision-making, and strengthens accountability, ultimately improving outcomes and creating a safer healthcare environment.

Overcoming challenges in teamwork requires a multifaceted approach, focusing on clear communication, defined roles, and a supportive culture. Key strategies enhance teamwork by addressing common challenges like miscommunication, role confusion, and interpersonal conflicts, ultimately improving team performance and patient safety.

Key Strategies to Enhance Teamwork and Patient Safety Include

1. **Effective Communication**: Encourage open, transparent communication channels where team members can share ideas, provide feedback, and express concerns without fear of judgment. Tools like regular team meetings, handovers, and communication platforms can facilitate this.
2. **Defined Roles and Responsibilities**: Clearly define each team member's role to avoid overlaps or gaps in tasks. Knowing who is responsible for specific duties ensures accountability and helps prevent confusion.
3. **Trust and Respect**: Building mutual trust and respect among team members is crucial. This can be achieved by promoting a culture where everyone's expertise and contributions are valued.
4. **Leadership Support**: Strong leadership is essential to guide the team, resolve conflicts, and create a positive environment. Leaders should provide direction, support, and recognition to keep the team motivated.
5. **Conflict Resolution**: Encourage constructive conflict resolution strategies to address disagreements. Addressing disputes promptly aids in avoiding the escalation of misunderstandings and supports team unity.
6. **Team Training**: Provide regular teamwork and communication training, focusing on skills like active listening, negotiation, and problem-solving, which enhance team dynamics.
7. **Shared Goals**: Align team efforts around common goals. When everyone understands the team's mission and how their role contributes, it fosters unity and collaboration.

The stages of teamwork and formation are commonly described by Bruce Tuckman's model, which outlines five key phases a team goes through in its development. Understanding these stages helps improve team performance and resolve challenges as the team evolves:

1. **Forming**: In this initial stage, team members come together and start to understand their roles and tasks. There is excitement, but also uncertainty, as individuals get acquainted with each other and their responsibilities. Leaders play a crucial role in providing direction and structure. Example: A hospital assembles a multidisciplinary team to improve infection control protocols. In the first few meetings, the doctors, nurses, pharmacists, and administrators are polite but hesitant, unsure of each other's roles. They start discussing the project goals and how they might contribute.
2. **Storming**: As team members start working together, conflicts and disagreements may arise due to differences in opinions, working styles, or personalities. This is a critical stage where leadership and conflict resolution are needed to help the team navigate through tensions, establish trust, and address power struggles. Example: As the infection control team begins planning specific interventions, disagreements arise over the best strategies. The doctors prioritize clinical guide-

lines, while the nurses emphasize patient education. Tension builds as team members debate whose approach is more effective. Leadership steps in to mediate and facilitate compromise.

3. **Norming**: Once conflicts are resolved, the team begins to develop stronger relationships and agree on shared norms and processes. Cooperation improves as roles are clarified, and team members start working more cohesively. Trust and collaboration become more evident during this stage. After several meetings, the team starts finding common ground. Example: Roles are clearly defined: doctors focus on treatment protocols, while nurses lead patient education efforts. The team agrees on a timeline, divides tasks, and begins collaborating smoothly, appreciating each other's strengths.
4. **Performing**: At this stage, the team functions at a high level, with established roles, strong communication, and effective collaboration. Team members are focused on achieving their goals and work independently with minimal supervision. Productivity and efficiency are maximized, and the team is now capable of handling challenges effectively. Example: Now that the team has settled into their roles, they are working efficiently to implement new infection control measures. The group communicates well, deadlines are met, and they achieve notable reductions in infection rates. Team members support each other and handle challenges with minimal supervision.
5. **Adjourning** (also called "Mourning"): This final stage occurs when the team completes its project or task, and members prepare to move on. There may be a sense of accomplishment, but also sadness as the team disbands. Reflecting on the team's journey and celebrating successes can help bring closure. Example: After the successful implementation of the infection control project, the team disbands, or the members change. Before they part ways, the team holds a final meeting to celebrate their achievements, review what worked well, and identify lessons for future projects. Members convey appreciation and contemplate their shared experiences.

Since quality improvement is an ongoing process, the team must remain active to assess performance, implement changes, and monitor outcomes over time. For example: Hospital Infection Control Team, Quality Improvement Team, Antimicrobial Stewardship Program (AMSP) Team, Cybersecurity Incident Response Team, Disaster Management Team. These examples show teams whose work is critical to the ongoing functioning and safety of their organizations, making perpetual operation essential. To avoid adjourning in perpetual teams such as Hospital Infection Control Committee, it's essential to maintain momentum and continuity through several strategies:

- **Evolving Goals**: Set ongoing, evolving objectives to keep the team focused on new challenges, ensuring they remain engaged and relevant over time.
- **Continuous Learning**: Encourage skill development and training to keep team members motivated and aligned with the latest practices and innovations.
- **Role Rotation**: Rotate roles or introduce new responsibilities to prevent stagnation and allow team members to grow within the team.

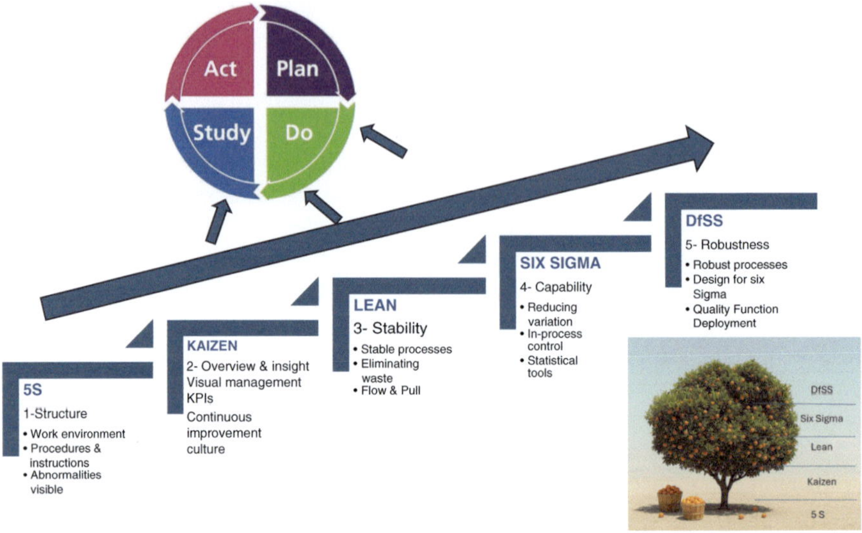

Fig. 13.5 Steps for selecting quality improvement approaches

- **Strong Leadership**: Ensure consistent leadership that fosters long-term vision, collaboration, and motivation, maintaining the team's purpose over time.
- **Celebrate Milestones**: Recognize and celebrate achievements along the way, reinforcing team cohesion and a sense of purpose without disbanding.

> Toyota's assumption is that if you make teamwork the foundation of the company and develop strong leaders, individual performers will give their hearts and souls to contribute to the team and make the company successful – **Jeffrey Liker**

Choosing the most suitable quality improvement approach depends on various factors, including the organization's goals, resources, culture, and specific challenges. Steps that can help to choose between various quality improvement approaches are shown in Fig. 13.5.

13.11 Where to Start

Selecting the Most Suitable Quality Improvement Approach

Steps that can help to choose between various quality improvement approaches:

1. **Identify Goals and Objectives**: Start by clearly defining the goals and objectives of the quality improvement initiative. Determine what you want to achieve, whether it's reducing defects, improving efficiency, enhancing customer satisfaction, or achieving regulatory compliance.
2. **Assess Current Situation**: Conduct a thorough assessment of the current state of processes, systems, and performance metrics. Identify areas for improvement, pinpoint bottlenecks, and uncover opportunities for enhancement.
3. **Understand Organizational Culture**: Consider the organization's culture, values, and readiness for change. Assess the level of employee engagement, leadership commitment, and willingness to adopt new approaches.
4. **Analyze Resources and Constraints**: Evaluate the resources available for quality improvement initiatives, including budget, time, expertise, and technology. Consider any constraints or limitations that may impact on the implementation of certain approaches.
5. **Select Appropriate Methodologies**: Based on the goals, current situation, organizational culture, and resources, select the most appropriate quality improvement methodologies. Options may include Six Sigma, Lean, Kaizen, Total Quality Management (TQM), Plan-Do-Study-Act (PDSA) cycles, or others.
6. **Consider Stakeholder Involvement**: Involve key stakeholders, including employees, leaders, customers, and suppliers, in the decision-making process. Seek their input and feedback to ensure alignment with organizational priorities and needs.
7. **Pilot Test and Evaluate**: Before implementing a quality improvement approach on a larger scale, consider conducting pilot tests or small-scale experiments to assess feasibility, effectiveness, and potential impact. Gather data and feedback to evaluate the results.
8. **Monitor and Adjust**: Continuously monitor progress and performance metrics throughout the implementation process. Adjust as needed based on feedback, lessons learned, and changing circumstances.
9. **Foster Continuous Improvement**: Regardless of the chosen approach, prioritize a culture of continuous improvement within the organization. Encourage ongoing learning, innovation, and adaptation to drive sustained excellence over time.

Selecting the Most Suitable Quality Improvement Approach and Self-Assessment

14

Contents

14.1 Self-Test

Some case-based scenarios that call for use of a particular quality tool or set of tools in quality.

Scenario 14.1: Hospital Emergency Department (ED) Wait Times
Case: The emergency department (ED) of a hospital is experiencing long wait times for patients seeking urgent medical care. This issue has led to patient dissatisfaction, overcrowding, and delays in treatment. Hospital management decides to apply quality tools to address this problem.

Solution: The hospital decides to implement the Lean Six Sigma methodology to improve ED wait times. They form a multidisciplinary team consisting of ED staff, nurses, physicians, administrators, and quality improvement experts.

1. Define: The team begins by defining the problem and setting specific goals. They collect data on current wait times, patient volumes, staffing levels, and process flow within the ED.
2. Measure: Using statistical tools, the team analyzes the collected data to identify bottlenecks and areas of inefficiency contributing to long wait times. They create process maps to visualize patient flow and identify opportunities for improvement.
3. Analyze: The team conducts root cause analysis using tools such as fishbone diagrams and Pareto charts to identify underlying factors contributing to long

S. Sharma, *Quality Demystified*, https://doi.org/10.1007/978-981-95-0339-1_14

wait times. They discover issues such as inadequate staffing during peak hours, inefficient triage processes, and delays in diagnostic testing.
4. Improve: Based on their analysis, the team develops and implements targeted interventions to address the identified issues. These interventions may include optimizing staffing levels, streamlining triage processes, implementing rapid diagnostic protocols, and improving communication between ED staff.
5. Control: The team establishes monitoring mechanisms to track the effectiveness of the interventions and ensure sustained improvement over time. They continue to collect data on wait times, patient satisfaction, and process metrics to identify any new issues and make further refinements as needed.

Explanation: In this scenario, the hospital utilizes the Lean Six Sigma methodology as a set of quality tools to systematically address the problem of long ED wait times. By following the Define, Measure, Analyze, Improve, and Control (DMAIC) framework, the multidisciplinary team is able to identify root causes, develop targeted interventions, and monitor the effectiveness of their efforts. This approach enables the hospital to improve patient flow, reduce wait times, enhance patient satisfaction, and optimize resource utilization in the ED. Through the application of quality tools, the hospital demonstrates its commitment to delivering high-quality, efficient healthcare services to its patients.

Scenario 14.2: Patient Discharge Process Optimization

Case: A hospital identifies delays and inefficiencies in the patient discharge process, leading to prolonged lengths of stay and decreased patient satisfaction. To address this issue, the hospital decides to utilize flowchart quality tools to optimize the patient discharge process.

Solution: The hospital forms a quality improvement team consisting of discharge planners, nurses, physicians, administrators, and quality improvement specialists to tackle the patient discharge problem.

1. Define: The team defines the problem by collecting data on the average length of stay, reasons for discharge delays, and patient feedback regarding the discharge process. They set specific goals to reduce length of stay and improve patient satisfaction related to discharge.
2. Flowchart: Using flowchart quality tools, the team maps out the current patient discharge process from the decision to discharge to the actual discharge of the patient. Each step of the process, including discharge planning, medication reconciliation, coordination of follow-up care, and patient education, is visually represented with decision points and interactions between healthcare providers.
3. Analyze: The team analyses the flowchart to identify potential sources of delays and inefficiencies in the patient discharge process. They identify issues such as delays in obtaining discharge orders from physicians, incomplete medication reconciliation, lack of communication between care team members, and inadequate patient education.

4. Improve: Based on their analysis, the team develops and implements targeted interventions to address the identified issues. These interventions may include implementing standardized discharge protocols, streamlining communication between healthcare providers, providing discharge planning training for staff, and utilizing electronic health record systems to facilitate medication reconciliation and documentation.
5. Control: The team establishes monitoring mechanisms to track the effectiveness of the interventions and ensure sustained improvement over time. They continue to use flowchart quality tools to periodically review and update the patient discharge process, incorporating feedback from frontline staff and monitoring length of stay and patient satisfaction metrics.

Explanation: In this scenario, the hospital utilizes flowchart quality tools to systematically analyze and improve the patient discharge process. By visually mapping out the current process and identifying areas for improvement, the quality improvement team is able to develop targeted interventions to streamline workflows, reduce delays, and improve patient satisfaction related to discharge. The flowchart serves as a valuable tool for communication, standardization, and continuous improvement, enabling the hospital to optimize the patient discharge process and enhance the overall patient experience. Through the application of flowchart quality tools, the hospital demonstrates its commitment to delivering high-quality, patient-centered care.

Scenario 14.3: Delay in Starting Operation Theatre

Case: A hospital's operating theatre (OT) frequently experiences delays in starting scheduled surgeries, leading to patient inconvenience, increased wait times, and resource inefficiency. To address this issue, the hospital decides to utilize the 5 Whys quality tool to identify the root causes of the delays.

Solution: The hospital forms a quality improvement team comprising surgeons, nurses, anesthesiologists, OR staff, administrators, and quality improvement specialists to investigate the causes of the delays using the 5 Whys quality tool.

1. Define: The team defines the problem by collecting data on the frequency and duration of delays in starting surgeries, reasons provided by OR staff, and the impact on patient care and hospital operations. They set specific goals to reduce the incidence of delays and improve the efficiency of the OT.
2. 5 Whys Analysis:
   (1) Why are surgeries frequently delayed?
      - Because the OR is not prepared on time.
   (2) Why is the OR not prepared on time?
      - Because the previous surgery ran longer than scheduled.
   (3) Why did the previous surgery run longer than scheduled?
      - Because the surgical team encountered unexpected complications during the procedure.

(4) Why were unexpected complications encountered?
   - Because the preoperative assessment did not adequately identify the patient's underlying health issues.

(5) Why did the preoperative assessment fail to identify underlying health issues?
   - Because there was insufficient communication between the surgical team and the patient's primary care provider, leading to incomplete medical history documentation.

3. Solution Implementation:
   - Improve communication between surgical team and primary care providers to ensure comprehensive preoperative assessments.
   - Enhance documentation processes to capture complete medical histories and identify potential risk factors prior to surgery.
   - Implement strategies to mitigate the impact of unexpected complications during surgeries, such as allocating additional OR time for complex cases.

Scenario 14.4: Cancellation of Operations

Case: A hospital is experiencing a high rate of cancelled operations, which disrupts patient care, increases healthcare costs, and affects staff morale. To address this issue, the hospital decides to utilize the 5 Whys quality tool to identify the root causes of operation cancellations.

Solution: The hospital forms a quality improvement team comprising surgeons, nurses, anesthesiologists, OR staff, administrators, and quality improvement specialists to investigate the causes of operation cancellations using the 5 Whys quality tool.

1. Define: The team defines the problem by collecting data on the frequency and reasons for operation cancellations, impact on patient care and hospital operations, and feedback from staff and patients. They set specific goals to reduce the incidence of cancelled operations and improve the efficiency of the OR.
2. 5 Whys Analysis:

(1) Why are operations frequently cancelled?
   - Because there is insufficient preoperative assessment.

(2) Why is there insufficient preoperative assessment?
   - Because patients do not adequately prepare for surgery.

(3) Why do patients not adequately prepare for surgery?
   - Because they lack clear instructions and education about preoperative requirements.

(4) Why do patients lack clear instructions and education?
   - Because there is limited communication between healthcare providers and patients during the preoperative period.

(5) Why is there limited or poor communication between healthcare providers and patients?
   - Because there are no standardized protocols for preoperative patient education and communication.

3. Solution Implementation:
   (1) Develop standardized preoperative protocols: Create clear guidelines and checklists for preoperative assessment, patient education, and communication to ensure consistency across all surgical specialties.
   (2) Improve patient education: Provide comprehensive preoperative information to patients through multiple channels, including written materials, videos, and one-on-one consultations with healthcare providers.
   (3) Enhance communication: Implement systems for regular communication between surgical teams, anesthesiologists, and patients to address any concerns, clarify instructions, and ensure optimal preparation for surgery.

Explanation: In this scenario, the hospital utilizes the 5 Whys quality tool to systematically analyze and identify the root causes of operation cancellations. By iteratively asking "why" and probing deeper into the underlying issues, the quality improvement team uncovers various contributing factors, including insufficient preoperative assessment, patient education, and communication. By addressing these root causes and implementing targeted interventions, such as developing standardized protocols, improving patient education, and enhancing communication systems, the hospital can reduce the incidence of cancelled operations and improve the efficiency of the OR. The 5 Whys analysis serves as a valuable tool for identifying systemic issues, facilitating interdisciplinary collaboration, and guiding evidence-based decision-making to enhance patient care and operational effectiveness in healthcare settings. Through the application of the 5 Whys quality tool, the hospital demonstrates its commitment to continuous quality improvement and patient-centered care.

Scenario 14.5: Root Cause Analysis of Medication Errors

Case: A hospital has been experiencing an increase in medication errors, leading to patient safety concerns and adverse outcomes. To address this issue, the hospital decides to conduct an RCA to identify the underlying causes of medication errors.

Solution: The hospital forms a multidisciplinary RCA team consisting of pharmacists, nurses, physicians, administrators, and quality improvement specialists to investigate the medication errors.

1. Define the Problem: The team defines the problem by collecting data on the types and frequency of medication errors, patient outcomes, and potential contributing factors. They set specific goals to reduce medication errors and improve patient safety.
2. Data Collection: The team gathers information on recent medication errors, including the medications involved, the individuals responsible, the circumstances surrounding the errors, and any contributing factors such as distractions, workload, or communication breakdowns.
3. RCA Investigation: The team conducts a systematic investigation to identify the root causes of the medication errors. They analyze the data collected and use

techniques such as the "5 Whys" or "Cause-and-Effect Diagram" to explore deeper into the underlying factors contributing to the errors.

4. Root Cause Identification: Through the RCA process, the team identifies multiple root causes contributing to medication errors, which may include:
   - Inadequate staff training and education on medication administration protocols.
   - High workload and staffing shortages leading to rushed or distracted medication administration.
   - Poor communication between healthcare providers during medication orders, transcription, and administration.
   - Deficiencies in medication reconciliation processes, leading to discrepancies in patient medication lists.
   - Lack of standardization in medication storage, labelling, and dispensing practices.
5. Interventions: Based on the identified root causes, the team develops and implements targeted interventions to address the underlying issues contributing to medication errors. These interventions may include:
   - Providing additional staff training and education on medication safety protocols.
   - Implementing double-checking procedures for high-risk medications.
   - Improving communication channels between prescribers, nurses, pharmacists, and patients.
   - Enhancing medication reconciliation processes and electronic prescribing systems.
   - Standardizing medication storage and labelling practices to reduce confusion.

Explanation: In this scenario, the hospital conducts a root cause analysis to systematically identify the underlying causes of medication errors. Through a collaborative and structured investigation process, the RCA team uncovers multiple root causes contributing to the errors. By addressing these root causes and implementing targeted interventions, the hospital aims to improve medication safety and reduce the incidence of errors. The RCA process serves as a valuable tool for identifying systemic issues, guiding evidence-based decision-making, and promoting a culture of continuous quality improvement in healthcare settings. Through the application of root cause analysis, the hospital demonstrates its commitment to patient safety and quality care.

Scenario 14.6: Analysis of Medication Errors in a Hospital

Case: A hospital's quality improvement team is tasked with addressing medication errors, which have been identified as a significant patient safety concern. To effectively prioritize and target their improvement efforts, the team decides to utilize a Pareto chart to analyze and visualize the frequency and types of medication errors occurring within the hospital.

Solution: The quality improvement team, consisting of pharmacists, nurses, physicians, administrators, and quality improvement specialists, collaborates to analyze medication errors using a Pareto chart.

1. Define the Problem: The team defines the problem by collecting data on medication errors, including the types of errors, frequency of occurrence, and their impact on patient safety and outcomes. They set specific goals to reduce medication errors and improve patient care.
2. Data Collection and Classification: The team gathers data on medication errors reported over a specific period, categorizing them into different types such as prescribing errors, dispensing errors, administration errors, and documentation errors. Each error type is further classified based on severity and potential harm to patients.
3. Pareto Chart Creation: Using the collected data, the team creates a Pareto chart to visually represent the frequency and distribution of medication errors by type. The error types are listed along the x-axis in descending order of frequency, and the corresponding frequencies are represented by bars on the y-axis (Fig. 14.1).
4. Analysis: The team analyzes the Pareto chart to identify the most common and significant medication errors contributing to the majority of adverse events. They focus on the top categories or "vital few" errors that account for the highest proportion of occurrences, as these represent the most critical areas for improvement.
5. Prioritization and Interventions: Based on the Pareto analysis, the team prioritizes their improvement efforts by addressing the most common and impactful medication errors identified. They develop targeted interventions and action plans to address the root causes of these errors, such as:

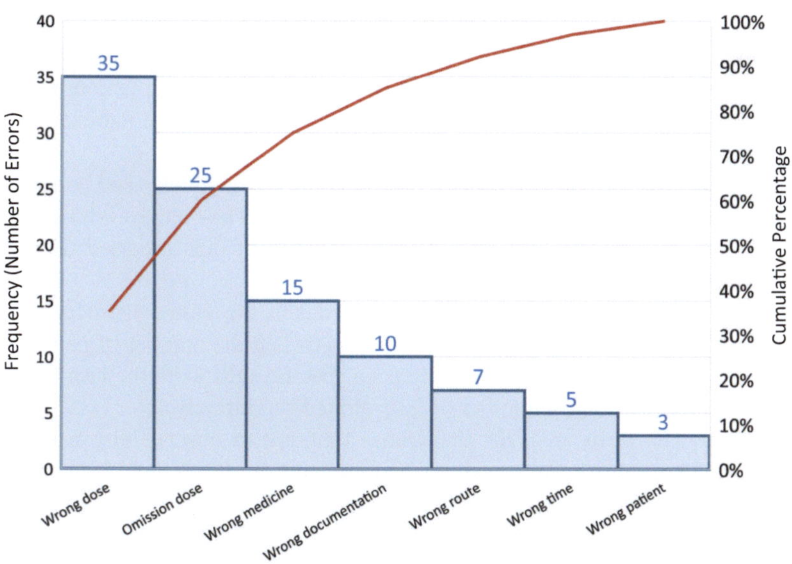

Fig. 14.1 Pareto chart of medication errors in hospital setting

- Implementing electronic prescribing systems to reduce prescribing errors.
- Enhancing medication reconciliation processes to minimize discrepancies.
- Providing additional staff training and education on medication safety protocols.
- Implementing barcode scanning systems for medication verification.
- Improving communication channels between healthcare providers involved in medication management.

Explanation: In this scenario, the hospital's quality improvement team utilizes a Pareto chart to identify and prioritize medication errors for intervention. By visually representing the frequency and distribution of errors, the Pareto chart helps the team focus their efforts on addressing the most critical and common errors that have the greatest impact on patient safety. Through targeted interventions aimed at reducing these key error types, the hospital aims to improve medication safety and enhance the quality of care provided to patients. The Pareto chart serves as a valuable tool for data-driven decision-making and resource allocation, enabling the hospital to efficiently allocate resources and prioritize improvement efforts to achieve meaningful and sustainable improvements in patient outcomes.

Scenario 14.7: Relationship Between Nurse Staffing Levels and Patient Falls
Case: A hospital is concerned about the occurrence of patient falls and wants to investigate if there is any correlation between nurse staffing levels and the number of falls. To explore this relationship, the hospital decides to use a scatter diagram.

Solution: The hospital's quality improvement team, comprised of nurses, administrators, and quality improvement specialists, collaborates to analyze the relationship between nurse staffing levels and patient falls using a scatter diagram.

1. Define the Problem: The team defines the problem by collecting data on nurse staffing levels (e.g., nurse-to-patient ratios) and the number of patient falls that occur over a specified period. They set specific goals to reduce patient falls and enhance patient safety.
2. Data Collection: The team gathers data on nurse staffing levels and patient falls from hospital records over a designated timeframe. They record the number of nurses on duty during each shift and the number of falls reported during the same period.
3. Scatter Diagram Creation: Using the collected data, the team constructs a scatter diagram with nurse staffing levels on the x-axis and the number of patient falls on the y-axis. Each data point represents a specific shift, with the nurse staffing level and corresponding number of falls plotted accordingly.
4. Analysis: The team analyzes the scatter diagram to observe any patterns or trends between nurse staffing levels and patient falls. They look for clusters or trends in the data points that may indicate a relationship between staffing levels and falls. For example, they may observe that shifts with higher nurse staffing levels tend to have fewer patient falls, while shifts with lower staffing levels have more falls.

5. Conclusion and Action Plan: Based on the analysis of the scatter diagram, the team concludes that there is a potential correlation between nurse staffing levels and patient falls. They use this insight to develop targeted interventions and action plans to address the issue. This may include:
    - Adjusting nurse staffing levels during high-risk periods or units with a higher incidence of falls.
    - Providing additional training for nurses on fall prevention strategies and patient safety protocols.
    - Implementing regular audits and monitoring systems to track nurse staffing levels and patient falls over time.

Explanation: In this scenario, the hospital utilizes a scatter diagram to explore the relationship between nursing staffing levels and patient falls. By visually representing the data on a scatter plot, the quality improvement team can identify any correlations or trends between the two variables. This analysis helps inform decision-making and resource allocation to improve patient safety and reduce the incidence of falls. The scatter diagram serves as a valuable tool for data visualization and analysis, enabling healthcare organizations to make evidence-based decisions and implement targeted interventions to enhance patient outcomes. Through the application of the scatter diagram, the hospital demonstrates its commitment to continuous quality improvement and patient-centered care.

Scenario 14.8: Improving Patient Satisfaction Scores in a Hospital

Case: A hospital's patient satisfaction scores have been consistently lower than desired, indicating a need for improvement in the quality of care provided (Figs. 14.2 and 14.3). To address this issue, the hospital decides to conduct a brainstorming session to generate ideas for enhancing patient satisfaction.

Solution: The hospital's quality improvement team, consisting of nurses, physicians, administrators, and patient experience specialists, organizes a brainstorming session to generate ideas for improving patient satisfaction.

1. Define the Problem: The team defines the problem by reviewing patient satisfaction survey results, identifying areas of concern, and setting specific goals to enhance patient satisfaction scores.
2. Brainstorming Session: During the brainstorming session, team members are encouraged to freely express their ideas and suggestions for improving patient satisfaction. The facilitator encourages creativity and ensures that all ideas are recorded without judgment or criticism. Figure 14.4 depicts various ideas generated during the brainstorming session for improving patient satisfaction.
3. Idea Generation: Team members generate a wide range of ideas for improving patient satisfaction, including:
    - Implementing bedside rounding practices to enhance communication with patients.
    - Providing additional amenities such as bedside entertainment options or complimentary services.

S no	Attributes	P1	P2	P3	P4	P5	P6	P7	P8	P9	P10	Average
1	Admission process	3	2	4	3	3	3	4	5	2	4	**3.3**
2	Staff communication & Behavior	4	4	3	4	4	4	5	4	3	5	**4**
3	Quality of Care	3	3	2	2	4	3	3	3	4	3	**3**
4	Nursing care	3	4	4	4	2	3	1	3	3	3	**3**
5	Hospital environment & Cleanliness	1	1	1	2	2	1	2	3	3	2	**1.8**
6	Food & Nutrition	4	3	2	4	3	2	2	2	4	2	**2.8**
7	Pain Management	3	3	3	2	2	2	1	2	2	3	**2.3**
8	Discharge process	4	5	4	4	3	5	4	5	4	4	**4.2**
9	Billing & Financial transparency	3	1	1	1	2	2	2	1	1	2	**1.6**
10	Post-discharge care & Follow-up	2	2	2	3	4	3	3	2	4	4	**2.9**
	Average	**3**	**2.8**	**2.6**	**2.9**	**2.9**	**2.8**	**2.7**	**3**	**3**	**3.2**	2.89

Sample size decided based on population size, confidence level, margin of error (Confidence interval), variability in responses, response rate, time and resources available

Fig. 14.2 Patient satisfaction score of IPD

S no	Attributes	P1	P2	P3	P4	P5	P6	P7	P8	P9	P10	Average
1	Access & Convenience	3	2	4	3	3	3	4	5	2	4	**3.3**
2	Timeliness of Care	4	4	3	4	4	4	5	4	3	5	**4**
3	Staff Behavior and Communication	3	3	2	2	4	3	3	3	4	3	**3**
4	Provider Competence & Expertise	3	4	4	4	2	3	1	3	3	3	**3**
5	Patient-Centered Care	1	1	1	2	2	1	2	3	3	2	**1.8**
6	Confidentiality and Privacy	4	3	2	4	3	2	2	2	4	2	**2.8**
7	Information & Education	3	3	3	2	2	2	1	2	2	3	**2.3**
8	Environment & Cleanliness	4	5	4	4	3	5	4	5	4	4	**4.2**
9	Billing & Payment Experience	3	1	1	1	2	2	2	1	1	2	**1.6**
10	Feedback & Complaints Handling	2	2	2	3	4	3	3	2	4	4	**2.9**
	Average	**3**	**2.8**	**2.6**	**2.9**	**2.9**	**2.8**	**2.7**	**3**	**3**	**3.2**	2.89

Sample size decided based on population size, confidence level, margin of error (Confidence interval), variability in responses, response rate, time and resources available

Fig. 14.3 Patient satisfaction score of OPD

Fig. 14.4 Ideas generated by brainstorming for improving patient satisfaction

- Enhancing the hospital environment with soothing color, artwork, and comfortable furnishings.
- Offering patient education programs to empower patients in managing their health.
- Implementing processes to reduce wait times and improve access to care.
- Enhancing staff training and communication skills to provide compassionate and patient-centered care.
- Implementing technology solutions to streamline processes and improve communication between patients and healthcare providers.

4. Idea Evaluation: After the brainstorming session, the team evaluates the ideas generated, considering factors such as feasibility, impact on patient satisfaction, and resource requirements. They prioritize the most promising ideas for further exploration and implementation.

5. Action Plan: Based on the ideas generated during the brainstorming session, the team develops a comprehensive action plan to address key areas of improvement in patient satisfaction. This may include implementing specific initiatives, assigning responsibilities to team members, setting timelines for implementation, and establishing metrics for monitoring progress.

Explanation: In this scenario, the hospital utilizes brainstorming to generate creative ideas and solutions for improving patient satisfaction. By tapping into the collective expertise and creativity of the quality improvement team, the hospital can identify innovative approaches to enhance the patient experience and address areas of concern. The brainstorming session serves as a valuable tool for fostering collaboration, promoting engagement, and generating actionable ideas for improvement. Through the application of brainstorming, the hospital demonstrates its commitment to delivering high-quality, patient-centered care and continuously improving the patient experience.

Scenario 14.9: Risk Assessment Using FMEA for Medication Administration
Case: A hospital's quality improvement team wants to assess the risks associated with the medication administration process to prevent medication errors and enhance patient safety. To accomplish this, they decide to conduct a risk assessment using Failure Mode and Effects Analysis (FMEA).

Solution: The quality improvement team, comprising nurses, pharmacists, physicians, administrators, and quality improvement specialists, conducts a risk assessment using FMEA for the medication administration process (Table 14.1).

1. Define the Process: The team defines the medication administration process, including the steps involved from prescription to patient administration. They identify key stakeholders involved in the process and establish specific objectives for the risk assessment.
2. Identify Failure Modes: The team identifies potential failure modes, or ways in which the medication administration process could fail, at each step of the process. This includes errors in prescribing, transcribing, dispensing, administering, and monitoring medications.
3. Determine Potential Effects: For each identified failure mode, the team determines the potential effects or consequences on patient safety and outcomes. This may include adverse drug reactions, medication errors, patient harm, or even death.
4. Assign Severity, Occurrence, and Detection Ratings: The team assigns severity, occurrence, and detection ratings to each failure mode based on their potential impact, likelihood of occurrence, and detectability. Severity ratings reflect the seriousness of the potential consequences, occurrence ratings estimate the likelihood of the failure mode happening, and detection ratings evaluate the chances of identifying the failure mode before it affects the patient.
5. Calculate Risk Priority Numbers (RPNs): The team calculates the Risk Priority Numbers (RPNs) for each failure mode by multiplying the severity, occurrence,

Table 14.1 FMEA risk assessment for medication administration process

Process step	Failure mode	Potential effect	Severity (S)	Potential causes	Occurrence (O)	Detection (D)	RPN (risk priority number)	Actions for improvement
Medication Ordering	Incorrect dose entered	Overdose or underdose	9	Miscommunication or error in transcription	5	3	135	Double-check by physician, use CPOE system
Dispensing	Wrong medication	Adverse drug reaction	10	Pharmacy misreads order	3	4	120	Barcode scanning at pharmacy, pharmacist verification
Medication Preparation	Wrong route selected	Ineffective treatment or harm	8	Lack of clear labeling	4	4	128	Labeling medication by route, training staff on routes
Administration	Wrong patient	Serious harm or death	10	Lack of patient verification	2	5	100	Two-patient identifier system, wristband scanning

Documentation	Failure to record	No evidence of administration, leading to repeated dose	7	Nurse forgets to chart	4	5	140	Implement electronic medication administration record (eMAR)
Monitoring after Admin	Failure to monitor side effects	Adverse reaction not identified	9	Lack of monitoring training	3	5	135	Training staff, automated reminders for monitoring

S = Severity (1–10, with 10 being the most severe)
O = Occurrence (1–10, with 10 being the most frequent)
D = Detection (1–10, with 10 being the hardest to detect)
RPN = Risk Priority Number (S × O × D), higher numbers indicate higher risk and priority for corrective action

and detection ratings. This approach helps identify and prioritize the failure modes that pose the highest risk, ensuring that those requiring immediate attention are addressed first.

6. Develop Mitigation Strategies: Based on the RPNs, the team prioritizes the highest-risk failure modes and develops mitigation strategies to reduce their likelihood of occurrence or mitigate their impact. This may include implementing double-checking procedures, barcode scanning systems, standardized protocols, staff training, or process redesign.
7. Implement and Monitor: The team implements the identified mitigation strategies and continuously monitors the medication administration process to assess their effectiveness. They track medication error rates, adverse drug events, and patient outcomes to evaluate the impact of the interventions and adjust as needed.

Explanation: In this scenario, the hospital utilizes FMEA to systematically assess the risks associated with the medication administration process and develop targeted mitigation strategies to enhance patient safety. By identifying potential failure modes, analyzing their potential effects, and prioritizing mitigation efforts based on risk priority numbers, the hospital can proactively address vulnerabilities in the medication administration process and prevent medication errors. FMEA serves as a valuable tool for promoting a culture of safety, improving process reliability, and reducing the likelihood of adverse events in healthcare settings. Through the application of FMEA, the hospital demonstrates its commitment to continuous quality improvement and patient-centered care.

Scenario 14.10: Addressing Hospital Readmissions Within One Month
Case: A hospital is concerned about the high rate of patient readmissions within one month of discharge, which not only impacts patient outcomes but also increases healthcare costs. The hospital's quality improvement team seeks to identify and address the underlying causes of readmissions to improve patient care and reduce healthcare expenditures.

Solution: The quality improvement team, comprising physicians, nurses, case managers, administrators, and quality improvement specialists, implements a multifaceted approach to address the problem of readmissions within one month.

1. Define the Problem: The team defines the problem by analyzing hospital data to determine the frequency, reasons, and characteristics of readmissions within one month of discharge. They identify specific patient populations, diagnoses, and contributing factors associated with readmissions.
2. Root Cause Analysis: The team conducts an RCA to identify the underlying causes of readmissions. They examine factors such as inadequate discharge planning, medication errors, lack of follow-up care, patient education deficiencies, and social determinants of health contributing to readmissions.

3. Targeted Interventions: Based on the findings of the RCA, the team develops targeted interventions to address the identified root causes of readmissions:
   - Enhanced Discharge Planning: Implement standardized discharge protocols, conduct comprehensive medication reconciliation, provide patient and caregiver education, and ensure appropriate follow-up appointments.
   - Transitional Care Programs: Establish transitional care teams to provide post-discharge support, coordinate care transitions, and conduct home visits or telehealth consultations to monitor patient progress.
   - Medication Management: Implement medication reconciliation processes, provide medication adherence support, and conduct medication therapy management to prevent medication-related readmissions.
   - Care Coordination: Improve communication and collaboration among healthcare providers, enhance care coordination between inpatient and outpatient settings, and ensure timely follow-up with primary care providers and specialists.
   - Addressing Social Determinants of Health: Identify and address social determinants of health (e.g., housing instability, food insecurity, lack of transportation) that may contribute to readmissions through community partnerships and referral programs.
4. Implementation and Monitoring: The team implements the targeted interventions and continuously monitors their effectiveness through regular data collection and analysis. They track readmission rates, patient outcomes, and adherence to intervention protocols to assess the impact of the interventions over time.

Explanation: In this scenario, the hospital employs a comprehensive approach to address the problem of readmissions within one month of discharge. By conducting a root cause analysis and implementing targeted interventions addressing discharge planning, transitional care, medication management, care coordination, and social determinants of health, the hospital aims to reduce readmissions and improve patient outcomes. This multifaceted approach acknowledges the complex factors contributing to readmissions and emphasizes the importance of collaboration among healthcare providers, patients, caregivers, and community organizations in addressing this issue. Through continuous monitoring and evaluation of the interventions, the hospital demonstrates its commitment to quality improvement and patient-centered care, ultimately leading to better outcomes for patients and reduced healthcare costs.

Scenario 14.11: Stratification of Critical Value Reporting in Laboratory Medicine

Case: A hospital laboratory is responsible for reporting critical values, such as abnormal lab results that require immediate clinical intervention, to healthcare providers for timely patient management. However, the laboratory has noticed variations in the timeliness of critical value reporting across different departments and shifts. To address this issue, the laboratory's quality improvement team decides to conduct a stratified analysis of critical value reporting to identify areas for improvement.

Solution: The quality improvement team, composed of laboratory technicians, pathologists, clinicians, administrators, and quality improvement specialists, conducts a stratified analysis of critical value reporting in the hospital laboratory.

1. Define the Analysis: The team defines the scope of the analysis, focusing on the timeliness of critical value reporting from the laboratory to healthcare providers. They aim to stratify the data based on variables such as department, shift, and type of critical value.
2. Data Collection: The team collects data on critical value reporting times from the laboratory information system, ensuring that the data are accurately stratified based on the chosen variables. They gather information on the department or unit where the critical value was generated, the time of day, and the type of critical value (e.g., abnormal hematology, chemistry, or microbiology result).
3. Stratification Variables: The team stratifies the critical value reporting data based on various variables, including:
   - Department or Unit: Emergency department, intensive care unit, medical-surgical floors, etc.
   - Shift: Day shift, evening shift, night shift.
   - Type of Critical Value: Abnormal hematology, abnormal chemistry, abnormal microbiology result.
   - Figure 14.5 illustrates the stratification of critical value reporting data according to shift by the quality improvement team.
4. Stratified Analysis: Using the collected data, the team conducts a stratified analysis of critical value reporting times for each stratum defined by the chosen variables. They calculate average reporting times and assess any variations or patterns between strata.

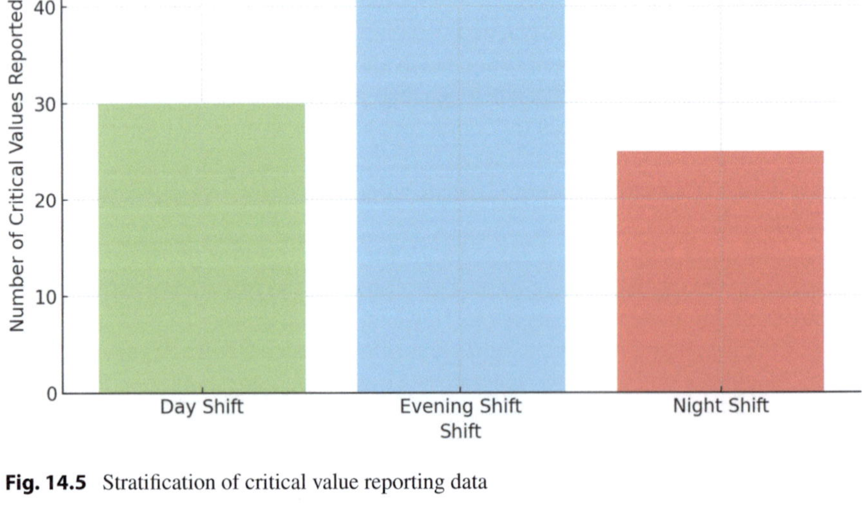

Fig. 14.5 Stratification of critical value reporting data

5. Interpretation: The team interprets the results of the stratified analysis to identify areas of improvement in critical value reporting timeliness across different departments, shifts, and types of critical values. They look for disparities or trends in reporting times between strata and investigate potential contributing factors, such as staffing levels, workload, or communication processes.
6. Tailored Interventions: Based on the findings of the stratified analysis, the team develops tailored interventions to address specific areas for improvement identified within different departments, shifts, or types of critical values. This may include:
   - Implementing automated alerts or notifications for critical values to ensure timely identification and reporting.
   - Providing additional training or resources for laboratory staff during high-volume periods or on certain shifts.
   - Streamlining communication processes between the laboratory and healthcare providers to facilitate timely reporting and action on critical values.

Explanation: In this scenario, the hospital laboratory utilizes stratification to analyze critical value reporting and identify opportunities for improvement in timeliness. By stratifying the data based on department, shift, and type of critical value, the laboratory can identify variations in reporting times and tailor interventions to address specific needs. Stratification helps ensure that critical value reporting processes are optimized for efficiency and effectiveness, leading to improved patient care and outcomes. Through the application of stratification, the laboratory demonstrates its commitment to delivering high-quality, timely patient care and continuously improving laboratory services.

Scenario 14.12: Addressing Challenges with Verbal Order Policy in Healthcare

Case: A hospital has been experiencing challenges with the implementation of its verbal order policy, leading to concerns about patient safety and compliance with regulatory standards. Healthcare providers have reported instances of miscommunication, delays in order transcription, and inconsistencies in documentation related to verbal orders. To address these issues, the hospital's quality improvement team decides to conduct a comprehensive analysis of the verbal order policy and develop targeted interventions for improvement.

Solution: The quality improvement team, comprising physicians, nurses, pharmacists, administrators, and quality improvement specialists, initiates a systematic approach to address challenges with the verbal order policy in healthcare.

1. Define the Problem: The team defines the problem by gathering input from healthcare providers and reviewing incident reports related to verbal orders. They identify specific issues such as miscommunication, transcription errors, and documentation inconsistencies, and assess their impact on patient care and safety.

2. Policy Review and Analysis: The team conducts a thorough review and analysis of the hospital's verbal order policy, including its content, procedures, and implementation guidelines. They assess compliance with regulatory standards, evidence-based practices, and best practices in verbal order management.
3. Stakeholder Engagement: The team engages key stakeholders, including physicians, nurses, pharmacists, and administrative staff, in discussions to identify barriers and challenges with the current verbal order policy. They seek input on areas for improvement and gather feedback on potential solutions.
4. Root Cause Analysis: Using techniques such as RCA, the team identifies the underlying causes contributing to challenges with the verbal order policy. This may include factors such as inadequate communication protocols, unclear documentation requirements, staffing shortages, and lack of standardized processes.
5. Interventions: Based on the findings of the policy review, stakeholder engagement, and root cause analysis, the team develops targeted interventions to address challenges with the verbal order policy:
   - Enhancing Communication Protocols: Implement standardized communication protocols for verbal orders, including the use of read-back and verification procedures to ensure accurate transmission of orders.
   - Improving Documentation Practices: Provide training and education on documentation requirements for verbal orders, emphasizing the importance of clarity, completeness, and timeliness in documentation.
   - Standardizing Processes: Develop standardized processes for receiving, transcribing, and documenting verbal orders, including clear roles and responsibilities for healthcare providers involved in the order management process.
   - Enhancing Technology Support: Implement electronic prescribing systems or computerized physician order entry (CPOE) systems to facilitate the electronic capture and transmission of orders, reducing the reliance on verbal orders.
   - Monitoring and Auditing: Establish monitoring and auditing processes to assess compliance with the verbal order policy, identify areas for improvement, and track outcomes related to order accuracy and patient safety.

Explanation: In this scenario, the hospital's quality improvement team takes a systematic approach to address challenges with the verbal order policy in healthcare. By conducting a comprehensive analysis of the policy, engaging stakeholders, identifying root causes, and developing targeted interventions, the team aims to improve communication, documentation, and processes related to verbal orders, ultimately enhancing patient safety and quality of care. Through the application of evidence-based practices and continuous monitoring, the hospital demonstrates its commitment to addressing challenges and continuously improving its policies and procedures to ensure safe and effective patient care.

Scenario 14.13: Standardizing Documentation of Allergy History in Patient Case Notes

Case: A hospital is experiencing challenges with the uniform documentation of allergy history in patient case notes, leading to inconsistencies in allergy identification and management. Healthcare providers have reported instances of missing or incomplete allergy information, which can compromise patient safety and treatment decisions. To address these issues, the hospital's quality improvement team decides to implement a standardized approach to documenting allergy history in patient case notes.

Solution: The quality improvement team, comprising physicians, nurses, medical records staff, administrators, and quality improvement specialists, initiates a systematic approach to standardize the documentation of allergy history in patient case notes.

1. Define the Problem: The team defines the problem by reviewing patient case notes and identifying instances of missing or incomplete allergy information. They assess the potential impact of incomplete allergy documentation on patient safety, treatment decisions, and regulatory compliance.
2. Stakeholder Engagement: The team engages key stakeholders, including physicians, nurses, medical records staff, and information technology (IT) personnel, in discussions to identify barriers and challenges with documenting allergy history. They seek input on current practices, challenges encountered, and opportunities for improvement.
3. Policy Development: Based on stakeholder input and best practices in allergy documentation, the team develops a standardized policy and procedure for documenting allergy history in patient case notes. The policy outlines clear guidelines for healthcare providers on how to obtain, record, and update allergy information accurately and consistently.
4. Training and Education: The team provides training and education sessions for healthcare providers on the new policy and procedures for documenting allergy history. Training sessions may include instruction on allergy assessment techniques, the importance of thorough documentation, and the use of standardized templates or EHR tools for recording allergy information.
5. Standardized Documentation Tools: The team develops standardized documentation tools, such as allergy assessment forms or templates within the EHR system, to facilitate the uniform recording of allergy history in patient case notes. These tools may include prompts for healthcare providers to document allergy type, severity, reaction, and date of onset.
6. Implementation and Monitoring: The team implements the standardized documentation policy and tools across all departments and units within the hospital. They monitor compliance with the new policy through regular audits of patient case notes, assessing the completeness and accuracy of allergy documentation. Feedback is provided to healthcare providers to reinforce adherence to the standardized approach.

Explanation: In this scenario, the hospital's quality improvement team takes proactive steps to standardize the documentation of allergy history in patient case notes. By engaging stakeholders, developing a standardized policy and procedure, providing training and education, and implementing standardized documentation tools, the team aims to improve the accuracy, completeness, and consistency of allergy documentation. Standardized documentation practices enhance patient safety by ensuring that healthcare providers have access to reliable allergy information for informed treatment decisions. Through continuous monitoring and feedback, the hospital demonstrates its commitment to quality improvement and patient-centered care.

Scenario 14.14: Challenges with Adhering to Real-Time Documentation of Safe Surgical Checklist

Case: A hospital's operating room (OR) is experiencing difficulties in adhering to real-time documentation of the safe surgical checklist, leading to incomplete or delayed documentation of critical safety measures during surgical procedures. Surgeons, nurses, and other OR staff have reported instances of forgetfulness, time constraints, and workflow interruptions that hinder their ability to complete the checklist in a timely manner. As a result, there are concerns about potential patient safety risks and compliance with regulatory standards. The hospital's quality improvement team decides to address these challenges and improve adherence to real-time documentation of the safe surgical checklist.

Solution: The quality improvement team, consisting of surgeons, nurses, anesthesiologists, surgical technologists, administrators, and quality improvement specialists, implements targeted strategies to address challenges with adhering to real-time documentation of the safe surgical checklist.

1. Identify Root Causes: The team conducts an RCA to identify the underlying reasons for the challenges with real-time documentation. Potential root causes may include:
   - Lack of standardized processes or workflows for completing the checklist.
   - High workload and time constraints in the OR, leading to distractions and interruptions.
   - Forgetfulness or oversight due to the complexity of surgical procedures.
   - Inadequate training or awareness about the importance of timely checklist documentation.
2. Workflow Optimization: Based on the findings of the RCA, the team collaborates with OR staff to optimize workflow processes and streamline documentation procedures. This may involve:
   - Establishing clear roles and responsibilities for completing and documenting checklist items.
   - Implementing standardized protocols for incorporating the safe surgical checklist into preoperative, intraoperative, and postoperative workflows.
   - Allocating dedicated time slots for completing the checklist before key milestones in the surgical procedure.

3. Training and Education: The team provides targeted training and education sessions for OR staff on the importance of real-time documentation of the safe surgical checklist and the revised workflow processes. Training sessions may include:
   - Reviewing the purpose and significance of each checklist item in enhancing patient safety.
   - Demonstrating best practices for completing and documenting checklist items efficiently and accurately.
   - Providing tips and strategies for overcoming common challenges and barriers to adherence.
4. Implement Reminders and Prompts: The team implements reminders and prompts within the OR environment to reinforce adherence to real-time documentation of the safe surgical checklist. This may include:
   - Using EHR systems or digital platforms to display checklist items and reminders at key points during the surgical procedure.
   - Incorporating audible alerts or visual cues to prompt OR staff to complete and document checklist items before proceeding with each stage of the surgery.
5. Continuous Monitoring and Feedback: The team establishes mechanisms for continuous monitoring and feedback to assess the effectiveness of the interventions and identify opportunities for further improvement. This may involve:
   - Regular audits of checklist documentation to evaluate adherence rates and identify areas for improvement.
   - Soliciting feedback from OR staff on the usability and effectiveness of the revised workflow processes and documentation tools.
   - Providing ongoing support and reinforcement to OR staff through coaching, mentoring, and recognition of achievements in checklist adherence.

Explanation: In this scenario, the hospital's quality improvement team takes proactive steps to address challenges with adhering to real-time documentation of the safe surgical checklist in the OR. By identifying root causes, optimizing workflow processes, providing training and education, implementing reminders and prompts, and establishing mechanisms for continuous monitoring and feedback, the team aims to enhance adherence to checklist documentation and improve patient safety during surgical procedures. Improving adherence to real-time documentation practices ensures that critical safety measures are consistently implemented and documented, leading to better outcomes for patients undergoing surgery. Through the application of targeted strategies and ongoing support, the hospital demonstrates its commitment to quality improvement and patient-centered care in the OR environment.

14.2 Self-Assessment Diagnostic Tool for Evaluating Safe, Reliable, and Effective Care

This diagnostic tool is designed to help organizations systematically evaluate their performance based on the various components for Safe, Reliable, and Effective Care. The framework, centered on patients and families, is built around two foundational domains—culture and the learning system—and nine interrelated components. It integrates strategic, clinical, and operational concepts essential for achieving safe, reliable, and effective care into a single, comprehensive model.

By using this tool, organizations can:

- **Assess Performance**: Evaluate how effectively they are addressing and implementing the framework's components, identifying strengths, gaps, and opportunities for improvement.
- **Leverage a Comprehensive Framework**: Utilize the framework as a roadmap, which succinctly brings together all the critical elements needed for achieving safe, reliable, and effective care.
- **Focus on Continuous Improvement**: Gain actionable insights to prioritize and address areas requiring development, fostering a culture of ongoing learning and enhancement.
- **Ensure Patient-Centered Care**: Align organizational efforts with the framework's core focus on partnering with patients and families to deliver respectful, reliable, and safe care.
- **Use Flexible Tools**: Apply the diagnostic tool both as a strategic planning guide and a resource for tracking progress and benchmarking performance over time.

This tool empowers organizations to systematically drive improvements in safety and reliability, ensuring care delivery is effective, patient-centered, and continuously evolving. Using this tool, organizations can establish clear goals, track progress, and build the capacity to achieve sustained improvements in the safety, reliability, and effectiveness of their care delivery.

Tables 14.2, 14.3, 14.4, 14.5, 14.6, 14.7, 14.8, 14.9, 14.10, 14.11 and 14.12 present a structured framework to assess and track progress in achieving across nine key components within healthcare organizations.

Table 14.2 Psychological safety for safe, reliable, and effective care

Focus area	Initial stage	Evolving: Making Progress	Advanced: Effective execution	Outstanding practice
Safe Environment	Staff feel hesitant to speak up due to fear of ridicule or negative reactions.	Some staff speak up, mainly in cases of imminent danger to patients.	Staff feel more comfortable asking questions and providing feedback but lack active encouragement.	All staff feel comfortable asking questions, providing feedback, and suggesting innovations as the norm.
Leadership's Role	Few leaders model behaviors that support psychological safety.	Some leaders encourage open communication, but practices are inconsistent.	Leaders and middle managers actively encourage psychological safety and address unsupportive behaviors.	All leaders model psychological safety behaviors, ensuring consistent support across the organization.
Feedback Mechanisms	Feedback is superficial, primarily during formal appraisals or after incidents.	Feedback quality improves but is often inconsistent or limited to select events.	Detailed feedback is regularly provided, and learning from incidents is increasingly shared.	Detailed and constructive feedback is routine, and learnings from adverse events are shared organization wide.
Empowering Innovation	Staff are reluctant to suggest innovations, fearing they won't be taken seriously.	Some suggestions are considered, but follow-through is inconsistent.	Innovations are regularly tested, and successful ideas are implemented more consistently.	Staff suggestions are routinely tested, implemented, and valued as part of continuous improvement.
Hierarchy Dynamics	Hierarchies discourage open communication and feedback sharing.	Hierarchies begin to flatten in some areas, promoting greater openness.	A semi-flat hierarchy supports constructive criticism and collaborative learning.	Flat hierarchies foster open dialogue, feedback, and collaborative problem-solving.

(continued)

Table 14.2 (continued)

Focus area	Initial stage	Evolving: Making Progress	Advanced: Effective execution	Outstanding practice
Organizational Learning	Limited sharing of lessons learned from adverse events.	Learning from events occurs but is not consistently shared across the organization.	Learning from incidents is shared across most departments and informs policy changes.	Routine sharing of lessons learned ensures continuous learning and improvement at all levels.
Culture and Expectation	Psychological safety is acknowledged but not practiced consistently.	Psychological safety practices are increasingly valued but lack standardization.	Psychological safety becomes an integral expectation but is not yet fully embedded.	Psychological safety is fully embedded as a cultural norm, with all levels actively supporting it.

Table 14.3 Accountability for ensuring safe, reliable, and effective care

Aspect	Just beginning	Making progress	Significant impact	Exemplary
Approach to Errors	Focus on punishing individuals involved in errors without considering system flaws.	Some understanding of system contributions, but focus remains on individual errors.	Systems contributions are prioritized in learning, but not all staff share this perspective.	All staff differentiate between individual and system issues, focusing on systems improvements.
Just Culture	No differentiation between individual and system contributions to harm.	Just culture approach adopted but not fully implemented across the organization.	Just culture principles are applied inconsistently; some middle managers trained in investigation methods.	Just culture algorithm fully implemented, creating a fair and transparent environment for all staff.
Training and Support	No training provided for middle managers in just culture practices.	Some managers trained, but practices are inconsistent across the organization.	Most managers trained in just culture algorithms, but application is uneven.	All managers are trained and consistently apply just culture methods to investigate and address errors.

(continued)

Table 14.3 (continued)

Aspect	Just beginning	Making progress	Significant impact	Exemplary
Accountability for Values	Staff are not held accountable for behaviors aligned with organizational values.	Some staff are held accountable, but the approach is inconsistent.	Many staff are held accountable, but the model is not uniformly applied across the organization.	All staff are accountable to their roles, responsibilities, and behaviors, aligned with organizational values.
Culture Measurement	Culture is not measured or assessed.	Culture measured every few years, but findings result in minimal action.	Culture measured annually, with plans drawn up for improvement, though follow-up is inconsistent.	Culture measured annually, with actionable plans implemented, leading to measurable improvements.
Staff Perception	Staff perceive accountability as punitive and unfair.	Perceptions of fairness improve, but variability exists across teams.	Many staff perceive the environment as fair and just, though not universally.	All staff perceive the environment as fair and supportive, fostering trust and collaboration.

Note the shift from punitive approaches to a fair, systems-focused, and supportive culture that aligns with organizational values

Table 14.4 Teamwork and communication for safe, reliable, and effective care

Focus area	Initial stage	Evolving: Making Progress	Advanced: Effective execution	Outstanding Practice
Communication Structure	Communication is unstructured, with frequent failures.	Some communication uses structured methods, but execution varies.	Most communication follows structured approaches, with rare failures.	All communication is structured and consistently executed, with exceptions being extremely rare.
Handovers and Briefings	Handovers and briefings lack structured processes.	Some handovers and briefings use standard formats, but not universally.	Most handovers and briefings follow standard processes.	All handovers and briefings are structured and standardized across the organization.
Behavioral Norms	Norms vary by seniority, profession, or personality, leading to inconsistency.	Behavioral expectations apply to all staff, but adherence is uneven.	Behavioral norms are largely standardized, with only occasional exceptions.	All teams adhere to agreed behavioral norms, with non-supportive behavior swiftly addressed.

(continued)

Table 14.4 (continued)

Focus area	Initial stage	Evolving: Making Progress	Advanced: Effective execution	Outstanding Practice
Team Identification	Staff do not consistently identify as part of a multidisciplinary team.	Some staff identify with multidisciplinary teamwork, but it is not the norm.	Multidisciplinary teamwork is the standard across most teams.	Multidisciplinary teamwork is fully embedded, with all staff contributing as equal team members.
Raising Concerns	Team members feel uncomfortable raising concerns.	A minority of staff feel comfortable speaking up.	Most staff are comfortable raising concerns, fostering a safer environment.	All team members feel psychologically safe and valued, regardless of seniority or role.
Team Dynamics	Teams are fragmented, and few are seen as positive workplaces.	Some teams are recognized as good workplaces, but this is inconsistent.	Most teams are high-functioning and supportive environments.	Teams exhibit exceptional dynamics, supporting psychological safety and valuing all opinions equally.
Communication Failures	Communication failures are common and often lead to issues.	Failures still occur but are less frequent.	Failures are rare, with communication being standardized and effective.	Communication failures are virtually eliminated due to rigorous processes and team collaboration.
Teamwork Training	No investment in teamwork or communication training.	Limited investment in training, with some staff receiving it.	Most staff have undergone teamwork and communication training.	All staff receive comprehensive teamwork and communication training, fostering consistent excellence.

Note the progression from fragmented teamwork and unstructured communication to exemplary practices that ensure psychological safety, consistent communication, and effective multidisciplinary collaboration

Table 14.5 Effective engagement and conflict resolution for safe, reliable, and effective care

Focus area	Initial stage	Evolving: Making Progress	Advanced: Effective execution	Outstanding practice
Engagement Skills	Few staff have the skills to effectively engage others in change.	Some staff are skilled in negotiation but lack consistency in applying it.	Many staff effectively negotiate improvements and process changes.	All staff are skilled in negotiation, focusing on collaborative, non-adversarial approaches.
Conflict Resolution	Conflicts are poorly managed, harming teamwork and agreement.	Conflicts are occasionally resolved, but behaviors often remain disruptive.	Conflicts are resolved in ways that benefit all parties and enhance patient care.	Conflicts are consistently resolved collaboratively, with staff reporting high levels of respect and harmony.
Focus on Interests vs. Positions	Staff focus on positions rather than underlying interests, leading to stalemates.	Some staff begin to differentiate between positions and interests in negotiations.	Most staff focus on interests, fostering productive discussions and resolutions.	All staff excel at differentiating positions from interests, using appreciative inquiry for mutual gains.
Impact on Teamwork	Poor negotiation skills undermine teamwork and collaboration.	Negotiation skills improve, but teamwork disruptions persist in some areas.	Effective negotiation supports stronger teamwork and agreement across teams.	Negotiation skills reinforce seamless teamwork, with minimal disruptive behaviors.
Patient-Centered Agreements	Differences rarely result in solutions that prioritize patient care.	Negotiations sometimes focus on patient-centered outcomes, but inconsistently.	Negotiations increasingly lead to agreements that improve patient outcomes.	All negotiations prioritize patient-centered agreements, improving care quality and stakeholder satisfaction.
Application in Improvement Work	Negotiation is rarely applied to improvement initiatives or process changes.	Some improvement initiatives involve negotiation, but practices vary.	Negotiation is widely used to align team members on quality improvement work.	Negotiation is fully integrated into quality improvement initiatives, achieving operational excellence.

Note the development of engagement or negotiation skills and their impact on teamwork, conflict resolution, and patient-centered care. It highlights the progression from ineffective practices to exemplary negotiation that fosters collaboration, respect, and operational excellence

Table 14.6 Continuous learning for safe, reliable, and effective care

Focus area	Initial stage	Evolving: Making Progress	Advanced: Effective execution	Outstanding practice
Learning from Defects	Defects are collected in reporting systems but rarely lead to shared learning.	Learning from improvement activities is shared in some areas but inconsistently.	Learning from improvement activities is shared broadly across most areas.	Learning is shared systematically both within and across organizations, covering the continuum of care.
Use of Learning Boards	Learning boards are absent in most areas.	Some care locations have learning boards, but they are not used consistently.	Most care locations use learning boards during daily huddles for reflection and planning.	Learning boards exist in all care locations and are used daily to drive planning and reflection.
Daily Huddles	Huddles occur sporadically, often depending on workload.	Huddles are routine in some care locations but not universally adopted.	Huddles are held consistently in most care locations and involve area leadership.	Huddles are conducted daily in all care locations, with leadership actively engaging in discussions.
Sharing of Learning	Learning from safety projects and root cause analyses is rarely shared.	Systems are in place to share learning but are not consistently utilized.	Learning systems are used broadly, but challenges remain in spreading learning across all areas.	Learning systems are fully operational, ensuring spread and adoption of learning across the organization.
Leadership Engagement	Leadership does not engage actively with learning systems or rounds.	Leadership participates in some learning activities, but engagement is inconsistent.	Leadership routinely uses learning boards and makes rounds to discuss and share learning.	Leadership actively drives continuous learning, ensuring system-wide improvement and engagement.
Organizational Outcomes	Errors, harm, and defects remain stable or increase over time, with little learning.	Some pilot areas show reduced errors, but recurrence remains a challenge.	Certain errors and defects are consistently reduced in specific areas.	Organizational data demonstrates significant reduction or elimination of errors, harm, and defects.
Cross-Organizational Learning	No system exists for sharing learning between organizations or across care settings.	Learning occurs within isolated units but is not shared across the continuum of care.	Some efforts are made to share learning across different care settings.	Learning is shared seamlessly within and between organizations, enhancing care continuity and outcomes.

Note the progression from minimal learning activities to an exemplary culture of continuous learning, where systems and leadership drive widespread adoption and improvement in safety and care quality

Table 14.7 Improvement for safe, reliable, and effective care

Focus area	Initial stage	Evolving: Making Progress	Advanced: Effective execution	Outstanding practice
Improvement Projects	Few improvement projects are underway, and they lack guidance from a framework.	Several projects are guided by an organization-wide framework and show measurable improvements.	Many projects achieve sustained improvements guided by a consistent framework.	Improvement is embedded in all areas, with sustained results across the continuum of care.
Use of Frameworks and Models	No consistent improvement framework or model guides efforts.	Projects are increasingly guided by an organization-wide framework or model.	Improvement framework is consistently applied, driving measurable and sustained improvements.	Frameworks are fully embedded and guide all improvement activities.
Team Involvement	Improvement efforts involve few teams, and multidisciplinary collaboration is rare.	Some projects involve multidisciplinary teams.	Multidisciplinary teams are routinely involved in improvement efforts.	Teams across the organization collaborate effectively, leveraging diverse expertise.
Training in Improvement Methods	Training is limited and inconsistent across the organization.	Improvement training is available to some staff but is not universally applied.	Training is provided broadly, enabling staff to apply improvement tools effectively.	Improvement methods are part of standard training, equipping all staff with necessary competencies.
Measurement and Analysis	Data for improvement is not routinely collected or analyzed.	Data collection and analysis occur for some projects but are not systematized.	Data is systematically used to track and sustain improvements across projects.	Data is routinely and consistently used to guide and sustain organization-wide improvement efforts.
Sharing and Spreading Learning	Learning from projects is rarely shared across the organization.	Some learning is shared, but systematic spread is limited.	Learning from improvement efforts is systematically spread across the organization.	Improvements are consistently shared across the organization, the continuum of care, and with stakeholders.

(continued)

Table 14.7 (continued)

Focus area	Initial stage	Evolving: Making Progress	Advanced: Effective execution	Outstanding practice
Sustainability of Improvements	Improvements are not sustained, and errors or issues often recur.	Some projects achieve sustained improvements, but these are isolated.	Sustained improvements are common, with learning embedded in most areas.	Improvements are consistently sustained and integrated into organizational practice.

Note the stages of progress in achieving excellence in organizational improvement, from initial efforts to an embedded, systematic approach that drives sustainable and organization-wide results

Table 14.8 Measurement for safe, reliable, and effective care

Focus area	Initial stage	Evolving: Making Progress	Advanced: Effective execution	Outstanding practice
Use of Data	Data is used to measure performance in a few areas but is rarely linked to improvement activities.	Data is used to measure performance and support many improvement projects.	Data supports almost all improvement projects and is integral to measuring performance.	Data is consistently used to drive all improvement measures at both system-wide and subsystem levels.
Data Systems	Limited systems exist for data collection and analysis, hindering broader communication.	Some data systems exist, allowing for limited cross-organizational measures.	Established data systems routinely share system-of-care performance across the organization.	Advanced data systems enable effective communication and action across all continuum partners.
Support for Improvement	Few projects are supported by data-driven insights.	Data increasingly informs improvement projects but is not fully integrated.	Data routinely informs improvement measures, driving sustained changes.	Data drives all improvement measures, ensuring alignment with shared organizational objectives.
Cross-Organizational Communication	Information sharing across systems is limited and inconsistent.	Some systems allow cross-organizational communication but not comprehensively.	Data systems routinely share information with key partners and stakeholders.	Data systems ensure seamless communication and action within and across all continuum partners.

(continued)

Table 14.8 (continued)

Focus area	Initial stage	Evolving: Making Progress	Advanced: Effective execution	Outstanding practice
Integration with Objectives	Data use is ad hoc and not aligned with organizational objectives.	Data use is partially aligned with objectives but lacks full integration.	Data use is well-aligned with organizational objectives, informing key decisions.	Data is fully integrated, informing all knowledge and actions needed to meet shared objectives.

Note the stages of progress in utilizing data and measurement to support safe, reliable, and effective care, moving from limited use to an exemplary, system-wide approach that drives improvement and aligns with organizational goals

Table 14.9 Reliability for safe, reliable, and effective care

Focus area	Initial stage	Evolving: Making Progress	Advanced: Effective execution	Outstanding practice
Standardization of Processes	Few standardized processes exist, and most evolve without deliberate design.	Processes are standardized but lack focus on human factors or reliability testing.	Processes are designed and include human factors considerations, but implementation varies.	All processes are intentionally designed, tested, and monitored for reliability and outcomes.
Staff Adherence to Processes	Staff rely on personal methods rather than standardized processes.	Some staff follow standardized processes, but deviations for personal preference are common.	Most staff follow standardized processes, with occasional deviations in patient interest.	All staff adhere to standardized processes, with rare deviations solely to meet patient needs.
Focus on Outcomes	Processes are not linked to outcomes, and reliability depends on individual effort.	Processes are sometimes linked to outcomes, but this is inconsistent.	Reliability and outcomes are monitored for key processes, though gaps remain in some areas.	All processes are linked to outcomes, and reliability is measured and continually improved.
Human Factors Integration	Human factors are not considered in process design.	Some processes are designed with human factors in mind, but this is not universal.	Staff are trained in human factors methodology, though implementation is inconsistent.	Human factors are fully integrated into all processes, ensuring reliability and patient-centered care.

(continued)

Table 14.9 (continued)

Focus area	Initial stage	Evolving: Making Progress	Advanced: Effective execution	Outstanding practice
Monitoring and Testing	Reliability is not tested or measured, and outcomes are highly variable.	Processes are occasionally tested but not routinely monitored for reliability.	Reliability of key processes is monitored over time, though some gaps in measurement remain.	All processes are regularly reviewed, tested, and updated based on learning and outcomes.
Flexibility for Patient Needs	Processes are rigid and do not account for patient preferences.	Standardized processes are applied but with minimal flexibility for patient needs.	Processes are standardized yet allow some flexibility for patient preferences.	Standardized processes are flexible and consistently incorporate patient preferences.

Note the progression of reliability in healthcare processes, from minimal standardization and monitoring to an exemplary state where processes are intentionally designed, patient-centered, and continuously improved for optimal outcomes

Table 14.10 Transparency for safe, reliable, and effective care

Focus area	Initial stage	Evolving: Making Progress	Advanced: Effective execution	Outstanding practice
Data Sharing	Data shared meets legal requirements or minimum requirements in terms of publishing data or communicating with a patient or family following an adverse event.	Data sharing occurs in some areas but is often outdated or superficial.	Data is actively shared with teams and used for learning and improvement.	Data is openly shared across the organization, fostering trust and driving improvement efforts.
Learning Boards	Learning boards are rarely used and are not regularly updated.	Some areas use learning boards, but updates are infrequent and often for leadership visits.	Learning boards are routinely used in many areas and updated regularly.	Learning boards are consistently used across all areas to share ongoing improvement processes.
Leadership Engagement	Leaders rarely visit the "shop floor" to engage with teams or review data.	Some leaders visit the "shop floor" occasionally but without regular engagement.	Leaders spend time monthly on the "shop floor" reviewing data and engaging with teams.	Leaders actively review learning boards and spend significant time supporting learning processes.

(continued)

Table 14.10 (continued)

Focus area	Initial stage	Evolving: Making Progress	Advanced: Effective execution	Outstanding practice
Patient and Family Involvement	Patients and families are not involved in adverse event investigations.	Patients and families are superficially involved in some adverse event investigations or improvement activities.	Patients and families are deeply involved in adverse event investigations and improvement activities.	Patients and families are fully engaged in improvement efforts and trust the organization's communication.
Use of Data for Improvement	Data is not displayed or used for learning and improvement.	Data is occasionally displayed but not used consistently for improvement.	Data is routinely displayed and linked to improvement efforts in most areas.	Data is central to all improvement efforts, ensuring visibility and clarity for all stakeholders.
Trust and Communication	Communication with patients and families is minimal and limited to legal requirements.	Communication is improving but lacks depth and consistency.	Patients and families feel more involved and receive transparent communication.	Patients and families trust the organization's communication and feel actively involved in its processes.

Note the development of operational transparency, progressing from minimal data sharing and engagement to an exemplary state where leaders, staff, patients, and families are fully involved in and trust the organization's improvement processes

Table 14.11 Leadership for improvement in safe, reliable, and effective care

Focus area	Initial stage	Evolving: Making Progress	Advanced: Effective execution	Outstanding practice
Organizational Goals	No clear organization-level goals for safe and reliable care.	Some organization-level goals for safe and reliable care are prioritized and monitored.	Most goals for safe and reliable care are actively monitored and supported by leadership.	All organizational goals for safe and reliable care are clearly defined, monitored, and fully supported by leadership.
Culture of Safety and Improvement	Culture of safety and improvement is not identified as a priority.	Improving culture is identified as a goal but not consistently supported across all areas.	Culture of safety and improvement is an organizational priority actively supported in most areas.	A strong culture of safety and improvement is embedded and supported across the entire organization.

(continued)

Table 14.11 (continued)

Focus area	Initial stage	Evolving: Making Progress	Advanced: Effective execution	Outstanding practice
Leadership Engagement	Leadership for care improvement is not coordinated across departments.	Some leaders are supported to coordinate improvement activities across services.	Most leaders are supported in coordinating improvement across departments and services.	All leaders are fully supported to coordinate and integrate improvement activities across the continuum of care.
Focus on System of Care	Leadership does not focus on system-level improvements or integration.	Leadership begins to focus on system of care but with limited integration efforts.	Senior leadership prioritizes the system of care and supports most leaders in integration efforts.	Senior leadership drives system-wide integration and improvement, ensuring seamless support and coordination.
Sharing Learning	Learning from safety projects is rarely shared across the organization.	Some learning is shared, but it is inconsistent and limited in scope.	Learning is shared across most areas and informs broader improvement activities.	Learning from safety and improvement projects is systematically shared and integrated across the organization.

Note the progression of leadership engagement in improving safe, reliable, and effective care, from minimal involvement to an exemplary state of active leadership, system-wide integration, and robust support for a culture of safety and improvement

Table 14.12 Person-centered care for safe, reliable, and effective care

Aspect	Just beginning	Making progress	Significant impact	Exemplary
Patient and Family Involvement	Patients and families are minimally involved in care discussions or improvement activities.	Patients and families are occasionally involved in discussions or improvement projects but not consistently.	Patients and families are actively involved in most care and improvement activities.	Patients and families are integral members of care and improvement teams, driving decisions and outcomes.
Care Alignment	Care is delivered based on clinical routines rather than patient preferences.	Care is sometimes delivered in agreement with patient and family views, but this is not systematic.	Care delivery is routinely aligned with patient and family views in most areas.	Care is fully aligned with patient and family preferences and incorporates their views at every level.
Empowering Patients	No infrastructure exists to support patients and families as active participants.	Some infrastructure exists, but it is limited and inconsistently utilized.	Infrastructure supports patient and family engagement in most areas of care and improvement.	Infrastructure fully empowers patients and families to take active roles in care and improvement teams.
Psychological Safety	Patient concerns and preferences are rarely solicited or acted upon.	Staff occasionally solicit patient input, but this is not a routine practice.	Staff routinely ask for and incorporate patient input into care planning and decisions.	Patients feel safe and empowered to voice concerns and preferences, shaping their care experience.
Teamwork and Communication	Communication with patients is unstructured and excludes them from team discussions.	Patients are sometimes included in team discussions, but communication is inconsistent.	Patients are included in most team discussions and are kept informed about their care.	Patients are equal partners in team discussions, with communication tailored to ensure clarity and inclusion.
Accountability	Staff accountability to patient-centered care is not clearly defined or enforced.	Some staff take responsibility for delivering patient-centered care, but this varies.	Accountability to patient-centered care is embedded in most areas, with clear expectations.	Staff are fully accountable for ensuring patient-centered care, with systems in place to monitor and support this.

(continued)

Table 14.12 (continued)

Aspect	Just beginning	Making progress	Significant impact	Exemplary
Negotiation	Care decisions are dictated by staff without consulting patients or families.	Patients and families are occasionally involved in care negotiations, but their input may be superficial.	Patients and families are routinely included in care decisions and negotiations.	Care decisions are fully collaborative, with patients and families taking active roles in the negotiation process.
Transparency	Patients and families are given minimal information about their care or outcomes.	Information sharing with patients and families occurs but is not comprehensive.	Patients and families are routinely provided with clear and transparent information about their care.	Full transparency is practiced, with patients and families actively involved in monitoring and improving outcomes.
Leadership for Improvement	Leadership does not prioritize patient involvement in care or improvement.	Leadership occasionally supports patient and family involvement in improvement activities.	Leadership actively fosters patient and family participation in most improvement efforts.	Leadership prioritizes and ensures patient and family involvement in all aspects of care and improvement.

Note the progression of person-centered care, emphasizing the involvement of patients and families in care delivery, decision-making, and improvement activities, while fostering a culture of empowerment and alignment with their values and preferences

The Unintended Outcomes of Quality Improvement and Myths 15

Contents

Overall, quality improvement initiatives in medical practice contribute to better patient care, informed decision-making, and continuous advancement in healthcare services. However, quality improvement efforts don't always succeed and can sometimes divert resources without yielding benefits. Continuous Improvement (CI) is about more than just metrics and key performance indicators (KPIs). It's about understanding the real-world impact of interventions and avoiding unintended consequences, a concept illustrated by the Cobra Effect. In this cautionary tale from British colonialism, a bounty for dead cobras led to entrepreneurs breeding them for profit, exacerbating the cobra problem when the bounty was removed. Similarly, poorly designed metrics in business can incentivize behaviors that undermine the

S. Sharma, *Quality Demystified*, https://doi.org/10.1007/978-981-95-0339-1_15

intended goals. Changes in one area can negatively impact another; for instance, improving early discharge might increase readmissions.

Also, the focus on compliance and bureaucracy in healthcare quality systems often overlooks the human element, leading to a disconnect between providers and patients. This rigid approach stifles creativity, diminishes engagement, and compromises the patient experience. To truly enhance healthcare quality, there's a need to shift toward a more human-centric approach that values the unique perspectives and insights of healthcare professionals and patients alike.

Leaders must anticipate and monitor these potential pitfalls, using balancing measures and adjusting initiatives as needed. To avoid breeding "worthless cobras," CI professionals should consider potential unintended consequences, engage with frontline operations (Gemba), monitor for unintended behaviors, and be prepared to adjust course as needed. Simply imposing rules from the top-down without understanding their impact on the ground can lead to unexpected challenges down the line. The scale and coordination of improvement efforts matter; isolated interventions may duplicate work or pose safety risks. A system-wide approach, considering care processes holistically, is more effective, allowing diverse skills to address improvement challenges. This involves fostering a culture of innovation, prioritizing continuous improvement, and actively incorporating patient feedback to create agile, patient-centered systems. Embracing this approach can transform quality initiatives from bureaucratic checklists into dynamic frameworks that genuinely deliver excellence in healthcare.

Quality is never an accident; it is always the result of high intention, sincere effort, intelligent direction, and skilful execution. – **John Ruskin**
Continuous improvement is better than delayed perfection – **Mark Twain**

15.1 Quality Assurance Myths Debunked

Historically, quality management in healthcare has been associated with the traditional perception that:

- Documentation processes are cumbersome
- Audits and assessments are resource-intensive audits
- Compliance-driven approaches that often feel disconnected from day-to-day patient care
- Periodic bursts of activity around accreditation periods, followed by lulls

This repetitive and frequently reactive approach has caused many healthcare professionals to perceive quality management as a burdensome task that distracts from their primary focus: patient care.

QA in healthcare is critical for ensuring patient safety, improving outcomes, and maintaining high standards of care. However, there are several myths surrounding QA in healthcare that can lead to misconceptions and challenges. Some examples of QA myths in healthcare:

15.1.1 Myth: Small Hospitals/Businesses Don't Need Formal Quality Management

Example, a small hospital, nursing home/primary care clinic may believe that because of its size, it doesn't need to invest in formal quality management systems or adhere to stringent regulations like larger hospitals. Implementing structured quality processes ensures that patient care is consistent, safe, and effective, regardless of the size of the healthcare facility. It helps in standardizing procedures, reducing errors, improving patient outcomes, and enhancing overall efficiency. Moreover, in today's healthcare landscape, where regulatory requirements and patient expectations continue to evolve, small businesses that prioritize quality management are better positioned to adapt, thrive, and provide excellent care to their patients. Therefore, irrespective of size, all businesses gain advantages from implementing structured quality processes.

15.1.2 Myth: The Quality Unit Purposely Finds Faults and Implements Corrective Actions

QA upholds standards to optimize healthcare. Despite misconceptions, QA doesn't just find faults; it safeguards patient safety and enhances care quality. Through audits and reviews, rather than being punitive, these activities are aimed at proactively identifying potential risks or deficiencies and implementing corrective actions to prevent harm to patients and improve overall care quality. QA identifies areas for improvement, fostering a culture of safety and excellence. Recognizing QA's vital role ensures optimal care delivery.

15.1.3 Myth: Quality Management Comes with High Costs and Does Not Yield Returns on Investment Such as Resource-Intensive Audits and Assessments

Over time, proficient quality management proves cost-effective by minimizing defects and waste. Despite initial investment concerns, implementing robust practices in healthcare yields long-term savings and better outcomes. For example, a hospital that invests in quality management initiatives such as staff training, process improvements, and technology upgrades may initially incur expenses. However, these investments can lead to reduced medication errors, fewer patient complications, and shorter hospital stays, resulting in cost savings, better patient outcomes, and enhanced reputation in the long run, which outweighs the initial investment.

15.1.4 Myth: The Quality Unit Just Wants Everything by the Book, Documentation ...

Some may perceive the Quality unit as overly focused on documentation and strict adherence to regulations, which can lead to frustration and resistance to QA efforts. Non-compliance can have disastrous consequences, and QA safeguards everyone's interests. In healthcare, QA ensures compliance with regulations to safeguard patient safety and quality of care. Despite perceptions of emphasis on documentation and regulations, QA activities like audits and inspections prevent errors and legal liabilities, ultimately protecting patients and healthcare providers.

15.1.5 Myth: The Quality Unit Is Only There to Tick Boxes and Adhere to Procedures

In healthcare, some may perceive the Quality Unit as solely focused on bureaucratic tasks and procedural compliance, leading to undervaluing their contributions. However, the truth is that the Quality Unit plays a crucial role in driving innovation, fostering a culture of continuous improvement, and optimizing resource utilization in healthcare organizations. For example, through initiatives such as Lean Six Sigma methodologies, the Quality Unit identifies inefficiencies, streamlines processes, and eliminates waste, resulting in cost savings and improved patient outcomes. Additionally, by promoting adherence to evidence-based practices and standards, the Quality Unit ensures the delivery of high-quality, safe, and effective care to patients. Therefore, while their emphasis on adherence to procedures may be perceived as bureaucratic, the Quality Unit's broader mandate encompasses driving innovation, improving efficiency, and enhancing overall quality of care in healthcare settings.

15.1.6 Myth: More Tests and Procedures Mean Better Quality Care

Example: Ordering unnecessary tests or procedures under the assumption that it demonstrates thoroughness or provides reassurance to patients. However, unnecessary tests can increase healthcare costs, expose patients to unnecessary risks, and contribute to overdiagnosis and overtreatment.

15.1.7 Myth: Compliance with Regulations Equals Quality Care

Example: Assuming that meeting regulatory requirements ensures high-quality care. While compliance with regulations is important, it does not guarantee optimal patient outcomes or safety. Quality care involves more than just meeting minimum standards; it requires continuous improvement and a focus on patient-centered outcomes.

15.1.8 Myth: Errors Are Always Due to Individual Incompetence

Example: Blaming healthcare professionals for errors without considering systemic factors that contribute to mistakes. In reality, errors often result from complex interactions between individuals, processes, and organizational factors. Addressing errors requires a systems approach that identifies and addresses underlying causes rather than blaming individuals.

15.1.9 Myth: QA Is Solely the Responsibility of the QA Department

Example: Assuming that QA is the responsibility of a separate department or team. In reality, QA is everyone's responsibility within a healthcare organization, from frontline staff to leadership. Effective QA requires collaboration, communication, and a commitment to continuous improvement across all levels of the organization. Organizations where everybody takes ownership of quality enjoy the highest operational performance.

15.1.10 Myth: High Patient Satisfaction Scores Equate to High-Quality Care

Example: Believing that positive patient satisfaction scores alone indicate quality care. Although patient satisfaction is a significant aspect of quality, it is not the only factor that determines it. Quality care encompasses factors such as clinical effectiveness, patient safety, timeliness, efficiency, and patient-centeredness, which may not always align perfectly with patient satisfaction scores.

Addressing these myths requires a comprehensive understanding of quality assurance principles, ongoing education, and a commitment to evidence-based practices that prioritize patient safety and outcomes above all else.

The Accreditation Process

16

Contents

The accreditation process for healthcare facilities is a structured evaluation of an institution's performance against defined standards, aiming to ensure patient safety, high-quality care, and continuous improvement. The process involves the formation of quality teams or committees, internal assessments, and external assessments. An overview of each phase is as under:

16.1 Formation of Quality Teams/Committees

Purpose
Quality teams or committees are responsible for driving the accreditation process. They coordinate efforts, implement changes, and ensure compliance with standards.

S. Sharma, *Quality Demystified*, https://doi.org/10.1007/978-981-95-0339-1_16

Steps for Formation

- **Identify Team Members:** Select a multidisciplinary team that includes physicians, nurses, administrative staff, and other key personnel. This ensures all areas of the facility are represented.
- **Assign Roles:** Appoint a Quality Coordinator to lead the process. Other members may take responsibility for specific areas such as infection control, patient safety, medication management, or documentation.
- **Set Objectives:** Define the team's goals and responsibilities. The primary aim is to prepare the facility for accreditation by addressing gaps in quality and safety.
- **Training and Capacity Building:** Ensure that the team receives adequate training on accreditation standards, quality improvement tools, and the use of relevant audits and checklists.

Key Responsibilities

- Monitor adherence to clinical and operational standards.
- Promote a culture of continuous quality improvement.
- Implement corrective actions and best practices across the institution.

16.2 Internal Assessment

Purpose

Internal assessments, or self-assessments, help the healthcare facility evaluate its readiness for accreditation. These assessments identify areas that meet the standards and those that need improvement.

The SaQushal Safety and Quality Self-Assessment Tool by NHSRC is designed to help healthcare facilities in India assess and improve their quality of care and patient safety practices. Developed in alignment with national and international standards, this tool enables health facilities to conduct internal evaluations, focusing on critical areas such as infection control, patient safety, medication management, clinical governance, and staff performance. By utilizing this tool, facilities can identify gaps, measure compliance with quality standards, and develop action plans for improvement. SaQushal promotes a culture of safety and quality by fostering continuous self-assessment and accountability, ultimately enhancing patient outcomes and organizational performance. The tool also supports facilities in preparing for external accreditation, ensuring they meet the necessary regulatory and quality benchmarks.

Process

- **Gap Analysis:** Conduct a detailed gap analysis to compare current practices with accreditation standards. This helps identify deficiencies and areas for improvement.

- **Internal Audits:** Use quality tools like checklists, clinical audits, and mock surveys to assess various aspects of care (e.g., infection prevention, patient safety, medication management).
- **Data Collection:** Gather data on key performance indicators (KPIs), patient outcomes, staff adherence to protocols, and patient feedback.
- **Corrective Action Plans:** Based on the findings, develop corrective action plans to address deficiencies. Assign timelines and responsibilities for implementing improvements.
- **Mock Assessments:** Conduct mock accreditation assessments to simulate the external assessment process, identify remaining gaps, and make final adjustments.

Tools Used in Internal Assessments

- **Audit Checklists:** Cover all departments including clinical, nursing, pharmacy, and administrative services.
- **Root Cause Analysis (RCA):** To investigate adverse events or near misses.
- **Plan-Do-Check-Act (PDCA) Cycles:** For continuous quality improvement.

16.3 External Assessment

Purpose

External assessments are conducted by accredited bodies to evaluate the facility's compliance with national or international standards, such as those provided by NABH in India, JCI, ISO standards, NQAS, QAI standards, etc.

Process

- **Pre-assessment or Preliminary Survey:** Some accrediting bodies offer a pre-assessment visit. This can provide feedback on areas needing improvement before the formal assessment.
- **Submission of Documents:** Submit required documentation, including policies, procedures, and evidence of internal audits and improvements.
- **On-site Visit by Assessors:** A team of external assessors conducts an on-site evaluation. The team may include clinical and non-clinical experts. Their task is to observe processes, review documents, interview staff, and interact with patients.
- **Interviews and Observations:** Assessors conduct interviews with healthcare professionals and management teams, observe patient care, and review facility infrastructure, hygiene, and safety protocols.
- **Assessment of Compliance:** The assessors evaluate the facility's compliance with various standards, such as patient safety, care delivery processes, infection control, medication management, patient rights, and more.
- **Feedback and Reporting:** After the assessment, the assessors provide a detailed report highlighting strengths, weaknesses, and areas for improvement.

16.4 Post-assessment Actions

Corrective Action
If the external assessment identifies any areas of non-compliance, the facility must take corrective actions within a specified timeframe. This may involve retraining staff, revising policies, or implementing new protocols.

Accreditation Decision
Once compliance is confirmed, the accreditation body grants accreditation status. This status is typically valid for a defined period (e.g., 3 years), after which re-accreditation is required.

16.5 Continuous Monitoring and Improvement

Accreditation is not a one-time process but a commitment to continuous improvement. Facilities must maintain compliance with standards through regular audits, ongoing training, and process improvements. Quality teams play a pivotal role in ensuring that the improvements made for accreditation are sustained.

The accreditation process, involving quality teams, internal and external assessments, enhances patient safety and healthcare quality. By embracing this structured evaluation and adhering to established standards, healthcare facilities can foster a culture of continuous improvement and deliver safer, higher-quality care.

16.6 Breaking the Cycle: Sustaining Quality Beyond Accreditation Surges

Quality management in healthcare has often been seen as a necessary evil—a burden to endure rather than a valuable asset to be utilized. This deeply entrenched perception within many healthcare institutions has limited the true potential of quality improvement efforts.

The path to accreditation is demanding, requiring hospitals to align with a comprehensive set of standards. In preparation, institutions often undergo significant transformations, including infrastructural upgrades, the adoption of advanced healthcare technologies, and the implementation of rigorous safety protocols. These efforts aim to secure a favorable assessment, temporarily showcasing heightened quality and operational excellence.

16.7 A Surge in Quality: Periodic Bursts of Activity Around Accreditation Periods, Followed by Lulls

In the journey toward hospital accreditation, the cycle often begins with an intense surge of activity. As accreditation deadlines approach, healthcare institutions frequently undergo dramatic transformations, scrambling to meet stringent standards.

This period is marked by a flurry of initiatives, from infrastructural enhancements to the rapid adoption of advanced healthcare technologies and the establishment of new safety protocols. The focus is singular: achieving accreditation and showcasing a temporary peak in quality and operational excellence.

16.7.1 The Reality Unfolds

However, these bursts of activity are typically followed by significant lulls. Once the accreditation is secured, the urgency dissipates, and with it, the rigor in maintaining the newly established standards. Resources that were heavily allocated during the accreditation preparation phase may be retracted, and the continuous monitoring necessary to sustain high-quality care often wanes and the pressure to revert to cost-saving measures may compromise quality. Additionally, there is often a tendency to ease the workload, which can undermine the standards initially put in place. The result is a cyclical pattern where hospitals experience intermittent periods of excellence followed by a gradual decline in standards.

16.7.2 The Re-surge

As the next accreditation survey approaches, hospitals once again enter a period of intense preparation. During the intervening period, key personnel may have changed, and trained staff may have moved on to better opportunities. As a result, the responsibility for quality often falls to a "Quality Manager," typically a junior staff member with little clinical background and minimal training. This individual faces the daunting task of maintaining quality with limited support from clinical teams, nursing staff, and other caregivers, who may view quality initiatives as a burden or necessary evil.

The results of the re-survey are predictable. Indicators often show dramatic improvements, with many metrics achieving either 0% or 100% compliance. In reality, the quality manager, lacking full cooperation from the clinical teams, may be managing the system just enough to present favorable results. Once the re-survey is completed, the cycle ends—until the next "festival" begins. This Yo-Yo effect, as it is often called, can persist for years, sometimes lasting for over a decade.

16.7.3 Implications of the "Festival": Undermined Patient Care

The Yo-Yo effect fundamentally undermines the purpose of accreditation. Although designed to ensure consistent, high-quality care, this cycle of short-term improvement followed by regression creates a false sense of security, ultimately compromising patient safety and the quality of care. This cyclical nature of quality management poses several risks. The intense focus on meeting accreditation criteria can create a superficial layer of compliance, masking underlying issues that resurface once the initial momentum fades. During the lulls, the absence of ongoing vigilance can lead to a reversion to old practices, cost-saving measures that compromise quality, and a general complacency that undermines the very purpose of accreditation.

16.7.4 Root Causes and Moving Forward

Addressing this phenomenon requires a cultural shift within healthcare institutions toward continuous improvement, embracing accreditation standards as a permanent operational philosophy. Accrediting bodies also have a role to play in this transformation. Accrediting bodies must also refine their assessment methodologies, to focus not just on the ability to meet standards at a point in time, but on the sustainability of those standards over the long term by incorporating unannounced visits and patient outcomes data into their evaluations. Incorporating unannounced visits, patient outcomes data, and measures of continuous improvement into the accreditation process can help ensure that hospitals remain committed to quality long after the surveyors have left.

16.7.5 Beyond the Mirage

The **Yo-Yo effect** in hospital accreditation exposes a cycle of temporary excellence that fails to maintain consistent high standards. Overcoming this challenge is crucial to genuinely enhancing patient care and ensuring that accreditation reflects true trust and excellence in healthcare.

Breaking free from these fluctuations requires a shift in how quality is perceived and managed within hospitals. While traditional inspections play a role, more effective strategies—such as *data-driven audits, real-time monitoring, patient feedback systems,* and *process-based indicators*—offer a continuous and holistic view of care quality. These approaches prioritize learning and improvement over mere compliance checks, fostering sustainable enhancements rather than periodic fault-finding.

Accreditation should not be a periodic assessment but a tool that drives continuous, high-quality patient care. Only through smarter, proactive quality management can healthcare institutions ensure lasting excellence. Figure 16.1 illustrates the DIKW hierarchy, which helps in understanding the relationship between data, information, knowledge, and wisdom. The hierarchy showcases the progression from raw data to valuable wisdom:

- **Data**: Raw, unprocessed facts and figures without context.
- **Information**: Data that has been organized or processed to provide meaning.
- **Knowledge**: Information that has been assimilated, understood, and utilized effectively.
- **Wisdom**: The ability to make sound decisions and judgments based on knowledge and experience.

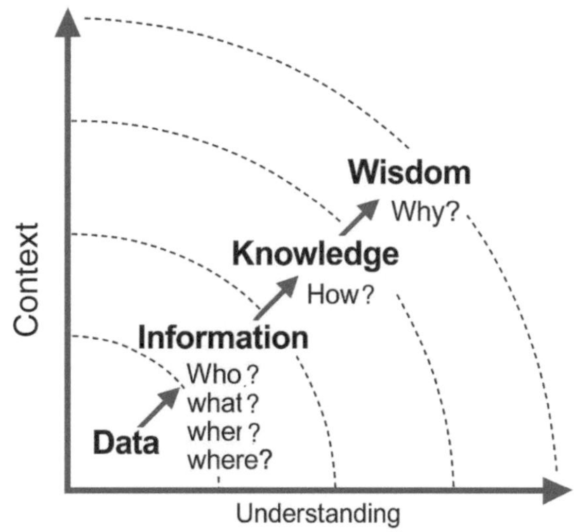

Fig. 16.1 The DIKW hierarchy: data, information, knowledge, and wisdom

Each level of the hierarchy builds upon the previous one, ultimately leading to insightful and informed decision-making. This progression is crucial for effectively managing and leveraging information in various contexts.

> Quality is not just the responsibility of the quality department. It is everyone's responsibility. – **W. Edwards Deming**

Part III

Clinical Audit

Clinical Audit

17

Contents

S. Sharma, *Quality Demystified*, https://doi.org/10.1007/978-981-95-0339-1_17

A clinical audit is an essential tool for improving the quality of care. Its primary objective is to stimulate interventions that improve clinical effectiveness and assess their impact. This section aims to provide healthcare staff with a comprehensive understanding of the concept and processes involved in clinical audit. It is intended for all individuals involved in or interested in conducting clinical audits.

The document offers the necessary knowledge to plan, design, and carry out a clinical audit, supplemented with resources such as examples and sample checklists that can be customized to suit the specific objectives of the audit at each facility.

Since its introduction in 1989, there have been various definitions of clinical audit. According to the National Institute for Health and Clinical Excellence (NICE), clinical audit is defined as a quality improvement process that aims to enhance patient care and outcomes through a systematic review of care against explicit criteria, followed by the implementation of necessary changes. It involves evaluating different aspects of care, including its structure, processes, and outcomes, against specific criteria. Where improvements are indicated, changes are implemented at the individual, team, or service level, and ongoing monitoring confirms the enhancement in healthcare delivery.

Another definition of clinical audit describes it as a quality improvement process that identifies areas for service improvement, develops and implements plans and actions to enhance medical practices, and verifies the impact of these changes through re-audit.

In essence, clinical audit is about ensuring adherence to best practices, identifying any gaps between established norms and actual practice, and making improvements to enhance patient care. The term "clinical audit" encompasses audits conducted by various healthcare professionals, including doctors, nurses, and paramedical staff. While medical audits primarily focused on doctors' practices, clinical audit recognizes the multidisciplinary nature of healthcare and evaluates the overall quality of care provided by the entire healthcare team.

It is crucial to emphasize that healthcare procedures typically involve multiple professional disciplines, and non-clinical staff, such as receptionists, secretaries, porters, and managers, also play a vital role in delivering quality service. Therefore, clinical audit is often a multidisciplinary and multisectoral activity, involving collaborations between health and social services, primary and acute care providers, and educational institutions.

17.1 Why Do Audit?

Clinical audit is an indispensable tool for assessing the quality of clinical care and identifying opportunities for improvement. It is widely recognized as a vital and essential component in enhancing the quality of care provided. Through clinical audit, we gain valuable insights into the effectiveness of current practices and can effectively demonstrate this to others, fostering confidence in the quality of care being delivered.

An exemplary success story highlighting the impact of clinical audit is the national audit of maternal death in developing countries, which has led to improved care for pregnant women and reduced maternal and child mortality rates.

Beyond its impact on patient care, clinical audit offers several benefits for healthcare professionals, including:

- Providing information to demonstrate the effectiveness and cost-effectiveness of services, thereby supporting their development.
- Serving as an educational activity that enhances participants' awareness of guidelines and procedures.
- Identifying training needs and creating opportunities for training and education.
- Assisting staff in monitoring their work, reviewing it, and seeking ways to improve.
- Ensuring better resource utilization and increased efficiency by directing staff time and resources toward practices that are proven to work.
- Offering a systematic approach to considering care objectives and outcomes.
- Providing an opportunity for increased job satisfaction among healthcare professionals.

Ultimately, the overarching goal of clinical audit is to enhance service user outcomes by improving professional practice and the overall quality of healthcare services delivered. It plays a crucial role in driving continuous improvement and ensuring the provision of high-quality care.

> Without data, you're just another person with an opinion.—**W. Edwards Deming.**

The Benefits for Patients and Society

- Ensures public safety and protects individuals from inappropriate, suboptimal, or harmful care.
- Ensures consistency in the delivery of care and treatment, promoting equitable access to healthcare and enhancing patient satisfaction.
- Reduces patient suffering and minimizes the risk of denial of available services or harm caused by excessive or inappropriate interventions.

The Benefits for Management

- Facilitates professional development and enhances working relationships, communication, and collaboration among staff, as well as between staff and service users and different agencies. Promotes multidisciplinary teamwork.
- Reinforces professional self-regulation within the healthcare system.
- Supports effective risk management, leading to a reduction in litigation and complaints.

- Strengthens the organization's ability to secure necessary resources.
- Improves cost-effectiveness and clinical effectiveness of services.
- Provides valuable insights that inform the need for organizational change and improvement.

17.2 Areas Which Can Be "Audited"

There are numerous topics which are suitable and relevant for clinical audit. Several ways of subdividing clinical audit topic areas have been devised, which can be categorized into three main areas according to Donabedian's framework (1966):

1. Structure (what is needed): This refers to the physical features and resources required for effective healthcare delivery. It includes factors such as the facility's infrastructure, the availability of staff with appropriate skills and training, clinic or hospital equipment, record-keeping systems, and the presence of essential resources for specialized units (e.g., neonatal or stroke units). It also encompasses the availability of critical equipment in emergency rooms, ambulances, or operation theatres, as well as the availability of materials in specialized areas like Cath labs.
2. Process (what is done): This category focuses on the actual activities and processes carried out in the delivery of healthcare services. It involves examining how resources are utilized and the adherence to established protocols and guidelines. Examples of process criteria include prescribing practices, referral patterns, utilization of laboratory investigations, discharge procedures, waiting times, post-operative pain management, and other aspects of care delivery. Evaluating process criteria is important as they directly influence the quality of care provided.
3. Outcome (what is expected to happen as a result): This category assesses the impact of healthcare activities on the health and well-being of service users. It involves measuring the results of healthcare interventions and determining changes in the patient's future health status attributable to the care received. Examples of outcome measures include blood sugar control, blood pressure management in response to therapy, rates of hospital-acquired infections, disease prevention, reduction in premature mortality, visual acuity after cataract surgery, and patient satisfaction with the care provided. Outcomes are considered the ideal indicators of care, but they can be challenging to measure accurately.

By examining these three dimensions (structure, process, and outcome), clinical audits can comprehensively evaluate the quality and effectiveness of healthcare services.

Other Processes Similar to and Related to Clinical Audit
There are several processes similar to clinical audit that are used to assess and improve the quality of healthcare services. However, none of these replace clinical audit and systematic clinical audit is the main way of assessing compliance of ongoing clinical care against evidence-based standards. These processes, along with clinical audit, contribute to continuous quality improvement in healthcare by evaluating performance, identifying areas for improvement, and implementing changes to enhance patient care and safety. Some of these processes include:

Quality Improvement Projects These are targeted initiatives aimed at improving specific aspects of healthcare delivery. They involve analyzing existing processes, identifying areas for improvement, implementing changes, and measuring the impact of those changes. Quality improvement projects can focus on various aspects of care, such as reducing medication errors, improving patient handoff processes, enhancing communication between healthcare providers, or streamlining workflow in a specific department.

Peer Review Peer review involves the evaluation of healthcare professionals' performance by their peers or colleagues (within or outside care setting) in their profession or field of practice with the intent to improve the quality of care. It is a process that helps ensure adherence to professional standards and guidelines. Peer review may involve case reviews, chart audits, or assessments of clinical decision-making. The purpose is to provide feedback, identify areas for improvement, and promote ongoing professional development. Often individual patient cases are discussed with peers to determine, with the benefit of hindsight, whether the best care was given. Peer review might include "interesting" or "unusual" cases rather than problematic ones.

Mortality and Morbidity (M&M) Mortality and morbidity reporting is a process in healthcare that involves the systematic collection and analysis of data related to patient deaths and significant adverse events or complications. The purpose of this reporting is to monitor and evaluate the outcomes of patient care, identify areas for improvement, and promote patient safety.

Mortality reporting focuses on the analysis of patient deaths. When a patient dies, relevant information such as the cause of death, contributing factors, and any associated issues or concerns is documented. This data is then reviewed and analyzed to understand the circumstances surrounding the death, identify any patterns or trends, and determine if there were any opportunities for improvement in the care provided.

Morbidity reporting, on the other hand, focuses on adverse events or complications that occur during patient care but do not necessarily result in death. This can include medical errors, infections, medication-related issues, surgical complications, or any other adverse outcomes. Similar to mortality reporting, data on morbidity events is collected, reviewed, and analyzed to identify underlying causes, assess the impact on patient outcomes, and develop strategies to prevent future occurrences.

The reporting of mortality and morbidity events serves several purposes:

1. Surveillance and monitoring: It allows healthcare organizations to track and monitor the occurrence of adverse events and deaths, providing valuable insights into patient safety and quality of care.
2. Quality improvement: Analysis of mortality and morbidity data helps identify areas for improvement in clinical practices, protocols, and systems. It enables healthcare professionals to implement interventions and changes aimed at reducing adverse events and improving patient outcomes.
3. Risk management: Reporting adverse events and deaths helps identify potential risks and vulnerabilities in the healthcare system, allowing for proactive risk management strategies to be implemented to prevent future occurrences.
4. Benchmarking and comparisons: Data from mortality and morbidity reporting can be used to compare performance across different healthcare providers or facilities, facilitating benchmarking and promoting a culture of transparency and accountability.
5. Education and training: Mortality and morbidity reporting provides valuable case studies for educational purposes, allowing healthcare professionals to learn from real-world experiences, identify best practices, and enhance their knowledge and skills.

Overall, mortality and morbidity reporting play a crucial role in promoting patient safety, quality improvement, and learning within healthcare organizations. By analyzing adverse events and deaths, healthcare providers can identify opportunities for improvement, implement evidence-based practices, and continuously strive to provide safer and more effective care to their patients.

Root Cause Analysis (RCA) RCA is a method used to investigate adverse events, near misses, or sentinel events in healthcare. It involves a systematic approach to identifying the underlying causes and contributing factors that led to the event. RCA aims to understand why an incident occurred and to develop strategies to prevent similar incidents in the future. It focuses on addressing system failures rather than blaming individuals.

Clinical Outcome Review Clinical outcome review involves the examination of patient outcomes and their correlation with the care provided. It assesses the effectiveness of interventions and treatments in achieving desired health outcomes. Clinical outcome review may involve the analysis of patient data, such as mortality rates, complications, readmission rates, functional outcomes, and patient-reported outcomes.

Compliance Monitoring Compliance monitoring involves assessing adherence to regulatory requirements, standards, policies, and protocols. It ensures that healthcare providers are following established guidelines and protocols in their practice. Compliance monitoring can involve regular audits, inspections, or assessments to identify any deviations from the expected standards and take corrective actions.

Confidential Enquiries

Confidential enquiries are different from processes based on standards or audits. Confidential enquiries are typically triggered by specific events, such as unexpected deaths or serious incidents, and are conducted to investigate the circumstances surrounding those events. These enquiries aim to identify any potential issues or system failures that may have contributed to the event and to make recommendations for improvement.

Confidential enquiries can be conducted at the national or local level. National confidential enquiries are often organized by government or professional bodies and involve collecting data from multiple healthcare organizations to gain a broader perspective on the issue. These enquiries may focus on specific areas of healthcare, such as maternal and child health, mental health, or surgical procedures.

On the other hand, local confidential enquiries are usually conducted within a specific healthcare organization or region. They are initiated to investigate incidents or events that have occurred within that particular setting. The purpose is to learn from the event, understand any contributing factors, and implement changes to prevent similar incidents in the future.

Unlike audits or standard-based processes, confidential enquiries do not primarily aim to assess compliance with predetermined standards. Instead, their focus is on investigating specific incidents, understanding the causes and circumstances surrounding them, and making recommendations for improvement to enhance patient safety and prevent future occurrences.

Confidential enquiries play a vital role in promoting a culture of learning from adverse events and driving improvements in healthcare systems. By maintaining confidentiality and focusing on the investigative aspect, these enquiries provide valuable insights and recommendations to enhance patient care and safety.

17.3 What Clinical Audit Is Not Or "Non-audit" Or "Quasi-Audit" Activities

Clinical audit is a specific process aimed at assessing and improving the quality of patient care through systematic review and evaluation. However, there are certain activities that may resemble clinical audit but do not fully meet its criteria. These activities are often referred to as "non-audit" or "quasi-audit" activities. **Not all "audit" that take place within the health service is clinical audit.** Some examples of what clinical audit is not:

1. Research studies: While research studies and clinical audits both involve data collection and analysis, they have different objectives. Clinical audit focuses on evaluating current practice against established criteria and implementing changes to improve care, whereas research studies aim to generate new knowledge and contribute to the existing evidence base.
2. Performance appraisal or performance management: Clinical audit should not be confused with individual performance appraisal or management processes. While audit findings may highlight areas for improvement at an individual level, the primary purpose of clinical audit is to assess and improve the overall quality of care provided by a healthcare organization or team, rather than evaluating individual performance.
3. Peer review: Peer review involves assessment of professional performance by colleagues within the same field. Although peer review may share some similarities with clinical audit, it typically focuses on individual professional practice rather than evaluating the entire care process or service delivery.
4. Complaints or incident investigation: While complaints and incident investigations are essential for identifying and addressing specific issues or incidents, they are not synonymous with clinical audit. Clinical audit is a broader process that aims to evaluate care across a range of patients or cases, identify systematic issues, and implement changes to improve overall quality.
5. Accreditation or certification processes: Accreditation and certification processes assess the overall quality and safety of healthcare organizations or services against predefined standards or criteria. While these processes may include elements of audit, they are comprehensive assessments that cover a wider scope than clinical audit alone.

In addition to clinical audit, there are several other forms of audit commonly conducted in healthcare organizations. These audits focus on different aspects of healthcare delivery and governance. Some examples are given below:

1. **Financial audit**: Financial audits assess the financial management and accounting practices of healthcare organizations. They ensure compliance with financial regulations, accuracy of financial statements, and proper use of resources.
2. **Compliance audit**: Compliance audits ensure that healthcare organizations adhere to legal and regulatory requirements, industry standards, and internal policies. These audits cover areas such as data privacy and security, healthcare regulations, licensing and accreditation standards, and ethical guidelines.
3. ***Internal audit:*** An internal mechanism that traces non-clinical activities and systems along "audit paths" to see if things happened the way they should have. For example, tracing a patient complaint from the initial letter of complaint through to a resolution to establish whether established grievance redressal system was followed appropriately.
4. **Operational audit**: Operational audits evaluate the efficiency, effectiveness, and performance of operational processes within healthcare organizations. They assess areas such as resource utilization, workflow management, patient flow, supply chain management, and information systems.

5. ***Organizational audit:*** An external, independent, and voluntary audit of the whole organization, based on a framework of explicit standards. Organizational audit looks at how well the organization is set up and runs daily.
6. **Information technology (IT) audit**: IT audits focus on the security, integrity, and effectiveness of information technology systems and infrastructure in healthcare organizations. They assess the confidentiality of patient data, data backup and recovery processes, system controls, and compliance with IT regulations and standards.
7. **Quality assurance audit**: Quality assurance audits examine the processes and practices in place to ensure and maintain the quality of healthcare services. They assess adherence to quality standards, patient safety protocols, infection control measures, and the effectiveness of quality improvement initiatives.
8. **Risk management audit**: Risk management audits identify and assess potential risks within healthcare organizations and evaluate the effectiveness of risk management strategies. These audits cover areas such as patient safety, adverse event reporting, risk assessment processes, and the implementation of risk mitigation measures.
9. **Human resources audit**: Human resources audits focus on the management of personnel within healthcare organizations. They assess areas such as recruitment and selection processes, employee training and development, performance management, workforce planning, and compliance with employment laws and regulations.

These different forms of audits play a crucial role in ensuring the overall effectiveness, efficiency, compliance, and quality of healthcare organizations. By evaluating and monitoring various aspects of healthcare delivery and operations, these audits help identify areas for improvement and support the provision of safe and high-quality care to patients.

- Routine monitoring of clinical outcomes: While clinical audit may involve the identification and measurement of clinical outcomes linked to the change process, it is important to note that routine monitoring of outcome data for performance monitoring purposes is not considered clinical audit. Clinical audit focuses on systematic review and improvement of care practices rather than ongoing monitoring of outcomes.
- Staff, patient, service user, carer surveys: Surveys conducted among staff, patients, service users, or carers are typically part of research projects or engagement activities. These surveys aim to gather opinions and feedback on treatment experiences and the quality of care provided, with the goal of identifying areas for improvement. However, it's important to note that conducting surveys alone does not constitute clinical audit. Clinical audit involves a more comprehensive process of reviewing and improving care practices based on established criteria and standards.

Key Points

- **Not all "audit" is clinical audit.** Be aware of the differences between project categories:

1. Clinical audit—audit against agreed standards of best practice. Clinical audit is a quality improvement activity alongside being a professional development activity that measures current clinical care and outcomes against explicit audit criteria (standards). The aim of clinical audit has always been quality improvement and is seen as a self-appraisal process.
   Clinical audit is a team endeavor.
   Clinical audit is now also a mainstream accountability and not solely clinician owned.
2. Research—aims to create new knowledge.
3. Service Evaluation—assesses the effectiveness of a service.

- Research discovers the right thing to do, i.e., the most effective practice. Audit is ensuring it is done right, i.e., if it is being practiced.
- **A service evaluation answers, "What standard does this service achieve?" A clinical audit answers "Does this service reach a certain standard?"**

17.4 Audit vs. Research vs. Surveillance vs. Service Evaluation

Audit, research, surveillance, and service evaluation are distinct activities within the realm of healthcare quality improvement, each serving different purposes and employing different methodologies. Table 17.1 depicts the key differences between clinical audit, research, and service evaluation.

Both clinical audit and research contribute to improving healthcare; they differ in their specific objectives, methodologies, and scope, and are two distinct activities in healthcare. Both audit and research answer questions about quality of care or service, involve sampling, data collection, and analysis of findings but there are key differences between these. Some key differences between audit and research:

Clinical Audit

- Objective: The primary objective of clinical audit is to evaluate and improve the quality of healthcare services provided within a specific organization or setting.
- Focus: Audit focuses on assessing current practices, identifying areas for improvement, and implementing changes to enhance the quality of care.
- Methodology: Audit involves systematic review of clinical practices against established criteria or standards. It often utilizes retrospective data analysis, clinical record reviews, and feedback mechanisms to evaluate performance and drive quality improvement.

Table 17.1 Difference between audit, research, service evaluation, and surveillance

Differentiation	Audit	Research	Service evaluation	Surveillance
Definition	Evaluate the degree to which best practices are being followed or enhanced.	Aims to establish Good Practice	Created and executed exclusively to evaluate or assess existing care.	Aimed at controlling outbreaks and assisting by recognizing and comprehending related risks.
Aim	Aims to improve services	The goal is to create new insights, assess current practices, or compare different methods to enhance a body of knowledge by questioning what actions should be taken.	Evaluates the existing service independently of any benchmark.	Systematic, statistical methods to allow timely public health actions
	Practice-based	Theory driven		
Measures	Measures against a standard	Usually involves testing of a hypothesis; What works and what is best	Designed to answer: "does this service reach a pre-determined standard"	The purpose is to determine the reason behind the outbreak.
Scope	Specific and local to one particular patient group and results are not transferable to other settings	Designed a way so that it can be replicated and so that its results can be generalized to other similar groups	Identifies the standards that the service achieves	
Interventions	Never includes trials or new therapies, regardless of whether they involve healthy volunteers or patients participating as volunteers.	Experiments or treatments may be conducted on human subjects, including patients, patient volunteers, or healthy volunteers.		It might include gathering personal information and samples to handle the incident or outbreak.

(continued)

Table 17.1 (continued)

Differentiation	Audit	Research	Service evaluation	Surveillance
Randomization	Patients are never assigned to different treatment groups through random allocation.	Patients can be placed into different treatment groups using random assignment.	No randomization	No randomization
	Never involves placebo treatment or control group	May involve placebo treatment or control group for the purpose of comparison and evaluation	Does not involve any intervention	Does not involve any intervention
Statistical analysis	Does not necessarily need a statistically valid sample size	Needs a statistically valid sample size		
	Basic statistical analysis usually suffices	Extensive statistical analysis is required		
Data collection	Involves analysis of existing data input from but may include administration of interview or questionnaire which help to determine whether standards have been met.	Collects additional data—treatment, samples, or investigations in addition to routine care	Usually involves analysis of existing data but may include administration of interview or questionnaire	The task may include evaluating current data or conducting interviews or surveys with individuals who have been exposed.
Next step	An ongoing process Responsibility to act on findings via an action plan	Often a one-off study No mechanism to act on findings	Entails examining existing data, though it might also involve conducting basic interviews or surveys.	
Ethics approval	Rarely requires ethics approval	Always require approval	Does not require approval	Does not require approval

- Scope: Audit typically focuses on evaluating the adherence to established protocols, guidelines, and best practices within a specific healthcare setting. The aim is to ensure that care delivery aligns with established standards.

Research

- Objective: The primary objective of research in healthcare is to generate new knowledge, expand understanding, and contribute to scientific advancements.
- Focus: Research focuses on exploring unanswered questions, testing hypotheses, and generating evidence to support the development of new interventions, treatments, or healthcare approaches.
- Methodology: Research involves design, implementation, and analysis of studies or experiments to gather data, test hypotheses, and draw conclusions. It often utilizes methodologies such as randomized controlled trials, observational studies, surveys, and interviews.
- Scope: Research typically aims to contribute to the broader scientific and medical knowledge base. It often involves larger sample sizes, longer timeframes, and generalizability to larger populations beyond a specific healthcare setting.

Surveillance

- Objective: The objective of surveillance is to monitor and detect patterns or trends in disease occurrence, public health risks, or other healthcare-related events.
- Focus: Surveillance focuses on continuous monitoring, data collection, and analysis of specific health indicators or events to detect and respond to potential threats or public health concerns.
- Methodology: Surveillance involves the ongoing collection of data from various sources, such as healthcare facilities, laboratories, or public health agencies. It includes data analysis, interpretation, and reporting to inform public health actions.
- Scope: Surveillance can be conducted at various levels, including local, regional, national, or global, depending on the nature of the health issue being monitored.

In summary, while audits primarily focus on evaluating and improving the quality of care within a specific healthcare setting, surveillance aims to monitor and detect health-related events or patterns to inform public health actions. Audit is more internally focused on quality improvement, whereas surveillance is broader in scope and often involves data collection and monitoring at a population or public health level.

Service Evaluation

- Objective: The objective of service evaluation is to assess the effectiveness, efficiency, and impact of healthcare services or interventions provided within a specific setting or organization. It addresses specific questions about the service concerned with inputs from patients, service users or carers and results are specific and local to a particular team or service and may lead to service redesign and reconfiguration in that area. The evaluation tool may, however, be used by more than one team or service.
- Focus: Service evaluation focuses on examining the overall performance and outcomes of a service or intervention, including its processes, outcomes, patient satisfaction, and resource utilization.
- Methodology: Service evaluation involves the collection and analysis of data, often through surveys, interviews, data extraction from records, and other evaluation methods, to assess the effectiveness and value of the service or intervention.
- Scope: Service evaluation can be conducted for a specific service, program, or intervention within a healthcare organization, aiming to inform decision-making, improve service delivery, and enhance patient outcomes.

While audit primarily focuses on assessing and improving the quality of care provided within a specific healthcare organization, service evaluation has a broader scope, examining the effectiveness and impact of healthcare services or interventions. For example, evaluating the implementation and impact of a new patient education program in a hospital setting. Assessing patient satisfaction through surveys and feedback to measure the quality of care and identify areas for improvement in a primary care clinic.

Audit is more focused on compliance with standards and guidelines, while service evaluation assesses the overall performance and outcomes of a service or intervention. Both activities contribute to quality improvement and enhancing patient care within healthcare organizations.

Examples of different types of studies in an in-patient unit for malnourished children:

1. Research Study: A study to examine whether children admitted to the unit benefit more from domiciliary or hospital management. This study involves comparing different treatment approaches to determine their effectiveness and gathering evidence to support decision-making.
2. Clinical Audit Project: A study to examine whether the unit is meeting a specific standard set by staff, such as ensuring that all children admitted to the unit receive a nutritional assessment within a day of admission to determine the appropriate management approach. This project aims to assess current practices against established criteria or standards and identify areas for improvement.
3. Service Evaluation: A study involving the collection of information about service users to assess whether the malnourished children who require the service

are being reached effectively. It also includes obtaining feedback from service users, referrers, and providers about various aspects of the service, such as accessibility, acceptability, and effectiveness. The purpose is to evaluate the overall performance and outcomes of the service to inform decision-making and improve service delivery.

Other Examples

While there may be overlaps in certain aspects, these activities have distinct purposes and methodologies. Clinical audit focuses on evaluating current practices and adherence to standards, research aims to generate new knowledge, surveillance is concerned with monitoring health events and trends, and service evaluation assesses the performance and impact of healthcare services. Each activity contributes to improving healthcare quality and informs decision-making in different ways.

Monitoring and Evaluation

Monitoring and evaluation (M&E) and audit are both important activities in assessing and improving the performance of organizations, projects, or programs. While there are some similarities between them, they have distinct purposes and approaches. A comparison between monitoring and evaluation and audit is discussed below:

Monitoring and Evaluation (M&E)

- Focus: M&E is an ongoing, continuous, and systematic process of gathering and analyzing data to monitor the progress, performance, and outcomes of a project, program, or organization.
- Purpose: The primary purpose of M&E is to provide regular feedback, ensure accountability, and support evidence-based decision-making for improving performance and achieving desired results.
- Scope: M&E assesses the implementation of activities, outputs, outcomes, and impacts, and helps identify strengths, weaknesses, and areas for improvement.
- Methods: M&E often involves a mix of qualitative and quantitative data collection methods, such as surveys, interviews, observations, and data analysis.
- Timeframe: M&E is typically conducted throughout the lifespan of a project or program to enable ongoing monitoring of progress and evaluation of results.

Audit

- Focus: Audit is a systematic and independent examination of financial records, processes, systems, or operations to assess compliance, accuracy, effectiveness, and efficiency.
- Purpose: The primary purpose of an audit is to provide assurance, verify financial integrity, identify risks, detect errors or irregularities, and ensure compliance with regulations, standards, or policies.

- Scope: Audit focuses on specific areas, activities, or functions within an organization, with a goal of assessing internal controls, risk management, and financial accuracy.
- Methods: Audit involves a structured and systematic approach, including sample testing, documentation review, interviews, and analysis of financial statements, reports, and records.
- Timeframe: Audits are often conducted periodically (e.g., annually, or biennially) or based on specific triggers (e.g., financial year-end) to provide independent verification and examination of financial or operational processes.

Although both evaluation and audit require the assessment of effectiveness and efficiency, audit differs from evaluation in orientation or objectives and the formulation of recommendations to promote improvement. Monitoring and evaluation are broader processes aimed at assessing overall performance, progress, and outcomes, while audit focuses more specifically on ensuring compliance, accuracy, and risk management in financial or operational aspects. Both M&E and audit contribute to organizational learning, improvement, and accountability, but they have different scopes, purposes, and methodologies.

Some examples are:

- Monitoring and Evaluation: What is the percentage of children/youth and/or families reached by the program? How effectively did the program reach children/youth and/or families beyond the target group? What is the level of satisfaction among participants regarding the program's delivery? What are the perceptions of staff, community partners, and referring agencies about the program? How participants came to know of the program?
- Monitoring: Were the program activities implemented according to the planned schedule (frequency, timing, location, duration)? To what extent was there adherence to the intended model of practice (program fidelity)? If there were deviations from the planned implementation, what were the limiting factors?
- Evaluation: Have the desired short-term, medium-term, or long-term outcomes been achieved by the program? Which outcomes are the most critical and challenging to attain?

Monitoring without evaluation is like navigating without a compass; it leads to waste.—**Anonymous**
Efficiency is doing things right; effectiveness is doing the right things.—**Peter Drucker**

Quality Improvement

Audit and quality improvement (QI) are two distinct but interconnected processes within healthcare. Quality improvement aims to improve the patient experience. Although audit is often more clinically orientated, QI can focus on more holistic issue.

Quality Improvement (QI)

Aims to proactively enhance processes and systems to achieve better outcomes and enhance patient care.

- Involves ongoing and iterative cycles of improvement, using data-driven approaches.
- Focuses on identifying problems, developing, and implementing solutions, and measuring the impact of interventions.
- Emphasizes continuous learning, innovation, and adaptation.
- It requires the active participation and collaboration of healthcare professionals and stakeholders to ensure effective implementation and outcomes.
- Implemented within the organization, with a focus on internal capacity building and improvement culture.

While both audit and quality improvement aim to enhance healthcare quality, audit focuses on assessing compliance with standards, while quality improvement focuses on driving positive change and continuous improvement in processes and outcomes. It can be achieved through the Plan-Do-Study-Act (PDSA) framework. PDSA cycles are iterative and have short durations, facilitating quick incorporation of improvements. These cycles in QI can occur on a weekly or even daily basis.

17.5 Important Don'ts

When conducting a clinical audit, there are certain "don'ts" or practices to avoid ensuring the effectiveness and integrity of the process. Some examples include:

1. Don't treat clinical audit as a punitive or blame-focused exercise: The purpose of clinical audit is to improve patient care and outcomes, not to assign blame or punish individuals. It should be approached with a positive and constructive mindset.
2. It should be conducted in a non-confrontational and non-judgmental manner—it's not a chance to name, shame, or blame. Audit should be transparent. Clinical audit should NOT be used for performance appraisal of staff, taking disciplinary actions, or creating competition between doctors. "Never judge the quality of professionals based on an audit—it's about enhancing care." It is neither a cost-cutting exercise nor needs assessment exercise.
3. Don't overlook the importance of stakeholder involvement: Engaging relevant stakeholders, including healthcare professionals, patients, and management, is

crucial for the success of a clinical audit. Don't exclude or disregard their input and perspectives.

4. Don't neglect patient confidentiality and data protection: Ensure that patient information and data collected during the audit are handled securely and in compliance with privacy regulations. Don't compromise patient confidentiality or breach data protection guidelines.
5. Don't rely solely on quantitative data: While quantitative data is important for measuring outcomes and identifying trends, don't overlook the value of qualitative data and the insights it can provide. Use a combination of both types of data for a comprehensive assessment.
6. Don't limit the audit scope to a single aspect: A clinical audit should cover multiple dimensions of care, including structure, process, and outcomes. Don't focus solely on one aspect while neglecting others, as this may provide an incomplete picture.
7. Don't ignore the feedback and recommendations: The purpose of a clinical audit is to drive improvement. Don't disregard the findings, recommendations, or feedback generated through the audit process. Act on the identified areas for improvement and monitor the effectiveness of implemented changes.
8. Don't view clinical audit as a one-time event: Clinical audit should be an ongoing and iterative process. Don't treat it as a one-off activity. Regularly review and reassess practices, monitor outcomes, and identify further areas for improvement.

Remember, clinical audit is a tool for quality improvement and learning, so it's essential to approach it with a positive mindset, engage stakeholders, protect patient confidentiality, and use a comprehensive approach to assess care.

17.6 Who Should Be Involved in the Clinical Audit Project?

A successful clinical audit project involves the collaboration and involvement of various individuals and groups. Key stakeholders who should be involved include:

1. Healthcare Professionals: Clinical audit requires the participation of healthcare professionals directly involved in patient care, including doctors, nurses, therapists, and other relevant staff members. Their expertise and insights are crucial for understanding current practices and identifying areas for improvement.
2. Patients and Carers: Involving patients and their caregivers in the clinical audit project is essential to gain a patient-centered perspective. Their experiences and feedback provide valuable insights into the quality of care and help identify areas where patient satisfaction can be enhanced.
3. Audit Team: An audit team, typically consisting of professionals with expertise in audit methodologies, quality improvement, and data analysis, should be involved in designing, implementing, and overseeing the clinical audit project. They provide guidance, ensure data accuracy, and drive the project towards its objectives.

4. Clinical Governance Leads: Clinical governance leads or managers play a crucial role in supporting and coordinating the clinical audit project within the healthcare organization. They provide oversight, allocate necessary resources, and ensure compliance with relevant policies and guidelines.
5. Quality Improvement Staff: Quality improvement personnel, such as quality managers or improvement advisors, bring their expertise in process improvement methodologies and facilitate the implementation of changes identified through the audit. They play a key role in driving the improvement process and monitoring outcomes.
6. Administrative Staff: Administrative staff, including managers and administrators, contribute to the clinical audit project by providing support in data collection, maintaining records, coordinating meetings, and ensuring smooth communication among stakeholders.
7. External Experts: In some cases, involving external experts or consultants with specific knowledge or experience in the clinical area being audited can bring valuable insights and a fresh perspective to the project.

It is important to have a multidisciplinary and collaborative approach, ensuring representation from different levels and departments within the healthcare organization. The involvement of various stakeholders promotes a comprehensive understanding of the current practices, encourages ownership of improvement initiatives, and enhances the likelihood of successful outcomes from the clinical audit project.

It is important to ensure the representation of as many relevant groups as possible on the clinical audit project team. The team should be multi-professional and consist of individuals from various cadres involved in delivering the service. During the planning phase of the clinical audit project, it is crucial to inform and involve the appropriate representatives from the beginning.

For example, if the chosen topic is "life-threatening obstetric hemorrhage," it would be beneficial to invite a member of the blood transfusion service to participate in the audit team. However, it is important to maintain a balance and avoid having an excessively large team. Each team member should be willing to dedicate time and contribute to the work.

In situations where individuals are unable to attend team meetings, they should still be consulted and kept informed about the clinical audit project throughout the entire process. Clearly identifying roles and responsibilities within the team, such as appointing an audit project lead and data collector, is essential for effective coordination and execution of the audit.

It is crucial to have at least one dedicated member of the clinical staff who can allocate sufficient time to guide the audit and present the results. While the actual data collection and analysis can be performed by nonmedical audit staff, such as medical record officers, there may be instances where additional resources are required to fund data extraction if it cannot be accommodated within existing staff capacity.

It is important to recognize that the priorities of those receiving care may differ significantly from those involved in delivering care. Therefore, when planning a clinical audit, it is beneficial to consider including service users in the process to gain insights into their experiences of receiving clinical care.

There are several common methods for involving service users in the clinical audit process, including:

- Gathering service user feedback, such as through letters of complaint.
- Analyzing comments made at service user forums.
- Conducting interviews with service users.
- Administering service user surveys.
- Organizing focus groups.
- Collaborating with expert user groups or non-governmental organizations (NGOs).
- Examining critical incidents from the perspective of service users.

By incorporating these methods, the clinical audit can benefit from the valuable perspectives and experiences of service users, enhancing the overall effectiveness and relevance of the audit process.

17.7 Stages for Stakeholder Involvement

The involvement of various stakeholder groups in the clinical audit process can be organized into different stages. These stakeholders can contribute in the following ways:

1. Topic and Objective Determination:
   - Participate in decision-making regarding the selection of clinical audit topics and objectives.
2. Methodology and Project Planning:
   - Contribute to and provide feedback on the clinical audit methodology, including the proposed criteria.
   - Provide support in creating and reviewing the project plan to ensure all elements are well-detailed and aligned with project goals.
3. Access and Support:
   - Grant permission to access the service user (patient) group for data collection.
   - Provide support and assistance to the clinical audit team throughout the process.
4. Data Contribution:
   - Serve as a source of data, providing relevant information for the audit.
5. Case Review and Analysis:
   - Review cases that do not meet the expected level of performance, especially when the stakeholder is an expert in the field.
   - Provide insights and explanations regarding the current care processes.
6. Analysis and Action Planning:
   - Contribute to the analysis of audit findings, including the identification of problems and areas requiring improvement.

- Assist in the development of action plans to address the identified areas for improvement.

7. Resource Allocation and Monitoring:
   - Help secure the necessary resources to support the implementation of change based on the audit findings.
   - Monitor the progress and implementation of agreed-upon actions.
8. Repeat Measurement and Analysis:
   - Contribute to the analysis of the findings from subsequent measurements, comparing them to the initial audit results.

By involving stakeholders at different stages of the clinical audit, their expertise, perspectives, and support can enhance the quality and effectiveness of the audit process and facilitate meaningful improvements in healthcare delivery.

17.8 Clinical Audit Topic Selection and Stages

Clinical audit topic selection involves a systematic process that ensures the chosen topic is relevant, feasible, and aligned with the goals of improving healthcare quality. Clinical audits play a vital role in assessing compliance with established national standards of care and contribute to the enhancement and development of these standards for future audits. Some clinical audits focus on monitoring the implementation of existing standards, while others aim to evaluate the introduction of new interventions and their adherence to guidance and service delivery standards. The process of clinical audit involves identifying areas of concern, creating action plans to address these issues, and measuring progress through subsequent audits.

It is the responsibility of healthcare professionals to conduct audits within their own areas of clinical interest to ensure the effectiveness and appropriateness of their practices. However, selecting a specific area of practice for audit can be challenging, necessitating the need for prioritization and careful consideration.

The following are the stages involved in clinical audit topic selection:

1. Identify the Purpose: Determine the overarching purpose of the clinical audit, such as improving patient outcomes, enhancing safety, or optimizing resource utilization.
2. Define Audit Criteria: Establish explicit criteria or standards against which the audit will be conducted. These criteria should be measurable and reflect best practice guidelines or evidence-based recommendations.
3. Prioritize Topics: Evaluate and prioritize potential audit topics based on their relevance, impact on patient care, potential for improvement, and alignment with organizational priorities. Consider input from stakeholders and experts to aid in the selection process.
4. Assess Feasibility: Evaluate the feasibility of conducting an audit on the selected topics. Consider factors such as data availability, accessibility, resources required, and the capability to implement changes based on audit findings.

5. Narrow Down Options: Refine the list of potential audit topics based on the relevance, feasibility, and impact. Select a manageable number of topics that can be effectively addressed within the available resources and timeframe.
6. Finalize Topic Selection: Make a final decision on the audit topic(s) to pursue based on prioritization, feasibility assessment, and alignment with organizational goals. Ensure that the selected topics have clear objectives and can generate actionable recommendations for improvement.

It is important to involve relevant stakeholders, such as healthcare professionals, administrators, patients, and quality improvement experts, in the topic selection process to ensure a comprehensive and collaborative approach. Their input can provide valuable insights, diverse perspectives, and support throughout the clinical audit journey.

17.9 Tools for Prioritizing Audit Topics

The organization should establish a framework for effective clinical audit that includes strategic planning and prioritization. All identified audit topics should be carefully considered, and a systematic approach should be used to prioritize them. It is essential for audit projects to align with the overall priorities of the organization and have a clear plan for improving patient care. However, it's important to recognize that resources are limited, including clinician time and support functions. This means that there is a practical limit to the number of audits that can be conducted within a given timeframe. Therefore, a systematic prioritization process is necessary.

There are various tools and approaches that can be used to prioritize audit topics. When prioritizing clinical audit topics, it is advisable to focus on areas that exhibit at least one of the following characteristics to maximize the potential for improvement:

1. **Clinical Significance**: Assess the potential impact of the audit topic on patient outcomes and the overall quality of care. Prioritize topics that have a significant clinical impact and where improvements can make a tangible difference such as causes of serious quality concern or problems associated with delivering care or a service, e.g., complaints and quality incidents or high complication rates, high hospital-acquired infection rate, adverse outcomes, or poor symptom control, etc.
2. **Burden of Disease/High Volume**: Consider the prevalence and severity of the condition or issue being audited. Topics that affect a large number of patients or have a high burden of disease should be given priority. Other topics could be such as most frequent reasons for referral, admission or treatment or frequent procedures performed. e.g., venepunctures, mandatory donor testing and donor selection, chest X-rays in the accident and emergency department.
3. **Risk Assessment**: Evaluate the potential risks associated with specific areas of practice. Focus on topics that pose a higher risk to patient safety or have a greater likelihood of adverse events if not properly addressed. For example, incorrect

clinical advice, late antenatal results, inappropriate donor selection or certain postoperative complications or high risk to staff needle stick injuries in an HIV unit.

4. **Caused Concern**: Topics that have raised concerns or have been associated with adverse events or suboptimal outcomes should be prioritized to address and mitigate these issues.
5. **High Cost**: Areas with significant financial implications or resource utilization should be considered, as improvements in cost-effectiveness can lead to substantial savings for the organization. For example, thrombolytic therapy with Tenecteplase in myocardial infarction or trastuzumab for breast cancer, HLA-matched platelets.
6. **Stakeholder Input**: Consult with relevant stakeholders, including healthcare professionals, patients, and other key individuals or groups involved in the care process. Gather their perspectives and insights to identify priority areas that align with their concerns and priorities.
7. **Evidence-Based Practice**: Review current evidence, guidelines, and best practices in the field. Identify areas where practice variation exists or where there is a gap between recommended guidelines and actual clinical practice.
8. **Organizational Priorities**: Consider the strategic priorities and goals of the healthcare organization. Focus on audit topics that align with the organization's mission, vision, and strategic objectives.
9. **Resource Availability**: Assess the availability of resources, including personnel, time, and financial resources, required to conduct the audit. Prioritize topics that can be feasibly audited given the available resources.

In addition to the above-mentioned factors, there are other important considerations when prioritizing clinical audit topics:

1. Key Professional or Clinical Interest: Is the topic of significant interest or relevance to the healthcare professionals involved? Topics that directly impact their practice or specialty should be given priority.
2. Availability of Reliable Data Sources: Are there reliable sources of data readily accessible for data collection purposes? It's important to ensure that the necessary data can be collected efficiently and effectively to support the audit project.
3. Feasibility of Data Collection: Can the required data be collected within a reasonable time period? Consider the practicality of data collection, including the time and resources needed to gather the data without causing significant disruptions to clinical workflows.
4. Amenable to Change: Is the identified problem or area of focus amenable to change? Prioritize topics where improvements can be made through changes in processes, protocols, or interventions, leading to measurable and meaningful outcomes.
5. Alignment with National or Local Initiatives: Is the topic relevant to national or local initiatives or priorities? Consider whether the audit topic aligns with broader healthcare objectives, guidelines, or quality improvement initiatives at the national or local level.

By prioritizing audit topics based on these characteristics, organizations can focus their efforts on areas that have the greatest potential for improvement and impact on patient care, while considering available resources and the overall strategic direction of the organization. By taking these additional considerations into account, organizations can further refine their selection of clinical audit topics, ensuring that they are meaningful, feasible, and aligned with professional interests and broader healthcare priorities.

When selecting audit topics, it is essential to prioritize areas or issues that can be measured against relevant standards and have the potential to drive improvements in healthcare outcomes and processes. It is not productive to audit areas where it is already believed that the practice is performing well. Therefore, it is advisable to choose projects that are focused, straightforward, and target areas where achievable changes can be made. To aid in the prioritization process, implementing a scoring system can be beneficial in ranking topics according to their importance and impact.

Criteria for Good Local Clinical Audit

1. Relevance: The topic chosen for the clinical audit should be relevant to the organization's objectives and priorities. It should address an area of practice that has the potential for improvement and can lead to meaningful changes in patient care.
2. Clear Objectives: The objectives of the audit should be well-defined and measurable with a quality improvement intent and should be part of a structured program. They should specify what is being audited, the expected outcomes, and the criteria for success.
3. Systematic: Topics chosen should be high risk; high volume or high cost or reflect national clinical audits. A clinical audit ought to evaluate both the process and the results of care.
4. Use of Standards: The audit should be based on established standards or guidelines that reflect the best practices in the field. These standards provide a benchmark for assessing the quality of care and enable comparisons with expected performance.
5. Multidisciplinary in Nature: Service users and those directly involved in the care process are part of the clinical audit process.
6. Stakeholder Engagement: Involving relevant stakeholders, such as healthcare professionals, patients, and management, throughout the audit process promotes buy-in, collaboration, and shared responsibility for improving care.
7. Data Collection: The audit should have a robust data collection process in place. This includes identifying the appropriate data sources, ensuring data accuracy and completeness, and maintaining patient confidentiality.

8. Analysis and Interpretation: The collected data should be analyzed thoroughly to identify patterns, trends, and variations in practice. The results should be interpreted in the context of the audit objectives and standards to draw meaningful conclusions.
9. Actionable Recommendations: The audit should provide clear and actionable recommendations for improving clinical practice. These recommendations should be feasible, practical, and aligned with the objectives of the audit.
10. Implementation and Follow-Up: It is crucial to have a plan for implementing the recommended changes and monitoring their effectiveness. Follow-up audits can be conducted to assess the impact of the implemented changes and identify further areas for improvement. Action plans need to tackle the local obstacles to change and pinpoint those accountable for enhancing services.
11. Continuous Learning: A good clinical audit fosters a culture of continuous learning and improvement. It encourages reflection on the audit findings, dissemination of lessons learned, and the integration of feedback into future audits and practice.
12. Accountability and Governance: Clear accountability and governance structures should be in place to ensure that the audit process is transparent, independent, and conducted with integrity. This includes adherence to ethical guidelines and appropriate oversight.

17.10 Time Required for Doing a Clinical Audit

The time needed to conduct a clinical audit can vary significantly based on the project's complexity and scope. Simple audit projects may involve data collection, analysis, and action planning within a short timeframe of an hour or two or a few weeks. However, more comprehensive audits may take several months or even years to complete. The key factor is to design the audit project in a way that ensures the generation of meaningful data while considering the available budget and time constraints.

It's worth noting that the implementation of necessary changes is often the most time-consuming aspect of a clinical audit project. This phase involves translating the audit findings into actionable recommendations and actively implementing them in practice. The duration of the implementation phase can vary significantly depending on the nature of the changes, organizational factors, and the level of stakeholder engagement.

Ultimately, the goal is to strike a balance between conducting a thorough clinical audit within a reasonable timeframe and ensuring the successful implementation of improvements in patient care. Effective planning, resource allocation, and ongoing monitoring are essential to manage the time required for each stage of the clinical audit project. What is important is to design a clinical audit project which will produce meaningful data which can be finished within the budget and time available. Implementing the necessary changes is often the most time-consuming part of any clinical audit project.

17.11 The Clinical Audit Cycle or Spiral

Clinical audit can be envisioned as a dynamic cycle or spiral aimed at enhancing the quality of care. This iterative process consists of several stages that work together to drive improvement (Fig. 17.1). These stages involve identifying best practices, measuring performance against established criteria, implementing actions to enhance care, and continuously monitoring progress to ensure sustained improvement. The spiral concept suggests that with each cycle, the aim is to elevate the level of quality, constantly striving for higher standards and better outcomes. It represents a continuous journey of refinement and advancement in the delivery of healthcare services.

Fig. 17.1 Clinical audit cycle of continuous improvement in patient care

17.12 Stage 1: Planning/Preparing for the Audit

Engaging Stakeholders

Effective stakeholder engagement is crucial in clinical audit. Since healthcare delivery often involves multidisciplinary teams, it is important to involve all relevant clinical and managerial disciplines related to the audit topic. Clear roles and responsibilities should be defined and agreed upon by all parties involved prior to the commencement of the audit. This ensures that everyone understands the purpose of the audit and their specific contributions to it. While the lead clinician sponsoring the audit may not directly participate in the audit activities, they should be informed about and supportive of the audit's progress and outcomes. Their approval is essential for maintaining the integrity and impact of the audit.

17.12.1 Identify Problem and Local Resources for the Audit

The selection of an audit topic is guided by the objectives of the audit and typically focuses on assessing adherence to healthcare processes that have demonstrated positive patient outcomes. It is important to involve all healthcare professionals who are involved in patient care to gather their perspectives. The clinical team plays a crucial role in prioritizing clinical topics. The following questions could help in selecting a topic:

- Is the topic a priority for the organization? Are there specific challenges or issues that have been identified in the relevant areas?
- Have patients recommended any topics? Patient perspectives may differ from those of clinicians, and their input can provide valuable insights. Earlier, we discussed strategies for involving patients in every phase of the audit, from design and data collection to implementing changes.
- Is there good evidence available to inform standards? This may include systematic reviews or national/state clinical guidelines that can serve as a foundation for establishing audit criteria and benchmarks.

By considering these factors, one can identify an audit topic that aligns with organizational priorities, incorporates patient perspectives, and is supported by evidence-based standards or guidelines.

17.12.2 Locate Relevant Information

To support a clinical audit, it is important to access relevant, reliable, and up-to-date information from various sources. Some sources for relevant information:

- Clinical Guidelines: Obtain clinical guidelines from reputable sources such as National Programmes, National Standard Treatment Guidelines (STGs), the

World Health Organization (WHO), National Institute for Health and Clinical Excellence (NICE), and the Scottish Intercollegiate Guidelines Network (SIGN). These guidelines provide evidence-based recommendations for best practices in healthcare.
- Local Hospital Guidelines: Refer to guidelines specific to your local hospital or healthcare organization. These guidelines may address local protocols and procedures relevant to the audit topic.
- Service Standards: Seek information on service standards from national service providers, such as health professionals, service users, and managers. They may have set measurable goals and standards for specific areas of care. For example, ensuring cashless services for pregnant women, mothers, and neonates under various healthcare schemes.
- National Health Systems Resource Centre (NHSRC): The NHSRC, or similar organizations, work to promote clinical standards through audits, guidelines, and related activities. They may provide valuable resources and support for clinical audits.

By accessing these sources of information, one can gather the necessary guidelines, criteria, and service standards to inform your clinical audit and improve the quality of care provided.

17.12.3 Planning the Delivery of Audit Fieldwork

To ensure a successful and focused clinical audit, it is essential for the audit team to have a clear understanding of the overall purpose and objectives of the audit. Without a well-defined purpose, the audit will fail to drive improvements in the quality and effectiveness of clinical care.

Objectives are specific statements that outline the aspects of quality that will be measured to demonstrate the achievement of the audit's purpose. Clearly stated objectives provide the audit team with:

- A clear definition of the audit's purpose and scope.
- Everyone involved in the audit has a common understanding of the audit's objectives and goals.
- Effective communication of the audit's purpose and scope to others.
- Guidance to keep the audit focused and prevent it from becoming too broad or losing direction.
- Consensus on realistic and achievable goals.
- Establish benchmarks to measure the audit's effectiveness.
- A focal point for writing the audit report.

The purpose of the audit can be expressed through aims and objectives, using verbs such as "improve," "ensure," "enhance," or "change." For example, in an audit of healthcare records, the overall aim could be:

To enhance compliance with healthcare records policies and procedures.

Detailed statements can be used to describe the specific aspects of quality that will be measured to fulfil the aim of the clinical audit. For instance, "*To* ***ensure*** *the content of the healthcare record provides an accurate chronology of events*."

Objectives of the audit of consent to treatment could include:

- Assessing whether consent forms are appropriately completed.
- Determining whether patients are offered a copy of the consent form.
- Establishing that consent is obtained by a qualified healthcare professional.

These objective statements provide a clear framework for measuring the different elements of quality and demonstrating how the overall aim of the clinical audit will be accomplished.

17.12.4 Determining the Necessary Skills and Personnel to Conduct the Audit

To ensure the success and effectiveness of a clinical audit, it is crucial to involve individuals with the appropriate skills and expertise from the very beginning. The identification of necessary skills and the selection of individuals possessing those skills should be given high priority.

A comprehensive assessment of the required skills will help in assembling a competent and capable audit team. This team should consist of professionals who possess the knowledge, experience, and competencies relevant to the audit's objectives and scope. It is essential to identify individuals who have a thorough understanding of the clinical processes being audited, as well as those who are experienced in data collection, analysis, and interpretation.

The following steps can assist in identifying the skills and individuals needed for the audit:

1. Define the audit objectives and scope: Clearly articulate the aim and purpose of the audit and determine the specific areas and processes to be audited.
2. Identify the required skills: Based on the audit objectives and scope, determine the specific skills and expertise that are necessary for conducting a successful audit. This may include knowledge of clinical practices, data analysis, research methodology, quality improvement techniques, and project management.
3. Assess the available resources: Evaluate the existing pool of professionals within the organization or healthcare setting to identify individuals who possess the required skills. Consider their availability, workload, and commitment to participate in the audit.
4. Seek external expertise, if needed: If certain skills are not available internally, consider collaborating with external experts or consultants who can contribute their specialized knowledge and experience to the audit.

5. Formulate the audit team: Select individuals who have the necessary skills and expertise and ensure that the team represents a diverse range of perspectives and roles within the healthcare setting. Clearly define the roles and responsibilities of each team member, including the audit lead, data collectors, analysts, and any other supporting roles.

By carefully considering the skills needed and involving the right individuals in the audit team, one can enhance the chances of conducting a successful clinical audit that generates meaningful insights and contributes to the improvement of healthcare practices.

Determining the Necessary Skills and Personnel Required to Conduct the Audit
To ensure the effective delivery of a clinical audit, it is important that all staff involved receive proper training and clear instructions regarding their roles and responsibilities. Each team member should possess the following:

1. Basic understanding of clinical audit: All team members should have a foundational knowledge of clinical audit principles, processes, and methodologies. This understanding will provide a common framework for conducting the audit.
2. Commitment to audit plans and objectives: It is essential that every team member comprehends and supports the goals and objectives of the audit. This commitment ensures a shared vision and promotes collaboration towards achieving the desired outcomes.
3. Clarity on expectations: Clear communication regarding the expectations of the audit team is crucial. At the beginning of the audit, it is important to establish and document the roles and responsibilities of each team member. This can be outlined in a terms of reference document, which clarifies the specific tasks and deliverables expected from each individual.

By ensuring that all staff members are appropriately trained and briefed, and possess a basic understanding of clinical audit, a cohesive and well-prepared audit team can effectively contribute to the success of the audit project.

17.12.5 Providing the Necessary Structures and Support

To ensure the smooth implementation of clinical audit work, it is necessary to establish appropriate structures and provide necessary support. This includes the following considerations:

1. Submission and review process: Completed forms, along with relevant standards, audit tools, and supporting documentation, should be submitted to the responsible

clinical lead, directorate, or governance committee for evaluation. This ensures that the proposed audit aligns with the service provider's requirements. Clear lines of accountability should be defined from the outset. Depending on local policy, formal decisions may be required before proceeding with the audit.

2. Timetable and planning: It is important to establish a timetable for designing and conducting the clinical audit. This helps in organizing and allocating resources effectively, ensuring that the audit progresses in a timely manner. Setting clear deadlines and milestones keeps the audit on track and enables efficient monitoring of progress.
3. Supportive tools and checklists: A simple clinical audit checklist can be a valuable tool in guiding the audit process. It provides a structured framework to ensure that all necessary steps and considerations are addressed during the audit. Additionally, providing appropriate templates and resources to the audit team facilitates consistent data collection, analysis, and reporting.

By establishing the necessary structures and support mechanisms, such as defined review processes, a well-planned timetable, and useful tools like checklists, the clinical audit can be executed effectively and in accordance with the requirements of the service provider.

17.13 Stage 2: Selecting Audit Review Criteria

When conducting a clinical audit on maternal death, it is essential to streamline the focus to ensure a manageable and effective process. Instead of attempting to cover all causes of maternal death, which could be time-consuming and overwhelming, a more practical approach is to concentrate on the major complications that result in death. Each of these complications can be considered as a separate audit topic.

To select appropriate audit review criteria, it is important to identify and define the best practices associated with the management of each specific life-threatening complication. These criteria should reflect the established guidelines and evidence-based practices for addressing and preventing such complications. Examples of criteria may include the timely administration of blood transfusions, appropriate use of antibiotics, and minimizing delays in reaching the operating theatre for necessary interventions.

By focusing on specific life-threatening complications and selecting criteria that align with best practices for managing each complication, the clinical audit can effectively assess the quality of care provided in relation to these critical areas. This targeted approach allows for a more focused and meaningful audit process that can lead to actionable recommendations for improving patient outcomes.

17.13.1 Define and Determine Standard and Criterion

Defining and determining standards and criteria is a crucial step in the clinical audit process. It involves establishing clear expectations for the quality of care to be achieved and developing measurable criteria to assess compliance with those standards. A standard outlines and specifies the level of care to be attained. Each standard has a title summarizing the area on which that standard focuses and explains the level of performance to be achieved. Standard criteria are explicit statements representing elements of care which need to be achieved in order for that standard to be reached. Measuring care, i.e., defined level or degree of expected compliance with audit criteria. In order to compare actual care with care that should be provided, each audit criterion should have an "expected level of performance" or "target" assigned to it usually expressed in percentage or proportion of cases.

Criteria is also called yard stick or measuring tool. For criteria to be valid and result in enhancements to service user care, they should adhere to SMART guidance, which means they must be Specific, Measurable, Achievable, Relevant, and Time-bound. Specific (explicit statements, not open to interpretation). The criteria should be explicit and focused, avoiding ambiguity or open interpretation. It is important to choose a specific area of interest or concern, keeping the scope well-defined. The problem statement for the audit should be concise and clear. In step 1, problem statement should be concise and limited to one short sentence.

Measurable—The criteria should be measurable against a set target or standard, either at the local or national level. Having measurable criteria allows for objective assessment and comparison between actual care and the desired level of performance. Clearly defining the standards and criteria is essential, using objective measures such as specific values or thresholds. Ensure that standards and criteria are clearly defined and objective. For instance, specify "Temp >38 degrees" instead of using the term "fever."

- Achievable (of a level of acceptable performance agreed with the stakeholder)—The chosen level of performance should be realistic and acceptable to stakeholders involved in the audit. It is advisable to limit the number of outcomes to one or two, ensuring that the data required for assessment are readily collectible. Consider the availability of necessary resources and access permissions before starting the audit process. Limit yourself to 1 or 2 outcomes. Ensure the data are easily collectible and that have all the necessary access before starting, such as computer logins. Additionally, verify that your time at the hospital is sufficient to conduct several reaudits.
- Relevant (related to important aspects of care)—The criteria should be directly related to important aspects of care and aligned with the audit objectives. It is important to consider whether the appropriate expertise and authority to conduct the audit in the chosen area is available. Selecting an audit topic that aligns with your knowledge and role will contribute to more accurate and meaningful results. Are the appropriate persons for this audit? It's often simpler for a student to evaluate four-hour waiting times in the A&E department than to measure the frequency of a specific technique being utilized in the operating theatre.
- Timely or theoretically sound—Consider the practicality and feasibility of the chosen criteria. Opting for criteria that can be assessed and audited quickly will maintain motivation and provide opportunities for re-auditing. Additionally, bas-

ing the criteria on current evidence and best practices ensures that the audit is grounded in sound theoretical principles.

- Theoretically sound or timely (evidence-based)—choosing something that can be done quickly will keep motivated and give more opportunity to reaudit.

By adhering to the SMART guidance when defining standards and criteria, the clinical audit process becomes more focused, effective, and conducive to generating actionable recommendations for improving the quality of care.

Criteria for clinical audits can be categorized into three main types: structural, process, and outcome criteria. Establishing explicit selection criteria ensures that the collected data are precise and relevant, focusing only on essential information. Clinical practice guidelines can be a valuable resource for developing criteria and standards. A few examples are:

1. ***Structural Criteria***:
   - Active management of the third stage of labor (AMSTL): Ensuring that oxytocin is administered at the time of delivery of the baby's shoulder in all deliveries.
2. ***Process Criteria***:
   - Asthma management: Requiring that all patients with asthma have their inhaler technique reviewed at least once a year to ensure proper usage and effectiveness of the medication.
3. ***Outcome Criteria***:
   - Myocardial infarction (MI) management: Ensuring that all patients with diagnosed MI receive aspirin within 30 min of hospital admission to reduce the risk of further complications.

These examples demonstrate how specific criteria can be developed based on clinical practice guidelines, aiming to improve the quality of care and patient outcomes. By implementing such criteria, clinical audits can effectively assess adherence to best practices and identify areas for improvement.

To Ensure That the Criterion Is Explicit and Measurable

In order to ensure that criteria are explicit and measurable, it is important to provide clear and specific statements that leave no room for ambiguity. Some examples of how to make the criteria more explicit, specific, and measurable are as below:

1. Clarifying Eligibility: Instead of stating, "All eligible women should have had a cervical smear," which lacks clarity, specify the age group and time frame: "All women aged 25-65 years who are eligible should have undergone a cervical smear within the past 5 years." This criterion makes the requirement explicit and measurable." Defining the Scope: Rather than using a broad statement like "The overall management of obstetric hemorrhage," which encompasses various scenarios, focus on specific indicators: "All women with per vaginum bleeding" or

"Perceived blood loss of more than 1000 ml within 24 hours of delivery" or "Patients presenting with clinical signs of shock (pulse >100/min, <100 mm/Hg systolic blood pressure)." These statements provide explicit, specific, and measurable criteria.
2. Specifying Management: Instead of using a vague statement like "The overall management of puerperal sepsis," which lacks clarity on the specific actions to be audited, focus on a particular aspect: "The use of antibiotics in the management of puerperal sepsis." This provides a more appropriate and measurable criterion.

Additional examples of specific criteria:

- "After diagnosis of the hemorrhage, was manual placenta delivery or uterine exploration effected in less than 15 minutes?"
- "Was oxytocin administered within 15 minutes of observation of uterine atony?"
- "All medicines in the crash cart should be in-date."

It's important to exercise caution and not attempt to audit too many criteria at once. Focusing on one or two criteria allows for a more focused and manageable audit process.

17.13.2 Define Ideal Standards

To ensure the usefulness of the criterion, it is necessary to define the ideal standards. The standard represents the desired level of care to be achieved for each specific criterion and is typically expressed as a percentage. **An audit standard serves as a minimum level of acceptable performance for that criterion**. Setting targets for compliance can be done in various ways, including:

1. Consensus Opinion: The audit team members and relevant stakeholders can engage in discussions and develop a consensus opinion regarding the appropriate standard to be met for the criterion. This collaborative approach ensures that multiple perspectives are considered in setting the target.
2. Benchmarking: Comparing the performance against national rates or established benchmarks can provide a reference point for setting the standard. By benchmarking, the audit team can align their target with recognized levels of performance in the field.

By defining the ideal standards through consensus or benchmarking, the clinical audit can establish clear expectations for the level of care to be achieved for each criterion. This allows for meaningful measurement and evaluation of performance against the set standards.

There are a number of ways to set targets for compliance, such as discussion and development of a consensus opinion among audit team members and relevant stakeholders and benchmarking against national rates.

Minimum vs. Optimum Standards

When setting targets for clinical audit, it is important to consider the distinction between minimum and optimum standards. An explanation of these terms and the factors to consider when setting targets is provided as below:

1. Minimum Standards: Minimum standards represent the lowest level of acceptable performance. They are used to differentiate between acceptable and unacceptable practices. Meeting the minimum standard indicates that the criteria have been met at a basic level, ensuring a certain level of quality and safety in healthcare delivery.
2. Ideal Standards: Ideal standards describe the level of care that would be provided under perfect or ideal conditions, without any constraints. While ideal standards may not be realistically achievable in all situations, they serve as a reference point for the best possible care. They represent an aspirational goal to strive for, even if they cannot be fully attained.
3. Optimum Standards: These standards represent a balance between minimum and ideal standards. They represent a realistic and achievable level of performance that is desirable in clinical practice. Optimum standards consider factors such as available resources, feasibility, and practicality while still aiming for a high level of quality and effectiveness.

A minimum standard represents the lowest level of acceptable performance, used to differentiate between acceptable and unacceptable practices. On the other hand, an ideal standard defines the level of care that could be provided under perfect conditions without any constraints. However, by definition, such a standard is rarely attainable. The expected level of performance or target can range from 0% (the criterion is something that must never be done) to 100% (the criterion is something that must always be adhered to). An optimum standard strikes a balance between the minimum and ideal standards, providing a realistic and achievable level of care under normal circumstances.

When setting targets, three factors should be considered and assessed to provide a balance between achievable performance and the highest possible quality of care:

Clinical Importance Targets should reflect the clinical significance of the criterion. It is essential to prioritize areas of care where achieving a higher level of performance will have a substantial impact on patient outcomes and overall quality of care. If a criterion is essential for ensuring the safety of service users, targets may be set at 100% or 0%. For instance, in a clinical audit concerning the safe administration of medication, a target of 100% might be set for the criterion that "medication is not administered to patients with a known allergy to the medication." All used syringes and needles should be discarded in the puncture proof container.

Practicability Targets should be practical and feasible to achieve within the given healthcare setting and available resources especially where clinical importance is not as significant. Unrealistic targets can lead to frustration and hinder progress, so it is important to set goals that are attainable. Resources required to fulfil the target performance level should be considered (manpower, equipment) and an acceptable performance level (one which is seen as both reasonable and attainable by those delivering and receiving care) should be identified. For example, in a clinical audit relating to the time frame within which service users should be seen in a particular outpatient clinic, a target of 90% may be deemed appropriate whereas a target of 100% may be deemed appropriate for shelf life of crash cart medicines "100% of the medicines in crash cart should be in-date." "100% of women in labor should be monitored using a partograph for charting.

Acceptability Targets should be acceptable to the relevant stakeholders, including healthcare professionals, patients, and regulatory bodies. Involving these stakeholders in the target-setting process helps ensure that the standards are meaningful, relevant, and widely supported. An optimum level of performance is set when the best care possible is identified given the resources available and normal conditions of caregiving. This will fall between the minimal acceptable level of care and the highest achievable level of care, which is possible only under ideal conditions without any resource constraints.

When choosing a standard for a clinical audit, it's vital to ensure it is both realistic and pertinent to the specific healthcare environment. Some considerations and examples for setting realistic standards:

1. Realistic Standard: The chosen standard should be achievable within the given healthcare setting, considering available resources, staff capabilities, and other practical constraints. It is essential to consider the feasibility of meeting the standard when setting targets.
2. Consensus and Discussion: Establishing an optimum standard involves careful judgment and collaboration with other members of the audit team. Engage in discussions to reach a consensus on the appropriate level of performance that balances clinical importance and practicability.
3. Direct Relationship and Timeframe: The standard should have a direct relationship to the audit criterion and be expressed within a suitable timeframe. This ensures that the standard is specific and measurable. For example:
   - "At least 80% of eligible women aged 25-65 should have had a cervical smear in the last 5 years."
   - "Blood transfusions should commence within one hour of the decision to transfuse in at least 80% of cases."
   - "Emergency surgery should be performed within 2 hours for at least 80% of patients suspected or diagnosed with uterine rupture."
   - "80% of patients admitted with a diagnosed stroke should receive thrombolytic therapy within 4.5 hours of symptom onset."

4. Clinical and Medico-legal Importance: The standard should reflect the clinical significance and medico-legal importance of the criterion. It is necessary to consider not only the overall performance, but also specific actions and interventions related to the criterion. For example, while 80% of women should have had a cervical smear, 100% of those with abnormal smears should have had appropriate follow-up or intervention.

Examples of Two Audit Criteria (with a Different Standard for Each Criterion) Are

1. Audit Criterion: Percentage of Patients with Septic Shock Who Were Given Anti-infective Treatment.
   - Standard: 100% Explanation: Since septic shock is a severe condition with a high mortality rate, it is crucial to ensure that all eligible patients receive timely and appropriate anti-infective treatment. Setting the standard at 100% reflects the critical nature of this criterion, aiming for optimal care and patient outcomes.
2. Audit Criterion: Percentage of Patients in the Emergency Seen Within Fifteen Minutes.
   - Standard: 80% Explanation: In the emergency department, timely access to care is vital. Setting a standard of 80% for patients to be seen within fifteen minutes acknowledges the potential variations in patient flow, acuity, and resource availability. While striving for prompt assessment and treatment for all patients, this standard allows for some flexibility, considering the complexity and unpredictable nature of emergency care.

It is important to note that standards may vary based on factors such as the severity of the condition, available resources, and the specific healthcare setting. Setting appropriate standards involves considering clinical importance, feasibility, and consensus among the audit team and relevant stakeholders. These examples demonstrate different standards tailored to the specific audit criteria and the associated clinical context (Table 17.2).

Table 17.2 Examples of standards and criteria: TYING IT ALL TOGETHER

Criteria	Standards
All children under 2 years should be immunized against tetanus and polio	90% of registered patients under the age of 2 years should have been immunized against polio and tetanus
All pregnant women should be immunized against tetanus	All registered pregnant women should have been received 2 doses of tetanus.
All notes of those patients with an allergy to penicillin should be marked	95% of patients with an allergy to penicillin should be clearly marked
All patients in the emergency should wait no longer than 30 min before a consultation	70% of patients in the emergency should wait no longer than 30 min before a consultation
All pregnant women should be recommended an HIV test	95% of pregnant women should have an HIV test before 20 weeks

> Without a standard, there is no logical basis for making a decision or taking action.—**Joseph M. Juran.**

17.14 Stage 3: Measuring Levels of Performance

17.14.1 Step 1. Develop a Methodology

Identify how the information/data is going to be collected in order to measure against the set criteria and standard.

17.14.1.1 Data Type

In order to measure against the set criteria and standards, it is important to identify how the information/data will be collected. The data can be classified into different types, each with its own characteristics and ways of description.

1. Quantitative Data: Quantitative data, also known as objective or "hard data," focuses on numerical or specific information. It includes measurable variables such as age, gender, length of stay, counts of specific actions or occurrences, frequency of events, and expected outcomes. This type of data is often linked to standards of best practice and allows for objective analysis and comparison.
2. Qualitative Data: Qualitative data, on the other hand, is descriptive and concerned with words rather than numbers. It is sometimes referred to as subjective or "soft data." Qualitative data collection methods aim to capture individuals' experiences, opinions, preferences, or narratives. Examples of qualitative data include comments provided on questionnaires, interviews, or open-ended survey responses. While qualitative data is typically descriptive, it can sometimes be represented in a quantitative form (e.g., using a rating scale or 8 out of 10 people…) or grouped into thematic categories to facilitate analysis (e.g., levels of satisfaction with care).

Depending on the nature of the audit topic, both quantitative and qualitative data may be collected. The selection of data types will depend on the specific objectives of the audit and the information needed to gain a comprehensive understanding of the topic under investigation. Integrating both quantitative and qualitative data can provide a more comprehensive and holistic view of the audit subject.

> What gets measured gets managed.—**Peter Drucker**:

When collecting qualitative data, it is important to use appropriate methods and techniques to ensure reliability and validity. These methods may include structured interviews, surveys, observations, or focus groups. For quantitative data, standardized measurement tools, checklists, or existing databases can be utilized to ensure consistency and accuracy in data collection.

When conducting an audit, it is essential to specify and agree upon the sources of data with the audit team. The chosen sources should provide the most accurate and complete data in a timely manner. The following are common sources of data for audits:

1. Health Facility Registers: Registers maintained at health facilities, such as admission and discharge logs, delivery registers, operating theatre registers, and emergency room registers, can be initial sources of case identification. Senior health professionals or designated staff often maintain these registers, making them valuable sources of information.
2. Patient Records and Clinical Documentation: Patient records and other routinely maintained clinical records, including nursing records, blood bank records, operating theatre logbook entries, and mortuary records, are primary sources of information. These records contain detailed information about patient care and outcomes and can provide valuable insights for the audit.
3. Computerized Information Systems: In modern healthcare settings, computerized information systems play a crucial role in data collection. Electronic health records, laboratory information systems, patient administration systems, and picture archiving and communication systems (PACS) are examples of computerized systems that can provide relevant data. Using these systems can streamline data collection and minimize duplication of effort.
4. Additional Data Sources: In some cases, additional data may be required to complement the routinely collected data. This could include laboratory reports, radiological images, or other specialized information. However, it is important to collect additional data only if absolutely necessary to avoid unnecessary duplication of information and workload.

It is worth noting that each data source may have its limitations (Table 17.3). Clinical records, for example, can be incomplete or have missing information. Overcoming these limitations often involves collecting data from multiple sources, such as cross-referencing clinical records with laboratory results or imaging reports, to ensure a more comprehensive data set.

The method of data collection can vary depending on the outcome being measured and the availability of computerized systems. It is crucial to determine the specific data needs, identify where the data can be found, and assign responsibility for data collection. Involving those who directly provide the service in data collection can help foster ownership and enhance the quality of the data collected.

Regularly reviewing the data with the data collectors before the completion of the clinical audit is important. This promotes their understanding of how the data are being used and allows for positive feedback on their efforts. It also provides an opportunity to address any data quality issues or discrepancies that may arise during the review process.

Table 17.3 Issues related to potential data sources for clinical audit standards

Data source	Issues	Possible actions to resolve
Patient care records, including inpatient, outpatient, primary care both paper and electronic	Records are lost, missing, or destroyed. Records are incomplete. The contents in the record are disorganized, making it difficult to find needed information. There is bias relating to what is or is not typically recorded. Information about the episode of care of relevance to the audit has not been filed in all the records. Handwriting is illegible. Authorization for access may be required, and it can take time to get authorization. Accessing electronic records requires technological skills.	Conduct a preliminary test before gathering data to detect possible issues. Monitor and clean data as they are received. Identify patterns that may cause bias, such as excluding patients whose records are frequently accessed. Provide feedback to data collectors. Revise the data collection forms or protocol as needed. If the audit's data is incomplete due to issues with the quality of data from the sources, include this in the clinical audit findings and implement measures to resolve the data source issues.
Departmental reporting systems, such as pathology and radiology reporting systems	The systems generally include accurate and complete records of results of investigations, directly linked to patient identifiers. Normally, the results are quality controlled before they are accessible to staff outside the service. Authorization for access may be required and it can take time to get authorization. Accessing electronic records requires technological skills.	Review and revise reporting systems to include aspects of quality of care Errors reporting Critical value reporting Lab utilization index

(continued)

Table 17.3 (continued)

Data source	Issues	Possible actions to resolve
Routinely collected data in local departmental and organizational information systems and national information systems	Routinely collected data: Tend to focus on processes of care or patient care transactions May overlook data pertinent to certain elements of quality of care. May not be complete and comprehensive Involve little if any additional cost to obtain and use Are not always complete, accurate and precise. Authorization for access may be required, and it can take time to get authorization.	Collection of data from several sources may overcome incomplete records. Promote efficient and accurate entry of data onto databases or spreadsheets for collation and analysis. Initiate the procedure for obtaining consent for data sharing at the earliest opportunity.
Specially collected data such as patient interviews or diaries	Inaccuracies or gaps in memory may occur for patients or caregivers, and information might not be documented fully or accurately.	Train data collectors Always pilot first. Retrain, if required
Forms completed by clinicians during a patient encounter	Because forms are not embedded in essential patient care records, they often are filled in after the encounter because of workload, which can provide unreliable or incomplete data, or not filled out at all. Form filling is seen as additional work that does not relate to normal care.	Simplify the process Keep carbon copy of the prescription
Postal survey questionnaires	Surveys can experience poor response rates. Clinicians or patients can be overloaded with survey questionnaires. It may be difficult to maintain anonymity of responses.	If postal questionnaire include pre-paid envelopes for respondents to use for returning questionnaires.

(continued)

Table 17.3 (continued)

Data source	Issues	Possible actions to resolve
Audio or video recording	Recordings of actual transactions: Affect privacy, confidentiality and anonymity, and therefore, require explicit consent Can produce a Hawthorne effect Can take considerable time to analyze.	Seek permission to make recordings and get consent for any use or disclosure Give patients adequate information about the purpose of the recording when seeking their permission. Ensure that patients feel free to decline permission for recordings. Cease recording upon patient request or if it negatively impacts the consultation or treatment. Refrain from participating in recordings against a patient's wishes. Safeguard the recording to maintain patients' privacy and dignity. Video recordings should be edited and anonymized by obscuring or "digitizing" identifying features. Patients should be given the opportunity to see it, and to withdraw consent for its future use or to specify the kinds of contexts in which it may be used Do not utilize recordings for any purposes beyond the original consent without acquiring additional consent. For anonymous recordings made during treatment which do not require separate patient consent are: Images from pathology slides (glass slides from microscopy) X-rays Laparoscopic images Images of internal organs Ultrasound images. Ensure that recordings are stored securely and appropriately. Avoid sending or recording identifiable images in ways that permit the recipient to forward it to others or use it for purposes beyond the original consent.

17.14.1.2 Data Collection Method/Strategy

The choice of a data collection method/strategy for a clinical audit depends on several factors, including the audit topic and objectives, feasibility in terms of resources and time frame, and the availability of relevant data sources and the likelihood of obtaining complete and reliable data. The three main data collection strategies are retrospective, concurrent, and prospective, each with its own merits and considerations:

1. Retrospective Data Collection: Retrospective data collection focuses on gathering evidence that has already been provided in the past. This strategy involves collecting data after the completion of treatment or care provided to service users. Retrospective data collection is often the easiest method as the data already exists and can be quickly gathered. However, it is important to note that retrospective data can become outdated, and there may be limitations in terms of data completeness and accuracy.
2. Concurrent Data Collection: Concurrent data collection focuses on gathering evidence that is being provided in the present, while treatment or care is ongoing. This strategy involves collecting data in real-time alongside the delivery of care. It is important to consider the potential for bias when collecting data concurrently with patient care, as the knowledge of an ongoing audit may influence clinical practice or behavior. Care should be taken to ensure that data collection does not disrupt or compromise the quality of care being provided.
3. Prospective Data Collection: Prospective data collection focuses on gathering evidence that will be provided in the future. This strategy involves planning and recording data on care that will be provided and gathering specific data for the audit. Prospective audits require careful planning and coordination to ensure that the necessary data is collected in a timely and systematic manner. While prospective data collection may be more time-consuming compared to retrospective or concurrent methods, it allows for the collection of specific data points that may not be routinely recorded.

Merits of prospective and retrospective data collection are shown in Table 17.4. It is important to note that the terms concurrent and prospective are sometimes used interchangeably. If the required data is not routinely collected, a prospective or concurrent audit should be considered, even though it may require more time and resources. For example, conducting a clinical audit on the effectiveness of providing special clinical diets to hospital in-patients may require collecting data as meals are delivered, especially if the exact meals delivered to patients on special diets are not routinely recorded.

The selection of the most appropriate data collection strategy should be based on careful consideration of the audit objectives, available resources, and the likelihood of obtaining complete and reliable data. For example, a clinical audit on the effectiveness of providing special clinical diets to hospital in-patients may require collecting data as meals are delivered, if the exact meals actually delivered to patients aren't routinely recorded for patients for whom a special clinical diet is requested.

Table 17.4 Merits of prospective and retrospective data collection

Prospective audit	Retrospective audit
Complete data set from each case	Predictable time frame
Carrying out audit may generate change in practice by raising awareness	Auditor's workload easy to organize
Meanings and abbreviations can be clarified	Number of cases can be achieved Results and changes available early
Time frame uncertain as incidence varies	Case notes may be missing; especially interesting or unusual cases
Workload harder to manage	Required data may not be recorded in notes
Some cases may be missed if not all staff aware of need to identify and audit cases	Guidelines may have changed since the time period chosen for audit

17.14.1.3 Data Collection Methods

Once the data collection strategy has been determined, it is important to design a suitable data collection tool. Various tools can be utilized for data collection in a clinical audit, including: The most commonly used tools are:

Questionnaire It is a set of questions and sent to the respondents through mail which should be returned back after answering. Questionnaire is used for survey method to get wider and more representative distribution of sample. Well-designed questionnaires can be incredibly valuable for gathering comprehensive and detailed information. Success of a questionnaire depends on its design; therefore, its physical appearance has to be attractive. Some tips for designing the questionnaire are given in Box.... This method is cheap and economical as no enumerator is required but a high non-response rate is a major limitation. Questionnaires do come with their own set of risks, such as incomplete or incorrect information. Bias due to non-response often remains indeterminate. This method could be used only for literate and cooperative respondents only.

Data Collection Forms/Proformas/Schedule This method of data collection is very much like the data collection through questionnaires, but the data is collected by the enumerator trained to collect such information. Response rate is high, and complete information is obtained as enumerator ensures the filling of all the questions, but success depends on honesty and competency of the enumerator. The form needs to be unambiguous, especially if several people are collecting the data. Forms from previous audits on similar topics can serve as valuable references.

Along with schedules, observation method can also be used but such a thing is not possible with questionnaires.

Clinical Case Review: Reviewing individual patient cases and extracting relevant data from their medical records or other clinical documentation.

Interviews: Conducting structured interviews with healthcare professionals, patients, or other stakeholders to gather information and insights related to the audit criteria.

Clinical Case Review: Reviewing individual patient cases and extracting relevant data from their medical records or other clinical documentation.

It is important to note that any structured surveys, whether administered via post, in hospitals, or through one-to-one interviews, should be approved by the appropriate Medical Audit committee or relevant ethical review board to ensure the protection of participants' rights and the ethical conduct of the audit.

The selection of the most appropriate data collection method or combination of methods will depend on the nature of the audit, the available resources, and the specific objectives of the audit.

Ensure Reliable and Valid Data Reliability

Reliability and validity are essential considerations in ensuring the quality of data collected during a clinical audit:

Reliability: It refers to the consistency and repeatability of the measurement process. It addresses the level of error or variation that may occur during data collection. To enhance reliability:

1. Standardization: Ensure that the data collection tool and methods are standardized and clearly defined, leaving no room for ambiguity or subjective interpretation. This helps to minimize variability in data collection.
2. Training: Provide adequate training and clear instructions to the individuals involved in data collection. This ensures that they understand the criteria, definitions, and procedures for data collection, leading to more consistent results.
3. Consistency Checks: Conduct consistency checks during data collection and data entry to identify and correct any errors or inconsistencies promptly. This helps maintain the reliability of the data.

Validity: Validity refers to the extent to which the audit measures what it intends to measure. To improve validity:

1. Clear Objectives: Ensure that the objectives of the audit are well-defined, specific, and aligned with the intended outcomes. This clarity helps to focus data collection efforts on the relevant aspects of care and ensures that the audit measures what it intends to assess.
2. Standardized Criteria: Establish clear and unambiguous criteria and standards for measurement. These criteria should be based on evidence-based guidelines or best practices to ensure that the audit assesses the desired aspects of care accurately.
3. Robust Data Collection Methods: Employ appropriate data collection methods and tools that align with the objectives of the audit and the nature of the data being collected. This ensures that the data collected are relevant, accurate, and aligned with the audit's purpose.

By considering reliability and validity throughout the audit process, from the design of the audit to the data collection and analysis stages, one can enhance the quality and credibility of the collected data and ensure that the audit findings are meaningful and reliable.

Tips for designing a questionnaire for an audit given in box…

Questionnaire Design Tips

When designing a questionnaire for an audit, consider the following tips:

- Include Audit Details: Clearly mention the audit project title and date on the questionnaire for easy identification and retrieval.
- Self-Explanatory and Brief: Ensure that the questionnaire is self-explanatory and easy to complete. Provide concise instructions directly in the questionnaire to assist respondents.
- Keep It Concise: Keep the questionnaire succinct and avoid overwhelming respondents with too many questions. Limit the questionnaire to around 20 questions to maintain participant engagement.
- Use Simple Language: Use plain and simple language that is easily understandable by patients. Avoid medical jargon or technical terms.
- Cover Letter: Include a cover letter with the questionnaire, providing contact details for any queries. Explain the purpose of the project, instructions for returning the questionnaire, assure confidentiality, and clarify how the collected data will be used.
- Ensure Anonymity: Develop completely anonymized questionnaires to maintain patient confidentiality.
- Sequential Ordering: Arrange the questions in the same sequence as the information appears in the data source to facilitate data entry and analysis.
- Avoid Double-Barreled Questions: Ask one question at a time and avoid combining multiple questions into a single item. This ensures clarity and prevents confusion in respondents' answers. For example, "How would you rate or evaluate the staff's efficiency and friendliness?"
- Clear and Unambiguous: Phrase questions clearly and unambiguously to avoid leading or biased responses. Use neutral language and provide response options that capture the intended information accurately. The questions need to be straightforward and without ambiguity. Occasionally, the wording of a question can imply the desired response. Avoid using leading questions. For example, the question "Do you think that the waiting times in the outpatient department are good?" It should be replaced with "How would you rate the waiting times in the OPD?" Tick box options could be used to respond to this question, e.g., Excellent, Good, Average, Bad, Very Bad, No opinion.

- Include Uncertainty Options: Provide response options for uncertainty, such as "not sure" or "no preference," to allow respondents to indicate when they are unsure about their answer. If these options are not provided, and the available choices do not align with the patient's views, they may either leave the question unanswered or provide a response in a free-text format that might be harder to analyze.
- Use Number Codes: Use consistent number codes throughout the questionnaire for recording decisions, for example, 1 is yes and 2 is no. Numeric codes are quicker to enter than alpha codes. Ensure consistency by using the same codes throughout the form.
- Use Filters: Incorporate filters or skip patterns to direct respondents efficiently through the questionnaire. This helps streamline the process and saves time for both respondents and data entry, e.g., if Yes, go to question 5′. Incorporating filters streamlines the process of completing the questionnaire, making it quicker and more efficient for respondents.
- Handle Missing Data: Use designated code (e.g., 9 or multiple 9 s) to represent missing data instead of leaving blank spaces.
- Additional Comments: Include a space for additional comments where respondents can provide extra insights or feedback that may be valuable for the audit.
- Improve Response Rate: If using postal questionnaires, aim for a response rate of 60% or higher. Encourage response by including a prepaid envelope for easy return.

By following these tips, you can design a well-structured and effective questionnaire that yields reliable and valid data for your audit.

Using Likert Scales for Question Responses

Limit the use of open-ended questions that require respondents to provide their own detailed descriptions. Analyzing and interpreting free text responses can be complex and time-consuming. Save open-ended questions for qualitative research methods like focus groups or interviews.

1. Create a dedicated questionnaire or proforma for each audit topic. Clearly define the questions and provide a small range of response options to ensure consistency and ease of analysis.
2. Yes/No or Limited Options: Use binary response options like "Yes/No" for certain questions that require straightforward answers. Alternatively, provide a drop-down list of limited options to capture the necessary information accurately.

For example:

1. Was an HIV test offered at ANC booking?
   - YES.
   - NO.
2. Did the woman have an HIV test at booking?
   - Test not offered.
   - Test offered but declined.
   - Test offered and accepted.
3. Likert Scale for Opinion-Based Questions: When seeking patient opinions or attitudes, Likert scales can be used. Offer response choices such as "Strongly agree/Agree/Undecided/Disagree/Strongly disagree" and ask respondents to circle or tick their chosen response.
4. Odd vs. Even Number of Choices: Be mindful of the number of response options provided. With an odd number of choices, respondents may tend to select the middle option, leading to less distinction in their responses. With an even number of choices, respondents are more likely to lean towards a positive or negative answer. Consider the desired level of granularity and balance when deciding on the number of choices.

By following these guidelines, a structured and standardized questionnaire can be designed that enables efficient data collection and analysis for your audit.

Summary data collection and designing a data collection form which is efficient, ethical, and focused on gathering the necessary information for the audit objectives.

Simplify and Automate: Streamline data collection forms and automate the process as much as possible to minimize errors and save time. Utilize electronic data collection tools or software that can facilitate efficient data capture.
Relevant and Specific Data: Collect data that is directly related to the objectives of the audit. Avoid gathering excessive data that is not essential to the audit question at hand. Focus on collecting relevant information to avoid an unnecessary burden on data collection.
Avoid Excess Identifiable Information: Ensure that data collection forms do not contain excessive patient or staff identifiable information. Respect confidentiality and privacy guidelines to protect sensitive information. Avoid using names or personally identifiable details of patients and healthcare professionals.

Consider Confidentiality and Ethics: Address issues such as confidentiality, information governance, information security, and ethical considerations when designing the data collection process. Seek approval from the local Research Ethics Committee for audits involving potentially sensitive topics.
Bias Awareness: Be mindful of potential biases that may influence the results. Design data collection forms and procedures in a way that minimizes bias, ensuring fairness and objectivity in data collection.
Pilot Testing: Before implementing the data collection process on a larger scale, pilot test the data collection forms. This allows for the identification of any issues, ambiguities, or areas for improvement, ensuring the reliability and validity of the data.

17.14.1.4 Deciding on a Population or a Sample

It is first necessary to define the population of concern (target population). A population can be defined as encompassing all the service users, events, cases, conditions, situations, or items that are the focus of the audit. A sample is a part of the population of interest. When conducting a clinical audit, the decision to target the entire population or use a sample depends on various factors. In order to determine which is suitable for the purposes of the clinical audit, the following should be considered: By considering these factors informed decision on whether to target the entire population or use a sample for the clinical audit can be made.

Inclusion/Exclusion Criteria Clearly define the criteria for including or excluding cases, patients, events, or situations in the audit. This helps in identifying the relevant population and ensuring that the data collected are precise and focused. Specify exactly the patients, cases, situations, circumstances, or events to be included in an audit (such as the specific diagnosis, condition, surgical procedure or special procedure, if the subject of the audit is care provided to patients with a diagnosis, condition, surgical or special procedure Or the specific referrals by reason, condition, source of referral or time period, if referrals relate to the subject of the audit or the exact events or circumstances and how they will be identified, if events or circumstances, such as patient falls, relate to the subject of the audit etc.). An example of a special situation for defining a patient population in a clinical audit on the effectiveness of completion of the consent process depends on being able to specify the procedures for which consent is required and then finding a "complete" list of patients who have had one of the procedures for which consent is needed.

The user group to be included, with any exceptions should be noted.

Timeframe Establish a suitable time period or interval for the audit. Consider whether the data will be collected from a specific timeframe in the past, present, or future. Align the timeframe with the objectives of the audit.

Size of the Population/Sample Size Evaluate the size of the population of interest. If the population is large, it may be impractical to audit the entire population, and sampling becomes more feasible. Sampling allows for a representative subset of the population to be studied.

Resources Evaluate the resources available for the audit, including time, budget, and personnel. Consider whether conducting the audit on the entire population is feasible given the available resources. Sampling can be more resource-efficient in certain cases.

Suitability Determine whether a sample or the entire population is more appropriate for achieving the objectives of the audit. Consider the level of precision required, the level of confidence needed in the findings, and the specific research question or objective of the audit.

Sample Selection Method If sampling is chosen, determine the appropriate sample selection method. Common methods include random sampling, stratified sampling, or purposive sampling, depending on the goals of the audit and the characteristics of the population.

17.14.1.5 Sample Size for Data Collection

Ideally, all patients who receive care can be audited. This includes every patient seen for a given condition over a specific period, every treatment administered, and every outcome achieved, to determine if their care met the established best practice standards. This is feasible if the total number of cases is small or if the required data is easy to collect.

However, if the number of patients in this population is too large this becomes impractical and thus instead one needs to look at a sample of the overall population representing the target population. Choosing a larger sample size than is unnecessary takes up extra time and resources without adding value and can mean that there is no time and energy left within the project team to address any issues of below-par practice and bring about improvement.

Sample sizes for data collection are often a compromise between the statistical validity of the results and pragmatical issues around data collection. Where performance may vary by clinical service, location, primary care center, staff shift pattern, or time of year, a single sample of 30 or 50 cases may produce biased results, particularly when the cases are consecutive. For scientifically valid outcomes optimum sample size is required to give a snapshot of what is happening, e.g., while for process-based clinical audit project approximately 20–50 cases may be appropriate but larger sample size is required for assessing outcome.

Determining the appropriate sample size for data collection in a clinical audit involves considering several factors. These factors include the size of the population, the desired degree of accuracy and confidence, and the expected incidence or compliance with the audit criteria. Some key points to consider include:

- Size of the Population: Assess the total number of cases, patients, or events in the population of interest. If the population size is small, it may be feasible to audit the entire population. However, for larger populations, sampling becomes necessary to ensure practicality.
  - **How difficult is it to find cases** that would be eligible for inclusion in the audit?
  - **How difficult is it to find the exact information needed** for the audit for each case that is to be included?
  - **How many cases** are required to be included in the audit **are there** in a given time period, such as a week, a month, or a year?
- Degree of Accuracy: Determine the desired level of accuracy for the audit findings. This reflects the acceptable margin of error or variation in the results. A smaller margin of error requires a larger sample size, while a larger margin of error allows for a smaller sample size.
- Degree of Confidence: Specify the level of confidence required in the findings. Commonly used levels of confidence are 95% or 99%. A higher level of confidence requires a larger sample size to minimize the risk of sampling error.
- Expected Incidence: Consider the expected compliance rate or incidence of meeting the audit criteria in the population. This information helps in estimating the sample size needed to detect significant variations from the expected performance.

> To understand God's thoughts we must study statistics, for these are the measure of His purpose.—**Florence Nightingale.**

The following example shows how this works in practice:

Another example is a secondary care team planning an audit of the care of patients with hypertension. There are 300 patients (size of population) being treated for the disorder, but the clinical audit team do not have time to review the records of all these patients. The audit criteria state that patients receiving treatment should have had their blood pressure checked and the result below 150–90 on three occasions in the past 12 months. The target for meeting this standard is set at 70%. However, the team are willing to accept 5% inaccuracy (degree of accuracy) due to sampling. In other words, if the findings give a level of 70%, on 95% of occasions (degree of confidence) the true value would lie between 65% and 75%. Epi Info software program calculated the sample size of 155 using the above parameters.

Sample size calculations can be performed using statistical software or formulas specific to the study design and objectives. These calculations aim to strike a balance between statistical validity and practical considerations, such as resource availability and time constraints. It's important to note that sample size calculations should be conducted for each audit criteria being addressed in the clinical audit project. This ensures that the sample size is appropriate for each specific objective of the audit.

By considering the size of the population, the desired accuracy and confidence, and the expected incidence, one can determine an optimal sample size that provides meaningful and reliable results for your clinical audit. Table 17.5 can help in choosing audit sample sizes and assumes an expected incidence of 50%, i.e., that standards will be met 50% of the time in order to be 95% sure (degree of confidence) that the results obtained from the sample will be within 5% (degree of accuracy) of the results that would have been obtained if data were collected from the whole population. Put another way, there is a 1 in 20 chance that your results will not be representative.

Using this table for an audit finding of 84% compliance with a clinical audit standard, using a sample size sufficient for a 95% level of confidence and a 5% range of accuracy, one could be 95% sure that the true value is 84% ± 5% or that the true value lies between 79% and 89% had whole population been looked at. In other words, one is 95% confident that the compliance with the audit standard in the entire population would be between 79% and 89%.

Note that sample sizes need to be proportionately smaller as the population size increases. For example, examining 357 out of 5000 patients provides the same level of certainty as assessing 44 out of a population of 50 patients. This is due to the reduced likelihood of unrepresentative results as the population size grows. Table 17.6 demonstrates how the sample size may vary based on the expected incidence for a population of 500.

Table 17.5 Sample size calculation

Population size	Sample size (95% confidence; ±5%)
50	44
100	79
150	108
200	132
500	217
1000	278
2000	322
5000	357

Table 17.6 Sample size for a population of 500 at different expected incidence

Confidence level (%)	Degree of accuracy (%)	Expected incidence ("best guess") (%)	Sample size
95	±5	50	217
90	±10	50	176
95	±5	40	213
95	±5	20	165
95	±5	5	64
95	±2.5	50	378
95	±2.5	5	185

17.14.1.6 Sampling Methods

Once a decision about sample size is taken, the next question is which cases will be included in the audit?

Choosing an appropriate sampling method is crucial for ensuring that the selected sample is representative of the target population in a clinical audit. There are several sampling methods available, and the choice depends on the specific characteristics of the population and the objectives of the audit. Some commonly used statistical sampling methods include:

Simple Random Sampling

In random sampling, each individual or case within the population has an equal chance of being selected for the sample. This method helps minimize bias and provides a representative sample when the population is homogenous. Random sampling can be achieved through various techniques each with its own advantages and disadvantages such as simple random sampling, stratified random sampling, convenience sampling, or cluster sampling.

A straightforward method for selecting these cases is to use a computer or random number table (Table 17.7). One can either take one number at a time, such as 2, 0, 1, 7, 4, etc., moving from left to right, or take two numbers at a time, reading down the table, like 04, 42, 20, 74, 88, 22, etc. These numbers then determine your sample, for example, the fourth, 20th, 74th, 88th, and 22nd patients from a list of all the patients in your population.

Table 17.7 Sample random table

2	0	1	7	4	2	2	8	2	3	1	7	5	9	6	6	3	8	5	1	0	2	1	0	9	6	1	0	5	1	5	5	9	2	5	2	4	4	2	5
7	4	4	9	0	4	4	9	0	3	0	4	1	0	3	3	5	3	7	0	2	1	5	4	4	7	8	6	9	4	6	0	9	4	4	9	5	7	3	8
0	4	7	0	4	9	3	1	3	8	6	7	2	3	4	2	2	9	6	5	4	0	8	8	7	8	7	1	3	7	1	8	4	7	8	4	0	5	4	7
2	2	4	4	8	9	6	5	6	8	9	5	3	2	5	2	3	8	3	7	1	5	1	2	5	4	0	2	0	1	3	7	5	6	8	7	6	5	8	9

Systematic Sampling or Quasi-Random Sampling

Systematic Sampling: Systematic sampling involves selecting every nth individual or case from a population list. For example, if the population size is 1000 and the desired sample size is 100, every tenth individual could be selected. This method is useful when the population list is readily available and ordered in a systematic manner.

To ensure that every patient in the audit population has an equal chance of being selected, the starting point needs to be picked randomly because otherwise the second patient would never have a chance of being selected if we always started with the first patient. In this instance the starting number must be between 1 and 4, i.e., auditing patients 1, 5, 10, 15, etc., or 2, 6, 11, 16, or 3, 7, 12, 17, etc.

Convenience or Consecutive Sampling

Convenience sampling involves selecting individuals or cases based on their easy accessibility or availability. It involves choosing the next, or last, e.g., the next or the last 50 patients, or alternatively, all patients seen over the course of the previous or next month. **Convenience Sampling—**Cases are selected for inclusion in the sample based on their accessibility. For instance, this might include patients visiting the heart clinic for a blood pressure check on a specific day. Though most convenient and practical way of selecting cases for a "snapshot" sample of the population but it is non-probability sampling method, i.e., the sample produced may differ in character from the overall population, therefore, the audit results may not be representative of the overall care that is given. This method is quick and convenient but may introduce bias, as it may not represent the entire population accurately. This method is often employed when time and resources are limited.

Consecutive Sampling Samples are included from an agreed start date, for example, all new heart patients from first September.

Stratified Sampling

Stratified sampling involves dividing the population into subgroups or strata based on specific characteristics and then selecting a proportional sample from each stratum. This method ensures representation from different subgroups within the population and can improve the precision of the audit findings. For example, if patient population is made up of 75% men and 25% women, taking a simple or quasi-random sample runs the risk of selecting only men. If certain aspects of care being audited pertain specifically to women, an adjust may be needed in the sample to ensure adequate representation. For instance, in a population of 500 patients, if the general ratio is 3:1 in favor of men, one would then have 375 men and 125 women. This way, one can focus on the relevant subgroup while maintaining a representative sample. This would result in the representative sample of 217 patients being split into 163 men:74 women. To select random sample, separate men and women into two groups and randomly select from both, i.e., 74 women from a population of 125, and 163 men from a population of 375.

Rapid-Cycle Sampling This approach is effective when there's an identified issue requiring prompt results. It involves conducting an audit with a small sample group, making necessary changes, and then re-auditing with another small sample to assess improvements. By using multiple small data sets, this method allows for continuous monitoring and accelerates the change cycle, making it faster to implement improvements in care.

Cluster Sampling: Cluster sampling involves dividing the population into clusters or groups, such as clinics or hospitals, and randomly selecting a few clusters to include in the sample. Within the selected clusters, all individuals or cases may be included, or a further sampling method can be applied. Cluster sampling is useful when it is impractical or costly to sample individuals directly.

Other Sampling Methods

Some other ways of choosing an audit sample are also there. However, care should be taken when using these as audit samples may end up less representative of the audit population:

Purposeful Sampling Purposeful sampling involves deliberately selecting individuals or cases that meet specific criteria relevant to the audit objectives. This method is useful when targeting a specific subgroup within the population or when seeking diverse perspectives. Purposeful sampling includes techniques such as quota sampling, snowball sampling, and maximum variation sampling.

Cases are selected specifically because they share characteristics, for example, diagnosis, age, gender, and ethnicity.

Quota Sampling Quota sampling is a non-random sampling method that aims to ensure representation of specific subgroups or quotas within the audit population. It is similar to convenience sampling in that individuals are selected based on their accessibility or availability. However, in quota sampling, steps are taken to ensure that subgroups are represented in proportions that reflect their presence in the population. Similar to convenience sampling, steps are taken to ensure that subgroups of the audit population are represented proportions, for example, diagnosis, age, gender, and ethnicity.

Reducing Bias

If the audit sample is not selected meticulously, it can lead to skewed results and inaccurate information. An unrepresentative sample may fail to capture the true picture, leading to erroneous conclusions and potentially misguided decisions. Therefore, it is important to choose the most appropriate method of sampling for the given situation, ensuring that the audit is as robust as possible and that reasonable inferences about the care given to the whole population can be drawn.

Consider daily, weekly, or seasonal fluctuations which may skew the data. For example, conducting an audit for a waiting time in OPD always in the early working hours or during festival days may not be representative of the waiting time during peak hours or peak load. Similarly, assessing availability of medicines in the first half of the month may not be representative of the availability of medicines in the rest of the month or year due to insufficient/irregular supplies.

In general, the risk of introducing bias is greater with the narrower time frame for data collection. The results thus obtained will not be representative of the population as a whole even if standards are met. To ensure that the results are the representative must take a sample across a longer time period.

Missing cases also may skew data; therefore, it is also important to make every effort to ensure that every case in the sample is included in the audit. For example, if a set of case notes or discharge summary cannot be located in file, they may be with complaints, legal services or held on to by the consultant because of problems in care. Not including such cases in the audit would indicate that care was better than it actually is. It is therefore important to make sure and find any missing notes.

Prospective audits often suffer from bias because of incomplete notes cases that met the audit criteria were not picked up and therefore not included in results. This may skew or distort results if missing cases are non-random. For example, if a particular staff member never completed the forms.

Reducing bias in the data collection process is crucial to ensure the accuracy and reliability of audit results. Some strategies to minimize bias include:

1. Random Sampling: Use random sampling methods to select participants or cases from the population. Random sampling helps ensure that each individual or case has an equal chance of being included in the audit, reducing selection bias.
2. Adequate Sample Size: Ensure that the sample size is large enough to capture a representative sample from the population. A larger sample size reduces the impact of random variation and increases the precision of the results.
3. Sampling Across Time and Locations: Consider conducting the audit over a longer time period and across different locations to account for daily, weekly, or seasonal fluctuations. This helps to capture a more comprehensive picture of the care provided and reduces bias introduced by specific time or location factors.
4. Avoiding Missing Cases: Make every effort to include all eligible cases in the audit. Missing cases can introduce bias if they are systematically different from the included cases. Take steps to locate and include any missing records or cases to ensure the sample is representative.
5. Consistency in Data Collection: Ensure consistent data collection practices among auditors or data collectors. Provide clear instructions and training to minimize inter-observer variability and bias in data collection.
6. Blinding: Consider blinding the auditors or data collectors to minimize observer bias. Blinding involves concealing certain information or characteristics of the cases being audited to prevent bias in data collection.
7. Data Validation: Implement checks and validation procedures to ensure the accuracy and completeness of the collected data. Verify the data against other sources or conduct independent data audits to identify and correct any errors or inconsistencies.

8. Monitoring and Quality Assurance: Regularly monitor the data collection process to identify and address any potential sources of bias. Implement quality assurance measures to ensure adherence to standardized protocols and minimize errors or biases.

By employing these strategies, the potential for bias in data collection can be minimized, leading to more reliable and valid audit results that accurately reflect the quality of care provided to the target population.

17.14.1.7 Sample Summary

- Begin by identifying the characteristics of the population under study.
- Determine the size of the target population.
- Decide whether to examine every case or select a sample based on the nature of the condition or objective of the audit. Options include random sampling, selecting all cases within a specific time period, sampling a fraction of cases, or selecting sequential cases. For common conditions the options would be:
  - A random sample of 5% of cases in 1 year, e.g., Miscarriages.
  - All cases in a short time period, e.g., all Caesarean sections in 1 month.
  - 1 in 5 cases, e.g., women attending for emergency contraception.
  - 30 sequential ectopic pregnancies.
  - All cases of pelvic inflammatory disease selected every Friday (convenience).
- Consider sample size calculations based on factors such as expected incidence, desired confidence levels, acceptable accuracy levels, and the type of audit (process-based or outcome-based).
  - Sample sizes may differ based on any of the following factors:
  - The expected incidence of the criteria being audited.
  - The required confidence levels. The confidence level does not always have to be 95%. It could be 90%, 99%, etc.
  - The acceptable level of accuracy which could be fixed at 10%, 5%, or 1%, etc.
  - Process-based clinical audit projects usually involve a "snapshot" sample of approximately 20–50 cases. Larger sample size is required for outcome-based projects.
- If a statistically representative sample size has been determined, select cases randomly to avoid selection bias.
- Take steps to minimize selection biases by ensuring all case notes are retrieved and calculating the proportion of missing cases to assess potential bias. Verify that selected cases meet the defined criteria and extract relevant information onto the audit form.
- Consult with colleagues involved in the audit to determine the number of cases they deem appropriate, considering time, resources, and the need for statistical validity.

Following these steps helps ensure a robust and representative sample for the audit, reducing biases and allowing for meaningful conclusions about the quality of care provided to the population of interest.

17.14.1.8 Duration of Data Collection

The time period chosen depends on the number of cases that are treated on a daily basis, the number needed to make a confident judgment of the care provided and inclusion and exclusion criteria. It should be the minimum required to collect only the essential information. Set a realistic deadline in order to keep the eagerness and adhere to timeline. The time period chosen depends on the number of cases that are treated on a daily basis, the number needed to make a confident judgment of the care provided and inclusion and exclusion criteria.

17.14.1.9 Who Collects the Data

Depending on the audit, data may be collected by more than one person or different people may be responsible for completing different data sets. There should be no confusion over terminology. It's essential to provide a clear definition for each data item to ensure consistent collection (inter-rater reliability). Additionally, all individuals involved in the data collection process should clearly understand their responsibilities, including what data to collect, how to collect it, and where to record it to ensure uniformity and accuracy in data collection, facilitating reliable and valid audit results.

17.14.1.10 Ensuring Data Quality

Data can be said to be of good quality when it does what it is needed to do. Clear definitions for each data item must be provided to ensure data collectors have a thorough understanding of what to collect, how to collect it, and when to collect it. Set up systems to track both the data collection process and the data to minimize the occurrence of reporting and input errors. Have a mechanism to report instances of missing data to the clinical audit group and approaches to analyze the data for any patterns as well as methods to account for the causes.

17.14.1.11 Pilot and Collect Data

Pilot First

Before proceeding with data collection for the full audit sample, it's crucial to conduct a pilot study. This helps test the reliability and validity of the data collection method. This involves collecting a small amount of data from approximately 10 cases to ensure that the data collection process is effective. Piloting helps identify any issues with the audit design, assess the feasibility of data collection, and estimate the required time and resources. Analyzing the pilot data allows for early identification of any necessary amendments, such as clarifying instructions, improving question clarity, or modifying the data collection form. The pilot data should be analyzed against the audit standards to ensure accurate measurement of current practice. It may also reveal opportunities to restructure open questions as tick-box options for easier completion.

Avoid Audit Fatigue

Consider selecting a population or sample that does not include individuals or organizations that have already undergone extensive audits and data retrieval. This helps prevent audit fatigue and ensures a balanced representation. It is important to collect a sufficient amount of data without overburdening the participants or data collectors. Strive to collect "just enough" data rather than collecting excessive or unnecessary information. For ongoing data collection, the amount of data collected can be reduced compared to the baseline and early repeat measurements. Whenever possible, integrate the data collection process with existing work activities or established data collection systems. It is generally recommended to study at least four topics within a year to ensure a comprehensive assessment.

17.14.2 Step 2. Data Analysis

Different Types of Data

There are two main types of data:

Quantitative data is concerned with definitive, numerical, factual, and specific data, for example, Yes or No; Age; Gender; Blood Pressure; Blood Groups. The analysis of this type of data is performed using mathematical techniques.

Qualitative data is usually descriptive and subjective based on personal values and opinions, for example, complaints, and narrative responses to questionnaires. This data needs to be analyzed differently and carefully due to the high levels of inference and interpretation that can be applied to the information.

Synthesize Data, Interpret, and Draw Conclusions

Once the data has been collected, the next step is to synthesize and interpret the data to draw conclusions. This involves collating all the collected data and transferring it onto summary sheets, spreadsheets, or databases for analysis and interpretation. It is important to ensure the anonymity and security of service user data, using unique identifiers and password protection when working with electronic files.

Data analysis involves calculating proportions or percentages of cases that met the criteria of good practice. This can be done manually or using computer software if a large number of cases are involved. Data coding may be necessary to organize the data for analysis. A coding system assigns numeric codes to different categories or responses for each variable measured, such as gender, age, or ethnicity. Consistent coding schemes should be used, including codes for missing or unavailable data.

For most audits, complex statistical analysis is not required. The analysis should be simple, clear, and easily understandable by all stakeholders involved in the provision of care. Quantitative data can be presented as averages, percentages, or other relevant summary statistics, depending on the type of data collected.

The purpose of data analysis and interpretation is to identify patterns, trends, and areas of improvement or concern in the care provided. It is important to link the findings back to the audit criteria and standards in order to assess the level of adherence to best practice. The conclusions drawn from the data should inform and stimulate change, with the aim of improving the quality of care and achieving better outcomes for patients.

Coding of Data

Correct coding of data is essential for accurate analysis and interpretation. A data coding sheet or coding manual can be used to assign numeric codes to all possible answers for each question in the dataset. Variables such as gender, age, and ethnicity may be measured for each subject, and a coding system should be established for variables that require it.

For example, a coding system for gender may assign the code 1 for "male" and 2 for "female." Similarly, a binary question with a yes or no answer may be coded as 1 for "yes" and 2 for "no." Ethnicity can have different codes assigned to each ethnic group.

It is important to maintain a clear and consistent coding scheme throughout the data collection and analysis process. This includes defining codes for missing or unavailable data. Code 9 is commonly used as a universal code for missing data, while 99 may be used if the options exceed 9.

When dealing with free text responses, it is beneficial to code the information into categories whenever possible. This can be done by grouping similar responses based on disease groups, therapeutic drug groups, or other relevant categories.

Analyzing Quantitative Data

When analyzing quantitative data in a clinical audit, there are several key steps to consider:

1. Data cleaning: Before analysis, it is important to review the data for any errors or inconsistencies. This includes checking for missing values, outliers, and data entry mistakes. Any necessary corrections or adjustments should be made at this stage.
2. Descriptive statistics: Descriptive statistics provide a summary of the data, offering insights into the central tendency, dispersion, and distribution of the variables. Common measures in descriptive statistics include mean, median, mode, standard deviation, and range. These statistics provide a snapshot of the data and help identify patterns or trends.
3. Comparisons: Depending on the audit objectives, one may need to compare different groups or variables. This can be done using statistical tests such as t-tests, chi-square tests, or analysis of variance (ANOVA). These tests assess the statistical significance of differences between groups and help determine if the variations observed are likely due to chance or if they represent meaningful differences.
4. Trend analysis: If data has been collected over multiple time periods, one can analyze trends to assess changes in performance or outcomes over time. This can be done using graphical representations such as line charts or by calculating annual or quarterly averages to observe patterns or improvements.

5. Benchmarking: Benchmarking entails evaluating the audit results by comparing them to established standards, guidelines, or best practices. This helps assess performance relative to external benchmarks and identifies areas where improvement is needed.
6. Subgroup analysis: In some cases, it may be relevant to analyze subgroups within your data. This could involve examining differences based on demographic factors (e.g., age, gender) or clinical characteristics (e.g., diagnosis, treatment modality). Subgroup analysis can reveal variations in performance across different populations and provide insights into potential disparities or areas for targeted interventions.
7. Interpretation: Once the data analysis is complete, it is important to interpret the findings in the context of the audit objectives and relevant clinical guidelines or standards. Identify key findings, trends, and areas of concern, and consider their implications for clinical practice. This interpretation should be communicated clearly and effectively to stakeholders, highlighting actionable recommendations for improvement.

Average is a measure that represents the central or middle value within a data set. There are three types of average:

Mean—the mathematical average. The mean or average is calculated by adding all the values in a data set and then dividing the sum by the number of values in the list.

Example 1: 4; 8; 12; 16.
Mean = 10 (The sum of the numbers) ÷ (the number of items in the list) = 40 ÷ 4.
Mode—the most frequently occurring item in a list.
Example 2: Male; Female; Male; Female; Female; Female; Male

Mode = Female.

The mode has the advantage that it can be used with non-numerical data while other averages cannot.

Median—The median is the middle value in a group of a sorted list of numbers. If the list has an even number of values, the median is determined by calculating the average of the two middle numbers.

Example 3: 1; 7; 3; 15; 13
Median = 7 (reorder the list to read 1; 3; 7; 13; 15. Remove the highest and lowest values until one is left. This is the median).

Example 4: 1; 7; 3; 13 Median = 5 (since there are two elements in this remaining list, the median is the mean of these two values – 3 + 7 = 10 ÷2 = 5)

Standard Deviation Standard deviation provides information about the dispersion of data around the mean. It is assessed relative to the mean. A large standard deviation indicates that the data points are widely spread out from the mean, while a small standard deviation signifies that the data points are closely clustered around the mean.

Standard deviation will influence your conclusions and findings when conducting clinical audits.

Example 5: The patients have a mean age of 32 years and a standard deviation of 1.2 years. Most patients are 32 years and those who are not will be a year or so older or younger.

Example 6: Patients have a mean age of 32 years with a standard deviation of 10.5 years. While most patients will be 32 years, the ages of those who are not are spread across a much wider range, up to 10.5 years older or younger.

Confidence Intervals—unless the audit population is small, a sample of subjects will be used. Confidence intervals indicate whether the information collected is the same as would have been collected had the whole subject population been audited. Standard deviation is used to indicate the reliability of an estimate. Calculating confidence intervals is a statistical technique and is unlikely to apply to the type of audits being considered here.

In order to calculate the average or percentage, it is necessary to identify both the total number of applicable cases for a criterion (the denominator) and the total number within the denominator group that met the criterion (the numerator). The percentage is then calculated by dividing the numerator by the denominator and multiplying the answer by 100.

Ensuring the correct denominator is identified is crucial in preventing errors when calculating the percentage of cases meeting the audit criterion.

17.14.3 Step 3. Compare Performance with Criteria

Comparing the collected data with criteria and standards is a crucial step in the interpretation stage of a clinical audit. The key points to consider are:

1. Comparison with criteria: Assess how well the collected data aligns with the predefined criteria or standards that were set for the audit. These criteria could be based on clinical guidelines, best practices, regulatory requirements, or internal benchmarks. Determine the extent to which the criteria were met for each variable or aspect being audited.
2. Concluding performance: Based on the comparison, draw conclusions about the overall performance. Determine whether the standards were met or not, and to what degree. This can involve calculating percentages or rates of compliance, identifying trends, and assessing variations across different subgroups or time periods.

3. Identifying reasons for non-compliance: If the standards were not met, it is important to investigate and identify the reasons behind the non-compliance. This involves analyzing the data and seeking input from the audit team, clinicians, and other stakeholders involved in the care process. Look for underlying factors, contributory issues, or systemic issues that may have influenced the results. This analysis aims to understand the root causes of non-compliance and identify areas for improvement.
4. Setting targets and priorities: Consider the targets associated with each criterion and compare the actual performance to these targets. Determine if the targets were realistic and achievable, or if they need to be revised based on contextual factors such as resource limitations or workload considerations. Prioritize areas for improvement based on the degree of non-compliance and the potential impact on patient care.
5. Review with stakeholders: Share the findings and conclusions with senior clinicians, administrative staff, and other relevant stakeholders. Engage in discussions to ensure a shared understanding of the results and the areas for improvement. Seek input from the team to gain additional insights and perspectives.
6. Root cause analysis: Conduct a thorough analysis of cases where unacceptable care was identified. Utilize tools such as process mapping, the "five whys" technique, or cause-and-effect diagrams (fishbone diagrams) to facilitate a root cause analysis. Identify the underlying factors contributing to the identified issues and use this information to develop effective solutions for improvement.

By systematically comparing performance with criteria, understanding the reasons for non-compliance, and engaging stakeholders in the analysis, targeted interventions and action plans can be developed to drive improvements in clinical care.

Comparison of Targets or Standard Clinical audit implies that the type and quality of care provided to individual patients are compared, not only with previously agreed criteria, but also with targets. An example of a criterion that has already been given is that "blood transfusion will begin within one hour of the decision to transfuse." The target associated with this criterion might be that it will be achieved in 100% of cases. But in recognition of, for instance, a low staffing level and a high workload in the blood bank or availability of blood, a lower target of 80% might be set at a particular facility.

In theory, if the 100% standard was not met that was set, there is potential for improving care. In practice, if the results are close to 100% of the standard, generally further improvement will be difficult to achieve and that other standards, with results farther away from 100%, are the priority targets for action. However, this decision also depends on the topic—in some life-threatening situations, it will be important to achieve 100% of the standard.

17.14.4 Step 4: Feedback Findings and Set Local Targets

The following steps can be taken:

- Present key findings: Share a summary of the audit findings with all staff members during hospital meetings. Encourage open discussion and elicit staff perspectives on the extent to which they believe the criteria are currently being met. Then, present the actual proportions determined from the audit data. This comparison can help create awareness and provide a stimulus for change by highlighting the gaps between current practice and desired targets.
- Report process and outcome indicators: In criterion-based audits, it is important to report both process indicators (e.g., percentages of assisted deliveries, cesarean sections, etc.) and outcome indicators (e.g., case-fatality rates, maternal and perinatal mortality ratios, etc.). This comprehensive reporting approach gives a more complete picture of the impact of the audit on the facility's quality of care and service utilization.
- Enhancing quality can result in lower mortality rates and better service utilization. Reporting on both forms of indicators can give a more complete picture of how audit is affecting their facility.
- Utilize visual presentations: Visual presentations using tables, charts, and graphs can enhance the effectiveness of conveying the audit findings. Tools like Microsoft PowerPoint can be used to create visually appealing and easily understandable presentations.
- Consensus on targets: Facilitate discussions among the staff to reach a consensus on the proportion of cases that should meet each criterion. Consider input from clinicians, administrators, and other relevant stakeholders. Based on this consensus, set realistic targets to be achieved in the next round of the audit. These targets should be ambitious yet attainable, considering the available resources and contextual factors.
- Identify changes in practice or service provision: Collaboratively identify and agree upon feasible and achievable changes in clinical practice or service provision that can help meet the set targets. This may involve process improvements, training initiatives, workflow modifications, or other interventions. Encourage staff engagement and ownership in the change process.
- Consider external expertise: If needed, invite outside experts to participate in this step of the process. Their specialized knowledge and experience can provide valuable insights and recommendations for improvement.

By engaging staff, setting realistic targets, and fostering a culture of continuous improvement, the clinical audit process can drive positive changes in clinical practice and service delivery, ultimately improving patient outcomes and satisfaction.

17.14.5 Step 5. Report Writing

When creating a report for clinical audit findings, it is important to consider a clear and organized layout that effectively communicates the information. Some suggestions for the report layout (Box):

Title Page

Title summarizes main topic of audit.

Date of audit. Audit Lead and team and the organization's logo.

Contents

Introduction

Include a paragraph about the service being audited, identifying the target group and its role within the organization. The introduction should also provide background information covering the issues/reasons for undertaking the audit such as national interest, local priority, safety issues.

Aims and Objectives

- Aim—the purpose of the audit—what the audit is intended to achieve.
- Objectives—the objectives breakdown the overall purpose (aim) into more detailed and specific components.

Methodology

- Describe the data collection methods, including the population studied, sample size, and data collection period.
- Explain the criteria used for assessing performance and any relevant standards or targets.
- This section also provides an explanation of the methods used for data collection, including:
  - Data Source—such as clinical notes, computer systems, and individual patients
  - Data type (State the audit type, criterion, indicator-based, patient survey, etc.)
  - Population size
  - Sample selection method
  - Sample size
  - Data collection tools (Description of questionnaire/proforma/telephone interviews)
  - Data collection method
  - Data collectors
  - Data Qualit
  - Time period.
  - Description of analysis method
  - Caveats (if required) Explain any caveats which may affect the results

Results

- Present the key findings in a clear and concise manner giving overall picture. For example, number of participants/sample studies, demographics, distribution in different groups, etc.
- Use tables, charts, and graphs to visually represent the data and facilitate understanding.
- Include both process indicators and outcome indicators, as appropriate and whether standards have been met, identifying areas for improvement and observing any trends that may assist in implementing improvements.
- Identify sections that demonstrate effective practices and those that require enhancement.

- Provide numerical values or percentages to quantify the performance for each criterion. Describe the overall observations that can be taken from the findings offering interpretations/explanations about the findings.
- Response rates and any deficiencies in the data are highlighted in this section.
- Analysis and Interpretation:
  - Analyze the findings in relation to the criteria and standards set.
  - Discuss the implications of the results and their significance for patient care and outcomes.
  - Identify any factors contributing to variations in performance or areas of non-compliance.

Recommendations:
- Based on the findings, propose specific recommendations for improvement.
- Focus on ranking the suggestions according to their possible impact and practicality.
- Provide supporting rationale for each recommendation and consider potential resources required.

Conclusion:
- Summarize the main findings and their implications.
- Emphasize the importance of taking action to address areas of improvement.
- Reinforce the value of ongoing clinical audit in driving quality improvement.

Appendices:
- Include any additional information that supports the findings, such as detailed data tables or charts.
- Provide explanations of the coding system used, definitions of terms, or other relevant documentation.

References (List of references used for the audit and those referred to in the audit report).

Remember to use clear and concise language throughout the report, avoiding unnecessary jargon. Consider the intended audience and ensure the report is accessible to a wide range of readers, including both clinical and non-clinical staff.

Additional Items to Be Included in a Re-audit Report

Additional items to include in a re-audit report are:

1. Background:
   - Provide a brief overview of the previous audit, including the date, objectives, and key findings.
   - Summarize the actions that were implemented as a result of the previous audit and their intended impact on improving care.
2. Action Plans:
   - Document the action plans that were developed based on the previous audit findings.
   - Evaluate the success or effectiveness of the implemented actions.
   - Discuss any challenges or barriers encountered during the implementation process.
3. Progress Report:
   - Present a progress report on the changes that have occurred since the previous audit.
   - Highlight any improvements or positive outcomes that have been achieved.
   - Discuss any areas where challenges or difficulties persist.
4. Comparison to Previous Audit:
   - Include a summary or table that compares the current audit findings with the findings from the previous audit.
   - Recognize trends or patterns that have developed over time.
   - Assess whether there has been an overall improvement or regression in performance.
5. Conclusion/Summary:
   - Provide a concise conclusion or summary of the re-audit findings.
   - Emphasize the key areas of improvement or areas that still require attention.
   - Discuss the ongoing importance of clinical audit and the need for continued monitoring and improvement efforts.

By including these additional items, the re-audit report provides a comprehensive overview of the progress made since the previous audit, highlights the effectiveness of implemented actions, and helps guide further improvement initiatives.

17.15 Stage 4: Making Improvements

Implement Changes in Practice Where Indicated

Data collection has no chance of making any impact unless it is followed up with the more difficult process of implementing changes. After a presentation and discussion on the audit results with the rest of the team and whoever else in the hospital the audit is relevant to, the team must agree on recommendations for change. Implementing changes in practice is a crucial step in the clinical audit process. Key consideratio\ns and actions to take when implementing changes are:

1. Agree on Recommendations for Change:
   - Engage in discussion with the audit team, relevant stakeholders, and healthcare professionals to identify recommendations for change based on the audit findings.
   - Use an action plan to document these recommendations, specifying the actions to be taken, responsible individuals, and deadlines.
2. Identify Reasons for Deficiencies:
   - Determine the underlying reasons for falling short of the desired targets or standards.
   - Assess if deficiencies are due to organizational factors, inadequate knowledge, insufficient skills, or inappropriate attitudes among healthcare providers.
   - Focus on addressing the specific reasons identified to ensure effective change.
3. Develop Tailored Strategies for Change:
   - Design strategies and interventions that target identified deficiencies.
   - Examples of strategies include developing clinical protocols, conducting training sessions, implementing regular ward rounds or review meetings, and providing feedback on performance.
   - Consider the specific needs and context of the healthcare setting when planning and implementing changes.
4. Address Attitudes and Behavior:
   - If inappropriate attitudes are identified as a barrier to providing quality care, address these attitudes through targeted interventions. Promote awareness, understanding, and empathy among healthcare providers to foster positive attitudes towards delivering optimal care. Attitudes are inferred as inappropriate if the other three possible reasons are excluded. For example, provider's reluctance to give appropriate care to women who are HIV positive, despite adequate protective resources such as surgical gloves and the necessary knowledge and skills, can by default be attributed to inappropriate attitudes.
5. Use Effective Implementation Methods:
   - Disseminate educational materials, such as guidelines or protocols, alongside selected implementation methods.
   - Implement strategies such as trainings, performance reviews, reminders, and feedback mechanisms to support behavior change and adherence to new practices.
   - Ensure ongoing monitoring and evaluation to assess the effectiveness of the implemented changes.
6. Set a Timeline for Re-evaluation:
   - Specify a timeline for re-evaluating the implemented changes to assess their impact on improving care.
   - Regularly review and analyze data to track progress and identify any additional areas for improvement.

By systematically addressing deficiencies, providing targeted education and training, and promoting changes in attitudes and behaviors, the implementation of recommended changes can lead to improved quality of care. Continuous evaluation and adjustment of interventions are essential to ensure sustained improvements in practice.

17.15.1 Writing an Action Plan

When creating an action plan for implementing recommended changes, it is crucial to include specific details and ensure accountability. Key elements to include in an effective action plan are:

1. Clearly Define the Change Needed:
   - Clearly specify what needs to change or improve based on the audit findings and recommendations.
   - Identify the specific area, process, or practice that requires modification or enhancement.
2. Outline Actions Required for Change:
   - Enumerate the precise steps required to accomplish the desired transformation.
   - Break down the actions into smaller, manageable tasks to ensure clarity and ease of implementation.
   - Be specific about the actions to be taken, such as developing protocols, conducting training sessions, or implementing new procedures.
3. Set Deadlines for Completion:
   - Assign a realistic deadline for each action item to be completed.
   - Ensure that the deadlines are achievable and allow sufficient time for implementation and any necessary adjustments.
   - Consider dependencies between actions and account for them when setting deadlines.
4. Assign Responsibility:
   - Clearly specify who is responsible for each action item.
   - Assign ownership to individuals or teams who will be accountable for ensuring the completion of the tasks.
   - Ensure that responsibilities are assigned to individuals with the necessary authority, skills, and resources.
5. Specify Evidence of Implementation:
   - Define the evidence or documentation required to demonstrate that the actions have been implemented.
   - Identify the types of documentation, reports, or data that will serve as evidence of successful implementation.
   - This may include records, completion reports, training attendance logs, or other relevant forms of documentation.

By including these essential elements in an action plan, a clear roadmap for implementing changes is provided, establish accountability among team members, and ensure that progress can be tracked and measured effectively. Regular monitoring and review of the action plan will help ensure that the desired improvements are achieved within the agreed-upon timeframe.

17.15.2 Evaluate Data Recording Over Time

Sometimes higher score attainment after initial audit may be attributable to better record keeping. Evaluate the completeness of case files over time if improvements in record keeping have occurred. Compare the completeness of case files before and after the initial audit to understand the impact of improved record keeping on data quality and interpretation. Consider the importance of accurate and comprehensive case documentation for assessing the quality of medical care.

Improvements in record keeping are a good thing, but if this occurs after the first audit, interpretations regarding improved medical care need to be nuanced against more complete case file recording.

17.16 Stage 5: Sustaining Improvements

This stage is critical for ensuring the audit's success: it evaluates the impact of the implemented changes and identifies whether additional improvements are necessary to meet the standards set in stage 2.

- Verify whether the changes implemented as a result of the initial audit have had a lasting effect.
- Assess if the improvements in practice have been sustained over time.
- Determine if further enhancements are required to achieve the standards identified in the previous stages of the audit.

17.17 Stage 6: Re-evaluate Practice and Feedback

To ensure that the cycle is completed and sustained re-audit would be required to assess the effectiveness of the implemented changes and measure current practice. The repeat audit will hopefully show that changes have been implemented, and improvements made.

Ideally, the follow-up audit will demonstrate that the changes have been implemented successfully and that significant improvements have been made. The activities in this step are essentially the same as those in Step 3 (Measure current practice).

To complete the audit cycle and ensure continuous improvement, the re-audit process is essential. The activities involved in this step are similar to Step 3, which is measuring current practice. Breakdown of the key activities involved in the re-audit process is given as follows:

1. Specify the Time Period:
   - Determine the time period required to collect a sufficient number of cases for the second round of audit.
   - Ensure an appropriate timeframe that allows for a comprehensive assessment of current practice.
2. Confirm Sources:
   - Identify and confirm the sources of data, such as registers and relevant records, to identify cases for the re-audit.
   - Ensure that the selected sources provide a representative sample for evaluation.
3. Retrieve and Assess Case Notes:
   - Retrieve the case notes of the identified cases from the sources.
   - Assess the proportion of cases successfully retrieved and evaluate the completeness of the case notes.
4. Ensure Compliance with Criteria:
   - Verify that the retrieved cases comply with the defined criteria and definitions used in the audit.
   - Ensure that the cases meet the eligibility requirements for evaluation.
5. Transfer Data to Audit Form:
   - Transfer the relevant information from the retrieved case notes to the structured audit form.
   - Ensure accurate and consistent data entry for analysis.
6. Analyze Data:
   - Analyze the data collected to determine the proportions of cases that meet the criteria of good practice.
   - Compare the results with the findings from the previous audit to assess the effectiveness of implemented changes and improvements.
7. Review Findings and Present Summary:
   - Review the findings with senior obstetric, midwifery, and administrative staff.
   - Prepare a summary of the key findings to present to all staff, fostering discussion and engagement.
8. Agreement on Further Changes:
   - Engage in discussions with the team to identify areas for further changes in clinical or administrative practices.
   - Determine if additional improvements are necessary based on the re-audit results.
9. Decide on Next Steps:
   - Evaluate the value of conducting further rounds of the audit cycle.
   - Determine if revision of the criteria or development of new criteria is needed for subsequent audits.
   - Consider the need for continuous evaluation and improvement.

By completing the re-audit process, healthcare providers can assess the sustained impact of implemented changes, identify areas for further improvement, and ensure the delivery of high-quality care over time. The continuous evaluation and feedback loop of the audit cycle contributes to ongoing improvement in clinical and administrative practices.

17.17.1 Develop Tools to Sustain Improvements

Once improvements have been implemented and sustained, it is beneficial to transition from conducting full audits to implementing monitoring systems. Monitoring allows for ongoing observation and assessment of key performance indicators to ensure that the improvements remain in place and continue to deliver the desired outcomes. This can involve the development of structures and systems that integrate the monitoring process into routine clinical and administrative practices.

However, if there is a decline or deterioration in performance indicators, it is important to reactivate the full audit process. Full audits provide a more comprehensive and in-depth assessment of current practice, allowing for a thorough analysis of the factors contributing to the decline and the identification of necessary corrective actions. The reactivation of the full audit serves as a mechanism to address the issues and make appropriate changes to restore and improve the quality of care.

Therefore, a combination of monitoring systems for sustained improvements and periodic full audits for performance reassessment is recommended to ensure continuous quality improvement in healthcare settings.

Completing a clinical audit cycle involves several stages that provide a snapshot of the audit process (Fig. 17.2). Key stages include:

1. Planning: Identify the objectives, scope, and criteria for the audit. Define the data collection methods and establish a timeline for the audit cycle.
2. Data Collection: Gather relevant data according to the defined criteria. This may involve reviewing patient records, conducting surveys or interviews, and collecting other necessary information.
3. Data Analysis: Analyze the collected data to assess current practice and determine the extent to which the criteria are being met. Applying the correct statistical methods and tools is essential for interpreting data effectively and uncovering patterns or trends.

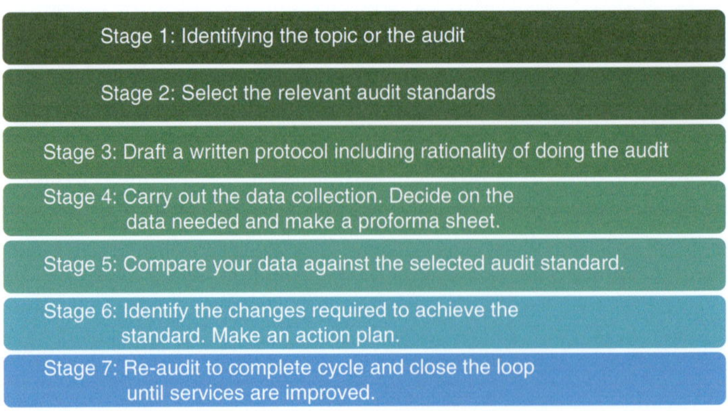

Fig. 17.2 Stages of clinical audit cycle for continuous improvement

4. Interpretation: Compare the audit findings with the established criteria and standards. Evaluate the level of compliance and identify areas where improvements are needed. Determine the reasons behind any deficiencies in meeting the criteria.
5. Implementation of Changes: Develop an action plan based on the audit findings and recommendations. Specify the changes required, assign responsibilities, set realistic timelines, and determine the evidence required to demonstrate the implementation of the changes.
6. Monitoring and Sustaining Improvements: Establish systems and structures to monitor the implemented changes and ensure their sustainability. Integrate monitoring processes into routine practices to track performance indicators and identify any potential deviations or decline in performance.
7. Re-evaluation and Feedback: Periodically re-evaluate the practice to assess the impact of implemented changes. Collect data to measure the sustained effectiveness of the improvements. Provide feedback to the healthcare team and relevant stakeholders on the progress made, areas of further improvement, and the need for ongoing quality assurance.

By completing these stages, healthcare providers can conduct a comprehensive clinical audit cycle that supports continuous quality improvement and enhances patient care.

17.18 Dysfunctional Consequences of Clinical Audit

While clinical audit is generally a valuable process for improving healthcare quality, there can be certain dysfunctional consequences that may arise if not managed effectively. Some potential dysfunctional consequences of clinical audit include:

1. Defensive Practice: Healthcare providers may become overly cautious and adopt defensive practices to avoid potential audit findings or negative feedback. This can lead to excessive testing, unnecessary procedures, or conservative decision-making, which may not always align with the best interests of the patients.
2. Target-Driven Care: Focusing solely on meeting audit targets and criteria may lead to a narrow focus on specific aspects of care, neglecting the overall patient-centered approach. Healthcare providers may prioritize achieving the audit goals rather than considering individual patient needs and preferences.
3. Data Manipulation: In an attempt to present favorable results, there is a risk of data manipulation or selective reporting. This can undermine the credibility and validity of the audit findings, making it difficult to identify areas that genuinely require improvement.
4. Resistance to Change: Healthcare professionals may resist changes recommended through the audit process, particularly if they perceive them as burdensome or conflicting with their established practices. Resistance can hinder the implementation of necessary improvements and impede progress.

5. Blame Culture: If clinical audit is used as a tool to assign blame or create a culture of punishment, it can have negative effects on the morale and motivation of healthcare providers. This can discourage open and honest reporting of errors or adverse events, hindering opportunities for learning and improvement.
6. Resource Burden: Clinical audit requires time, resources, and effort to collect and analyze data, implement changes, and monitor progress. If the burden of audit activities becomes overwhelming, it may divert healthcare professionals' attention and resources away from direct patient care.

To mitigate these dysfunctional consequences, it is crucial to approach clinical audit with a focus on continuous learning, collaboration, and quality improvement. Encouraging a supportive and non-punitive environment, involving healthcare professionals in the audit process, and providing adequate resources and training can help address these challenges and promote positive outcomes.

These actions can help reduce potential dysfunctional consequences of clinical audit:

1. Multidisciplinary Approach: Involve staff at all levels who are affected by the clinical audit, ensuring their active participation or providing relevant information about the audit process. A multidisciplinary approach fosters a sense of ownership and shared responsibility among team members.
2. Structured Program and Local Leadership: Integrate the clinical audit into a structured program with a designated local lead. This helps provide clear guidance, coordination, and accountability throughout the audit process.
3. Manageable Number of Standards: Limit the number of standards in a clinical audit to a manageable level. Overwhelming clinicians with too many standards can lead to decreased engagement and dilution of focus.
4. Evidence-Based Guidelines: Base the audit standards on good quality evidence-based guidelines. This increases the acceptability and credibility of the standards among healthcare professionals.
5. Balanced Selection of Standards: Ensure that the audit standards cover a balanced selection of care processes, outcomes, and patient satisfaction. Including standards of interest to clinicians increases their engagement and relevance.
6. Flexibility in Using Standards: Be flexible and careful in how the audit standards are used. Recognize that context and individual patient needs may require adaptation or exceptions to the standards while maintaining the overall goals of the audit.
7. Expert Interpretation: Seek expert interpretation of the audit standards, both from local experts and external, independent sources. This helps ensure accurate understanding and application of the standards.
8. Valid and Timely Feedback: Provide feedback on audit findings that is perceived as valid, credible, timely, and derived from reliable data sources. Effective feedback can motivate clinicians to initiate change and improvement.

9. Repeat Audits: Conduct repeat audits to assess whether improvements in care have been implemented as a result of the initial clinical audit. This helps track progress and identify areas that may require further attention.
10. Monitor and Sustain Improvements: Develop mechanisms and systems to monitor and sustain service improvements once the audit cycle is completed. This ensures that changes are integrated into routine practice and that improvements are maintained over time.

Further, to reduce potential dysfunctional consequences of clinical audit, the following actions can be taken:

1. Promote a Learning Culture: Emphasize that clinical audit is a learning and improvement process rather than a punitive exercise. Foster a culture that encourages open communication, collaboration, and reflection on both positive and negative findings.
2. Stakeholder Engagement: Involve all relevant stakeholders, including healthcare professionals, administrators, and patients, in the audit process. Encourage their active participation, input, and ownership of the audit findings and improvement initiatives.
3. Clear Communication: Ensure clear and transparent communication about the purpose, goals, and process of the clinical audit. Provide regular updates on progress, share findings, and explain the rationale behind any changes or recommendations resulting from the audit.
4. Training and Education: Provide training and educational opportunities to healthcare professionals on clinical audit methodologies, data collection, and analysis. This will enhance their understanding of the audit process and increase their engagement and confidence in implementing changes.
5. Emphasize Patient-Centered Care: Remind healthcare providers to prioritize patient-centered care alongside meeting audit criteria. Encourage them to consider individual patient needs, preferences, and shared decision-making in the context of audit recommendations.
6. Utilize Intrinsic Motivation: Tap into the intrinsic motivation of clinical staff by fostering a sense of performing well for the common good and aligning with their self-image and professional values. Highlight the positive impact of the audit on patient care.
7. Data Integrity and Accuracy: Establish mechanisms to ensure the integrity and accuracy of data collected during the audit. Implement quality control measures, such as regular data audits and validation processes, to minimize the risk of data manipulation or selective reporting.

   - Resource Allocation: Allocate adequate resources, including time, personnel, and technology, to support the clinical audit process. Ensure that healthcare professionals have the necessary support and tools to effectively collect, analyze, and utilize audit data.

- Continuous Improvement: Encourage a continuous improvement mindset by regularly reviewing and updating audit criteria, processes, and targets based on emerging evidence, best practices, and feedback from stakeholders. This will help maintain relevance and prevent stagnation in the audit process.
- Monitor and Address Resistance: Proactively monitor and address any resistance or reluctance to change among healthcare professionals such as lack of motivation, incomplete data collection, or other underlying causes. Develop appropriate strategies to overcome these challenges and provide necessary support. Engage in open dialogue, address concerns, and provide support and incentives to encourage participation and implementation of recommended improvements.
- Use Incentives: Plan strategies that incorporate implicit incentives to encourage participation and improvement. Recognize and reward individuals and teams for their involvement in the audit process, such as providing positive feedback, local awards, or opportunities for presenting findings at conferences or in journals.
- Celebrate Successes: Recognize and celebrate achievements and positive outcomes resulting from the clinical audit. Acknowledge the efforts of healthcare professionals and teams in driving improvement and encourage a sense of pride and motivation in their work.
- By implementing these actions, healthcare organizations can minimize the potential dysfunctional consequences of clinical audit and create a positive environment for continuous learning, quality improvement, and ultimately, better patient outcomes.

Ethical, Confidentiality, Data Protection, and Legal Issues

18

Contents

18.1 Ethical Issues

Clinical audits and quality improvement (QI) projects focus on evaluating and improving healthcare practices, processes, and outcomes without introducing additional interventions beyond standard clinical care. This is because their primary goal is to assess and enhance existing practices rather than to test new treatments or introduce changes that could impact patient care directly. Key reasons why these projects avoid patient-specific interventions:

- **Purpose of Evaluation**: Clinical audits and QI projects aim to measure adherence to established standards, guidelines, or protocols and identify areas for improvement in care delivery. They assess current practices rather than introduce new ones.

S. Sharma, *Quality Demystified*, https://doi.org/10.1007/978-981-95-0339-1_18

- **Focus on Process Improvement**: The focus is typically on improving healthcare systems, protocols, and professional behaviors rather than individual patient outcomes through new or experimental treatments.
- **Risk Avoidance**: Introducing new interventions could potentially impact patient safety or introduce risks. To protect patients, clinical audits and QI projects maintain their scope within routine care practices and rely on data collected as part of standard care.
- **Ethical Considerations**: Without introducing changes to patient treatment, clinical audits and QI projects avoid the ethical and regulatory challenges of interventions. This distinction ensures that these projects remain non-invasive and do not require additional consent beyond routine care.
- **Compliance with Guidelines**: Clinical audits are often designed to check compliance with established clinical guidelines, ensuring that patients receive evidence-based care. The goal is to identify deviations and improve adherence without altering patient management directly.
- **Data Collection Approach**: The data used in these projects is typically derived from routine records, observations, or administrative data, avoiding interference with patient treatment plans.

Unlike research, clinical audit or quality improvement projects by definition do not involve intervention to patients other than their usual clinical management. Therefore, these do not usually require formal ethical approval. Clinical audits play a pivotal role in enhancing general clinical standards. However, several ethical issues can still arise during their conduct. Key ethical considerations include:

- **Informed Consent and Patient Autonomy:** Clinical audits and QI projects often rely on patient data collected during routine care. Even though these projects do not typically introduce new interventions, and therefore, formal ethical approval may not be required for clinical audits, there is still a responsibility to respect patient autonomy. Patients should be informed when their data is being used, and in some cases, explicit consent might be needed, especially if identifiable or sensitive information is involved.
- **Privacy and Confidentiality**: Maintaining the confidentiality of patient information is crucial. During audits or QI projects, there may be a risk of unauthorized access or inappropriate handling of sensitive data. In India, healthcare data protection is governed by regulations like the IT Rules, 2011 and the pending Personal Data Protection Bill, 2019, which mandate consent, secure storage, and accountability. The ICMR Ethical Guidelines, 2017 emphasize privacy in research, while the proposed DISHA aims to regulate digital health data. Ethical guidelines like the Declaration of Helsinki and CIOMS stress privacy and confidentiality in research, ensuring patient rights and data security worldwide. Regulations play a crucial role in safeguarding digital health by ensuring the privacy, security, and ethical use of patient data (See below)
- **Data Security**: Audits and QI projects involve reviewing large sets of clinical and operational data. Ensuring that this data is securely stored and transmitted,

and accessible only to authorized personnel, is crucial to avoid data breaches or misuse.

- **Conflict of Interest**: If the audit or QI project is led by individuals with vested interests, such as financial or professional benefits from specific outcomes, there is a risk of bias. This can compromise the integrity of the findings and the proposed recommendations.
- **Impact on Healthcare Delivery and Patient Care**: Although these projects aim to improve care, the implementation of audit procedures or data collection activities can inadvertently disrupt patient care or impose additional burdens on healthcare staff. Ethical considerations should include minimizing disruptions and ensuring that patient care is not compromised.
- **Equity and Fairness**: If QI projects or audits target specific patient populations or clinical services, they must be designed to ensure fair representation and avoid introducing disparities. The findings and improvements should aim to benefit all patient groups equally.
- **Transparency in Use of Audit Results**: The results of an audit or QI project may reveal poor practices or suboptimal outcomes that could lead to reputational risks for individuals or institutions. Ethically, the results should be reported transparently and objectively, without unfairly assigning blame or overlooking key findings.
- **Accountability and Responsibility**: Ethical dilemmas can arise regarding who is accountable for implementing recommendations from audits and QI projects. It's crucial to clearly define responsibilities for action and follow-up to prevent neglect of identified issues.
- **Professional and Organizational Ethics**: The culture within an organization can influence how findings from audits or QI projects are received and acted upon. Healthcare professionals involved in clinical audit should adhere to professional standards and ethical guidelines. They should conduct audits with integrity, ensuring accuracy, honesty, and responsible use of data. There might be pressure to underreport negative findings or resist recommendations, raising ethical concerns about transparency and accountability.
- **Harm vs. Benefit**: While these projects aim to improve care, there may be situations where changes based on project outcomes could inadvertently lead to negative consequences for some patients or healthcare providers. Ethical deliberation should weigh the benefits of the proposed changes against potential risks.
- **Continuous Improvement**: Ethical considerations should also include a commitment to ongoing evaluation and improvement of the audit process itself, ensuring that it remains aligned with ethical principles and best practices.

While clinical audit may not require the same level of formal ethical approval as research, clinical audit and QI projects ethically require safeguarding patient rights, maintaining transparency, protecting privacy, ensuring fairness, and upholding the integrity of the process and findings, and ensure the overall quality of healthcare provision. Not following these principles may have far-reaching consequences for clinicians, patients, healthcare providers, and purchasers if these principles are not followed. All clinical audits must be conducted within an ethical framework and obtaining approvals from institutional review boards (IRBs) when needed, and being transparent about methodologies and findings can help mitigate these ethical concerns especially regarding confidentiality and disclosure of audit results. All staff and volunteers must adhere to a duty of confidentiality.

The starting point in any consideration of ethics is to remember that the audit project selected should benefit patients and not do any harm. There may also be times when it is uncertain where research ends, and audit begins. Pre-audit investigation may be performed to measure baseline performance if there are no standards and no prior knowledge of best practice on which to base standards. However, if at any time a comparison of treatments or introducing new treatments is involved then it becomes a research project. For details see the section on clinical audit.

- Research is concerned with discovering the right thing to do: Audit is ensuring it is done right

Privacy is an ethical consideration that is important for both families and healthcare workers. On the one hand, the deceased and her family have the right to privacy, but on the other it is almost impossible to investigate a maternal death and maintain complete privacy. Relatives and healthcare workers need to be assured that, as much as possible, their privacy will be maintained, and that the identities of the women whose deaths are being investigated, their families, and the healthcare professionals involved in their care are kept confidential, known only to those doing the actual investigation. Data collection forms, case summaries, review meetings, and any reports or dissemination of results should not contain personal identification. For example, this is particularly necessary when the death has occurred as a result of an illegal abortion.

In some cases, such as confidential enquiries, complete *anonymity* is the rule. However, in others, such as facility- and community-based case reviews, it is usual to identify both the deceased, and the health workers involved in the care. When caregivers are identified in any approach, their willingness to cooperate in subsequent investigations will be greater if confidentiality is maintained within the group carrying out the review.

- These approaches should be used as a tool to discover why patients/women are dying and to make changes to reduce these events. The audit findings should not be used to blame individuals or institutions, nor to punish persons or groups. Investigations that are carried out in a manner seeking to attribute blame for an adverse event inhibit the willingness of people to cooperate with the process.
- Conversely, health workers must be held accountable for their actions. Accountability can be encouraged by carrying out any approach in a way that

seeks to improve care by educating both the caregivers and the community. However, it will occasionally be necessary for appropriate persons (e.g., supervisors, licensing boards, medical councils) to discipline health workers who are persistently negligent, despite efforts to encourage and educate.

18.2 Confidentiality and Legal Issues

Clinical audit is an educational, confidential, non-punitive exercise. Patient records and other routinely maintained clinical records, such as nursing records, blood bank records, and operating theatre logbook entries, are primary sources of information. The confidentiality of the extraction process is crucial, usually with no mention being made of the names of the professionals providing the care.

As it is a confidential exercise, patients and relatives will not see the results of a clinical audit. This should reassure those with concerns that the endorsement of criteria may increase the risks of litigation. The confidential nature of audit is thus central to its continuing existence, with neither patients nor providers being individually identified.

Measures to protect confidentiality:

- Data Protection: Clinical audit involves the collection, storage, and analysis of patient data. It is essential to comply with applicable data protection laws including digital health data, and strict confidentiality guidelines when conducting a clinical audit to safeguard patient confidentiality and privacy. This includes obtaining informed consent for data collection and ensuring data security and confidentiality throughout the audit process.
- Anonymization and De-identification: To protect patient confidentiality, personal identifiers should be removed or anonymized from audit data. This helps prevent the identification of individual patients during data analysis and reporting. The key to these ID codes must be kept separately in a secure place; for example, it should not be kept in the same folder as the data collection sheets. Destroy the key once the audit is reported.
- Access Control: Only authorized personnel involved in the clinical audit should have access to patient data. Implementing proper access controls, such as restricted user accounts and secure storage systems, can help prevent unauthorized access or disclosure of sensitive information. The patient id code should only be decoded in case one might want to identify which data collection form corresponds to which patient. Sometimes, patient identifiable information may be required on the data collection form. For example, during a prospective data collection exercise, the data collection form might have to follow a patient through their pathway of care. In such cases record only the hospital ID number as the only patient identifiable information.
- Secure Data Storage & Transfer: Implement robust security measures to protect patient data during storage. This includes using secure servers or databases with access controls, encryption, and firewalls. Regularly updating and patching soft-

ware systems is essential to address any vulnerabilities. When sharing patient data with other parties, such as external consultants or collaborating organizations, use secure methods of data transfer. Encrypt the data during transit and use secure file transfer protocols or encrypted email services to prevent unauthorized access.

- Audit Trail: Maintain an audit trail that records activities related to patient data access, modifications, or sharing. This helps to track any unauthorized access or breaches and facilitates accountability and investigation if necessary.
- Confidentiality Agreements: Healthcare professionals and audit team members should sign confidentiality agreements or adhere to confidentiality policies to ensure that they understand their responsibilities in safeguarding patient information.
- Legal Basis for Data Processing: It is important to establish a legal basis for processing patient data in the context of clinical audit. This may include obtaining patient consent, fulfilling a legal obligation, or pursuing legitimate interests as defined by applicable data protection laws.
- Compliance with Health Records Laws: Clinical audit should adhere to local laws and regulations regarding health records management, retention, and disclosure. To ensure compliance, familiarize yourself with the specific legal requirements in your jurisdiction.
- Reporting and Disclosure: When reporting audit findings, it is crucial to present aggregated and anonymized data to prevent the identification of individual patients. Avoid disclosing sensitive or personally identifiable information unless explicit consent has been obtained.
- Audit Governance and Oversight: Establishing clear governance structures and oversight mechanisms for clinical audit can help ensure compliance with legal and ethical requirements. This may involve assigning responsibility to a designated audit lead, establishing audit committees, or seeking legal advice when necessary.
- Regular Review and Updates: Regularly review and update your policies and procedures related to patient confidentiality to stay aligned with evolving privacy regulations and best practices. Stay informed about any changes in privacy laws that may impact on your clinical audit practices.
- External Regulations and Standards: Apart from data protection and privacy laws, clinical audit may need to adhere to additional regulations and standards specific to the healthcare sector or the audit's focus area. Familiarize yourself with relevant regulations, such as those related to clinical research or quality improvement initiatives.

18.2.1 Global and Indian Regulations Protecting Digital Health

Global regulations play a crucial role in safeguarding digital health by ensuring privacy, security, and ethical use of patient data. Prominent among these is the Health Insurance Portability and Accountability Act (HIPAA) in the United States, which establishes standards for the protection of sensitive patient information. In the European Union, the General Data Protection Regulation (GDPR) provides comprehensive guidelines on data privacy, giving individuals greater control over their personal health data and imposing strict requirements on organizations handling such information. Additionally, international frameworks like the International Medical Device Regulators Forum (IMDRF) set standards for the safety and performance of digital health technologies globally.

In India, the regulatory landscape for digital health is evolving to address the unique challenges of its vast and diverse population. The Digital Personal Data Protection Act (DPDPA), inspired by GDPR, aims to protect individuals' personal data, including health information, by mandating consent, data minimization, and stringent security measures. Furthermore, the National Digital Health Blueprint (NDHB) outlines a comprehensive strategy for digitizing healthcare in India, emphasizing interoperability, data privacy, and security standards to ensure seamless and secure health information exchange. The Telemedicine Practice Guidelines, issued by the Ministry of Health and Family Welfare, provide a regulatory framework for the delivery of healthcare services through digital platforms, ensuring that telehealth practices adhere to ethical and legal standards.

18.2.2 Digital Personal Data Protection Act in India

The Digital Personal Data Protection Act (DPDPA) in India is a landmark legislation aimed at regulating and safeguarding personal data in the digital realm, with significant implications for the healthcare sector. The Act mandates that organizations, known as data fiduciaries, obtain explicit consent from individuals before processing their personal data, ensuring that patients maintain control over their sensitive health information. It enforces data localization requirements, requiring certain categories of sensitive data, including health records, to be stored within India's borders to enhance data security and sovereignty.

A key feature of the DPDPA is its emphasis on transparency and accountability. Healthcare providers and digital health platforms must implement robust data protection measures, conduct regular audits, and adhere to strict compliance protocols to prevent data breaches and unauthorized access. The Act grants individuals the right to access, correct, and delete their personal data, fostering trust and accountability within the healthcare ecosystem. Non-compliance can result in substantial fines and penalties, highlighting the government's commitment to enforcing data protection standards.

In the digital health context, the DPDPA impacts EHRs, telemedicine services, and health informatics systems by ensuring that patient data is managed securely and legally. This alignment with global standards like the EU's GDPR enhances India's position in the global digital health market. However, implementation challenges include the need for significant investments in data infrastructure, training for compliance, and ensuring interoperability while adhering to localization mandates. Small- and medium-sized healthcare providers may find these requirements particularly burdensome.

18.2.3 Challenges of Data Sovereignty in Digital Health

Data sovereignty—the principle that data is subject to the laws and governance structures within the nation where it is collected and stored—poses significant challenges in the realm of digital health. As healthcare systems increasingly rely on digital technologies for patient records, telemedicine, and health informatics, ensuring compliance with diverse and often stringent data sovereignty laws becomes complex and resource-intensive.

One major challenge is the fragmentation of international regulations. Different countries have varying requirements for data storage, processing, and transfer. For instance, the European Union's General Data Protection Regulation (GDPR) imposes strict rules on data protection and mandates that personal health data of EU citizens be stored within the EU or in countries with adequate data protection measures. In contrast, other nations may have more lenient or entirely different regulations. This disparity complicates the operations of global digital health platforms, which must navigate and comply with multiple legal frameworks simultaneously.

Data localization requirements further exacerbate these challenges. Some countries, like Russia and India, mandate that certain types of sensitive data, including health records, be stored within their borders. While intended to enhance data security and protect national interests, these requirements can increase operational costs for healthcare providers and technology firms. Establishing and maintaining local data centers or partnering with regional service providers demands significant financial and logistical investments, which may be particularly burdensome for smaller healthcare organizations.

Cross-border data transfers present another layer of complexity. Healthcare providers often collaborate internationally for research, telemedicine services, and patient care, necessitating the movement of data across borders. Ensuring that such transfers comply with both the originating and receiving countries' data sovereignty laws can be challenging. Legal mechanisms like Standard Contractual Clauses (SCCs) or Binding Corporate Rules (BCRs) under GDPR aim to facilitate compliant data transfers, but they require meticulous implementation and ongoing management to remain effective.

Security and privacy concerns are paramount in digital health, where data breaches can have severe consequences for patient privacy and trust. Adhering to data, sovereignty laws necessitate robust cybersecurity measures tailored to the specific legal requirements of each jurisdiction. This includes encryption, access controls, and regular security audits, which can be technologically demanding and costly. Moreover, differing standards across regions may lead to inconsistencies in data protection, potentially leaving gaps that malicious actors could exploit.

Technological limitations also pose significant hurdles. Integrating disparate data systems to comply with various sovereignty laws can be technically challenging. Interoperability issues arise when healthcare IT systems must communicate and share data securely across different regulatory environments. Developing and maintaining such integrated systems require advanced technical expertise and substantial investment, often out of reach for smaller healthcare entities.

The **impact on healthcare providers and patients** cannot be understated. Providers must navigate a maze of regulations to ensure compliance, diverting resources away from patient care and innovation. Patients may experience delays in accessing their health information or benefits from advanced digital health services due to the complexities of data sovereignty. Additionally, stringent data localization can limit the scalability of digital health solutions, reducing their accessibility and effectiveness in addressing public health needs.

Economic implications also play a role. Compliance with diverse data sovereignty laws can increase operational costs for healthcare organizations and technology companies, potentially leading to higher costs for patients. These financial burdens may stifle innovation and slow the adoption of cutting-edge digital health technologies that could otherwise enhance patient outcomes and streamline healthcare delivery.

By harmonizing data protection, enhancing security protocols, and promoting ethical standards, these regulations ensure that the benefits of digital health advancements are realized while minimizing risks to patient privacy and safety.

18.3 Measures to Protect Clinical Staff Confidentiality

Protecting staff confidentiality is also essential in clinical audit. It is also important to remember that clinical audit reports and presentations should not include identifiers such as names of clinicians or staff unless they have given their consent. Clinical audit reports and presentations should not identify clinicians by name unless the people concerned have agreed to this. This should be anonymized using unique identification, (ID), code in the same way as patient information, for example, Consultant 1, Consultant 2, etc. This unique ID code should be used on data collection sheets, spreadsheets, and databases. This maintains the essential spirit of audit as a non-punitive educational tool for improving the quality of care. Some measures to safeguard staff confidentiality are:

- Staff Awareness: Educate all staff involved in the clinical audit about the importance of maintaining staff confidentiality. Emphasize the ethical and legal obligations to respect the privacy and confidentiality of fellow colleagues.
- Limited Access: Limit access to staff information to authorized individuals who have a legitimate need to know. Implement role-based access controls and ensure that only authorized personnel can access and handle staff data.
- Anonymization and De-identification: Remove or modify personally identifiable information from staff records to ensure anonymity. Replace names, contact details, and other identifying information with unique identifiers or codes.

- Confidentiality Agreements: Require all individuals who have access to staff information to sign confidentiality agreements or data protection agreements. These agreements should outline their responsibilities in maintaining staff confidentiality and the consequences of any breaches.
- Secure Data Storage: Implement secure storage systems for staff data, such as password-protected databases or encrypted files. Ensure that access to these systems is limited and protected by strong passwords and other security measures.
- Training and Education: Provide training to staff members on staff confidentiality and privacy. Ensure that they understand the importance of keeping staff information confidential and the procedures in place to protect it.
- Non-disclosure Policies: Implement clear policies regarding the non-disclosure of staff information to unauthorized individuals. This includes restrictions on sharing staff information with external parties unless explicitly authorized or required by law.
- Data Transfer: When sharing staff information with other parties, use secure methods of data transfer. Encrypt the data during transit and use secure file transfer protocols or encrypted email services to prevent unauthorized access.
- Regular Review and Updates: Regularly review and update policies and procedures related to staff confidentiality to stay aligned with privacy regulations and best practices. Stay informed about any changes in privacy laws that may impact on staff confidentiality.
- Monitoring and Auditing: Implement monitoring and auditing mechanisms to ensure compliance with staff confidentiality policies. Regularly review access logs and conduct audits to identify any potential breaches or unauthorized access to staff information.

By implementing these measures, healthcare organizations can ensure that staff confidentiality is respected and protected throughout the clinical audit process. This promotes a culture of trust, privacy, and professionalism within the organization.

18.4 Measures to Protect Confidentiality Involving Patient or Staff Identifiable Data

- Always consider whether real data is needed to identify patients or staff. There may be situations where it is necessary to record patient identifiable information on the data collection form, such as during a prospective data collection exercise. In such cases, it is important to adhere to the following principles for handling patient identifiable information:
- Justify the Purpose: Clearly articulate the purpose(s) for using patient identifiable information and ensure that it aligns with the objectives of the clinical audit. The use of confidential information should be necessary and serve a legitimate purpose.
- Minimum Necessary: Only collect and use the minimum amount of patient identifiable information that is required to fulfil the objectives of the clinical audit.

Avoid collecting unnecessary personal details and use anonymized or de-identified data whenever possible.

- Strict Need-to-Know Basis: Access to patient identifiable information should be restricted to individuals who have a legitimate need-to-know for the purpose of the clinical audit. Implementing appropriate access controls is vital to ensure that only authorized personnel can access sensitive information.
- Responsibility and Awareness: All individuals involved in the clinical audit must understand their responsibilities in handling patient-identifiable information and maintaining confidentiality. Provide training and awareness programs to ensure compliance with ethical and legal requirements.
- Compliance with the Law: Familiarize yourself with the relevant laws, regulations, and guidelines governing the handling of patient identifiable information in your jurisdiction. Ensure compliance with applicable data protection and privacy laws to safeguard patient confidentiality.

By adhering to these principles, healthcare professionals can effectively handle patient identifiable information during a clinical audit while minimizing the risks to patient confidentiality and upholding ethical standards. Some other data protection considerations to keep in mind:

- Lawful and Transparent Processing: Process the data in a fair and lawful manner, ensuring that patients are aware of the purpose and scope of the clinical audit. Transparency should be maintained regarding the processing and use of patient data.
- Limited Purpose: Ensure that the data collected is used solely for the purposes of the clinical audit and not for any unrelated activities. Avoid using the data for research or other purposes without appropriate consent or legal basis.
- Data Minimization: Collect only the necessary and relevant data required for the clinical audit. Avoid collecting excessive or unnecessary information that is not directly related to the audit objectives.
- Secure Data Storage: Protect patient data by applying robust security measures, including the following actions:
- Ensure that paper documents containing patient or staff identifiable information are stored in a secure location, such as a locked room or drawer, at your workplace.
- Keep sensitive documents out of sight; do not leave them unattended on desks or other visible areas.
- Restrict removal of sensitive documents from the workplace.
- Dispose of data securely once the audit report is finalized.
- Ensure all electronic data is protected by passwords.
- Store electronic data on a secure server rather than on individual computer hard drives.
- Avoid sending data to email addresses outside the organization.
- Delete information promptly when it is no longer necessary.
- Prevent removal of patient-identifiable data from the workplace on devices like USB drives or laptops.

- Make sure reports and presentations do not include identifiable details of patients or staff.
- Patients and staff should not be identifiable in reports and presentations.
- Data Retention: Retain patient data only for as long as necessary to fulfil the objectives of the clinical audit. Once the audit is complete, securely dispose of the individual data collection forms while retaining aggregated or summarized data for future comparisons or re-audit purposes.
- Data Transfers: If there is a need to transfer data to other organizations or countries, ensure that adequate protection measures are in place. Comply with relevant data protection laws and regulations to maintain the confidentiality and security of the data during transit.
- By adhering to these data protection principles, healthcare professionals can ensure the privacy and confidentiality of patient information throughout the clinical audit process. It is important to stay informed about applicable laws and regulations in your jurisdiction to ensure compliance with data protection requirements.

18.5 Digital Health

Artificial intelligence (AI) and machine learning (ML) are transforming digital health by enabling advanced diagnostics, personalized treatment plans, and predictive analytics, which necessitates robust regulatory frameworks to ensure their safe and ethical deployment. Companies like Google, Microsoft, and Amazon are at the forefront of integrating AI into various aspects of healthcare, from diagnostics to personalized treatment plans. For instance, Google's DeepMind has developed AI systems capable of accurately diagnosing eye diseases from retinal scans, enabling earlier and more precise interventions. Microsoft's AI-driven tools assist in predicting patient deterioration by analyzing vast amounts of clinical data, allowing for timely interventions that can save lives. Amazon Web Services (AWS) provides scalable cloud-based AI solutions that support medical research, streamline administrative tasks, and improve operational efficiencies within healthcare facilities. Additionally, these tech giants are investing in machine learning and natural language processing to enhance EHRs, facilitating better data management and interoperability across different healthcare systems. However, the integration of AI in healthcare also presents significant challenges. Data privacy and security remain paramount concerns, as the handling of sensitive patient information must comply with stringent regulations like HIPAA. Additionally, integrating AI systems with existing healthcare infrastructure can be complex and costly, requiring substantial investment and collaboration between tech firms and healthcare providers. There is also the risk of algorithmic bias, where AI systems may inadvertently perpetuate existing disparities if not properly trained on diverse datasets. To address these challenges, large technology firms are implementing robust data encryption and security measures to protect patient information and ensure compliance with regulatory standards. Collaborations with healthcare professionals are being strengthened to facilitate the seamless integration of AI technologies, ensuring that these tools complement rather than disrupt existing workflows.

Globally, regulatory bodies such as the U.S. Food and Drug Administration (FDA) and the European Medicines Agency (EMA) have been actively developing guidelines to oversee AI-driven medical devices and applications. The European Union's proposed AI Act aims to classify AI systems based on risk levels, imposing stringent requirements on high-risk applications like those used in healthcare to ensure transparency, accountability, and data integrity. Similarly, the FDA has introduced a precertification program for AI software, emphasizing continuous monitoring and real-world performance evaluation to maintain safety and effectiveness.

In India, the regulatory landscape for AI in healthcare is evolving to address the unique challenges of its vast and diverse population. The DPDPA plays a crucial role in governing the use of AI by setting strict standards for data privacy and security, particularly for sensitive health information. Additionally, the Ministry of Health and Family Welfare has issued guidelines for the ethical use of AI in healthcare, focusing on ensuring that AI systems are transparent, unbiased, and inclusive. The NDHB also outlines strategies for integrating AI technologies into the healthcare infrastructure, emphasizing interoperability, data protection, and the establishment of a regulatory sandbox to test and validate AI innovations before widespread implementation.

Blockchain technology is increasingly recognized for its transformative potential in managing and securing Personal Health Records (PHR), addressing critical data security challenges within the healthcare sector. PHRs are comprehensive records maintained by individuals that encompass their medical history, diagnoses, medications, treatment plans, immunization dates, allergies, radiology images, and laboratory test results. Unlike EHRs, which are managed by healthcare providers, PHRs are controlled by patients themselves, empowering them to take an active role in their healthcare management. Integrating blockchain with PHR offers several significant advantages in enhancing data security. Firstly, blockchain's immutable nature ensures that once health data is recorded, it cannot be altered or deleted without consensus from the network, guaranteeing the accuracy and reliability of PHRs and preventing unauthorized modifications. Secondly, the decentralized control and ownership inherent in blockchain eliminates reliance on a single centralized authority, thereby reducing vulnerabilities to data breaches and cyberattacks. Patients retain ownership and control over their PHRs, deciding who can access their information through secure, permissioned access protocols. Additionally, blockchain employs robust encryption and privacy measures, utilizing advanced cryptographic techniques to protect data. Each PHR entry is encrypted and linked to previous entries, ensuring that sensitive health information remains confidential and accessible only to authorized parties, thereby safeguarding against unauthorized access and enhancing patient privacy.

Blockchain provides transparent and traceable access, as every access or modification to a PHR is recorded on the blockchain, creating an audit trail that allows patients and healthcare providers to monitor who has accessed the records and when, enhancing accountability and reducing the likelihood of unauthorized data sharing. Smart contracts, which are self-executing contracts with terms directly written into code, can automate access permissions based on predefined rules. For example, a patient can grant temporary access to a specialist for a specific treatment, and the smart contract ensures that access is revoked once the treatment is completed, maintaining strict control over data sharing. However, the integration of blockchain with PHR also presents several challenges. Scalability is a major concern, as managing large volumes of health data on a blockchain can be resource-intensive, necessitating solutions such as off-chain storage and layer-two scaling technologies. Interoperability is another significant hurdle, as integrating blockchain with existing healthcare systems and ensuring compatibility across different platforms is crucial for widespread adoption. Additionally, regulatory compliance poses challenges, as blockchain-based PHR systems must adhere to healthcare regulations like HIPAA in the United States or GDPR in Europe to ensure legal and ethical data management. User adoption and education are also critical, as patients and healthcare providers need to be informed about the benefits and functionalities of blockchain to foster trust and encourage its adoption.

Responsible innovation involves continuous assessment of the ethical implications, fostering collaboration between technologists, healthcare providers, and regulatory bodies, and maintaining a steadfast commitment to safeguarding patient privacy and promoting equitable access to digital health solutions. By addressing critical security challenges through decentralization, encryption, and transparent access controls, and by embedding ethical considerations into the development and deployment processes, AI can pave the way for a more secure, efficient, and responsible healthcare ecosystem. As technology continues to evolve, ongoing collaboration and a balanced approach to innovation will be essential to fully realize the potential of AI in safeguarding personal health information and improving patient outcomes.

Despite these challenges, several real-world applications demonstrate the promise of blockchain in enhancing PHR data security. Projects like MedRec by MIT aim to create decentralized record-keeping systems that enhance patient control and data security, while blockchain platforms such as Guardtime and Medicalchain are developing solutions that enable secure, interoperable, and patient-centric health record management. Importantly, the integration of blockchain with PHR exemplifies the principle of innovation with responsibility. This means not only advancing technological capabilities but also ensuring that such innovations are implemented ethically, transparently, and in ways that prioritize patient rights and data integrity.

18.6 Is Ethics Committee Approval for Audit Required?

The requirement for obtaining ethics committee approval for a clinical audit can vary depending on the country, institution, and the nature of the audit itself. In some cases, clinical audits may be exempted from formal ethical review because they do not involve interventions or experimental procedures on patients and are primarily focused on quality improvement. Even if formal ethical approval is not required, it is recommended to seek guidance from relevant institutional or professional guidelines and consult with local ethics committees or review boards. They can provide valuable insights and ensure that the audit adheres to ethical standards and meets the expectations of patient confidentiality and data protection. In contrast, research always requires ethical approval. Mislabeling research as a clinical audit does not remove this requirement. When determining if ethical approval is necessary, consider the following factors:

1. Nature of the Audit: Determine if the project qualifies as a clinical audit or if it is actually research. If the audit includes interventions, modifications to standard care, or experimental procedures, ethical approval is likely required. Audits solely focused on quality improvement or retrospective data analysis may be exempt, but confidentiality and informed consent should still be addressed. Clinical audits aim to answer questions like "Are we following best practices?" or "What impact are these practices having on patients?" They do not involve:
   - Introducing new treatments or practices.
   - Using control groups or placebos.
   - Disturbing patient care beyond standard clinical management.
   - Randomly assigning patients to different treatments.
   - Audits may, however, involve patient input at various levels.
2. Informed Consent: Evaluate if obtaining informed consent from patients is necessary or if it can be appropriately waived. If patient data is being collected prospectively or if identifiable information is involved, obtaining informed consent may be required to ensure transparency and respect for patient autonomy.
3. Professional Guidelines: Consult relevant professional bodies or associations that provide guidance on clinical audits. They may have specific recommendations or ethical frameworks that can help inform your decision-making process.
4. Applicable Regulations and Guidelines: Review the regulations, laws, and guidelines specific to your country or region regarding clinical audits. Different jurisdictions may have varying requirements for ethical approval, and it is crucial to understand and comply with the relevant regulations.
5. Institutional Policies: Check with your institution or organization to determine if they have specific policies or requirements regarding ethical approval for clinical audits. Some institutions may have their own guidelines in place, even if it is not mandated by external regulatory bodies.

Assess these factors carefully and seek guidance from relevant authorities, such as ethics committees or institutional review boards, when in doubt. They can provide expert advice and help ensure that ethical considerations are adequately addressed in the planning and conduct of the clinical audit.

18.6.1 Audit Project Involving Patient Survey

Patients may be invited to participate in surveys to assess whether clinical standards are being met. These surveys could cover sensitive topics, so ethical considerations are crucial. Surveys should be designed to minimize disruption for patients, and all surveys—whether for staff or patients, and administered by post, in the hospital, or in one-on-one interviews—should be reviewed for appropriateness. All structured surveys, staff or patient, administered by post, in hospital, or via a one-to-one interview, undertaken for clinical audit should be reviewed. When undertaking a survey, it is important to consider the patient's, service users', or carers' rights, dignity, and time.

Ethical questions that one should consider are:

- Is a survey appropriate? Is the survey necessary, or can the information be sourced elsewhere? Is one group of patients being excessively targeted? Surveys should focus on gathering information that isn't available from other sources and is directly related to care processes or outcomes. Essentially, they should assess whether the standards of best practice are being adhered to and the impact of these practices on patient care. Will there be value addition to the clinical audit project? Carrying out a survey unnecessarily, or asking more questions than necessary, does not add value to your project.
- Are the questions worthwhile asking? Keep your questionnaire succinct. Limit the questionnaire to meaningful questions, ideally 20 or fewer. Avoid overwhelming patients, as lengthy surveys may deter participation.
- How to ensure that the patients are alive? Check your sample, particularly in case of postal survey, check against the latest hospital records, as sending questionnaires to deceased patients is a frequent error that can cause distress to relatives. Similarly, inaccurate recording of patients' names or other personal details can be offensive. These errors may convey a lack of care and attention to detail, undermining trust and rapport.
- Are the numbers involved sufficient?
- How to ensure patient confidentiality and communicate this to the patient? Safeguard patient confidentiality by anonymizing responses where possible. Include a cover letter explaining the survey purpose, contact information for questions, a confidentiality statement, and instructions on questionnaire return, a statement on how patient's confidentiality will be protected, ideally this should be achieved through developing completely anonymized questionnaires, and what one intends to do with the data collected.
- Will the survey cause minimal disturbance? Keep it simple and don't collect more data than necessary. Ensure surveys do not interfere with patient treatment or cause patients to question their care, potentially impacting their treatment outcomes.
- How will you make sure that the patient is aware that their participation is entirely voluntary and their decision to take part or not will not affect their treatment? How will you ensure that patients consented freely for taking part in the project?

Patients must not be coerced into taking part and must have the option to decline or withdraw without affecting the care that they receive in any way detriment or antagonism Clearly state that participation is voluntary and that the decision to participate or not will not affect treatment. It's important to remind participants that they have the right to withdraw from study at any time, without needing to provide a reason.
- Is your survey addressing a sensitive topic? When addressing sensitive health matters, frame questions carefully to avoid unintended distress or discomfort for participants. Sensitive topics should be handled respectfully to prevent any potential harm.

When conducting patient surveys as part of a clinical audit, there are several ethical concerns to consider:

1. Informed Consent: Obtain informed consent from patients before including them in the survey. Patients should be fully informed about the purpose of the survey, how their responses will be used, and any potential risks or benefits associated with participating.
2. Confidentiality and Anonymity: Ensure that patient responses are kept confidential and anonymous. Use unique identifiers or codes instead of personal identifying information to protect patient privacy. Protect the data and restrict access to it, ensuring only authorized personnel can view it.
3. Voluntary Participation: Participation in the survey should be voluntary, and patients should not face any pressure or coercion to participate. They should have the right to decline or withdraw from the survey at any time without any negative consequences to their care.
4. Data Security: Implement appropriate measures to secure the survey data, both during data collection and storage. This includes using secure platforms, encrypting data if necessary, and preventing unauthorized access.
5. Data Use and Reporting: Clearly communicate how the survey data will be used and reported. Ensure that the data will be used solely for the purpose of quality improvement and enhancing patient care. Avoid disclosing individual patient information in any public reports or publications.
6. Ethical Review: Depending on the nature and scope of the survey, it may be necessary to seek ethical review and approval from an ethics committee or institutional review board. This is particularly important if the survey involves sensitive topics or vulnerable populations.
7. Cultural Sensitivity: Take cultural considerations into account when designing the survey and interpreting the results. Ensure that questions and response options are culturally appropriate and sensitive to diverse patient backgrounds.
8. Transparency and Feedback: Provide feedback to patients regarding the survey results and any actions taken as a result of the survey. Maintain transparency throughout the process and demonstrate respect for patient perspectives and experiences.

By addressing these ethical concerns, patient surveys as part of a clinical audit can be conducted in a responsible and respectful manner, ensuring patient rights and promoting trust between healthcare providers and patients.

18.7 Publications of Audit Project Results

Publications of clinical audit projects serve as a means to disseminate the findings and outcomes of the audit to the broader healthcare community. They provide an opportunity to share knowledge, promote transparency, and contribute to the overall improvement of healthcare quality and patient outcomes. Clinical audits are often published because their topics or methodologies can interest a broader audience. They can showcase how an audit cycle, initially yielding poor results against standards, was effectively adopted by making necessary changes and later demonstrated significant improvement in practice through a re-audit. Although clinical audit projects can sometimes be published without ethical approval, it is important to note that some journals may refuse to publish these quality improvement reports if there are ethical concerns or if ethical approval has not been obtained. If publication on results of the project is the focus of the publication, that means it is a research activity, rather than a clinical audit project, which requires ethical approval. These issues should be considered at the beginning of the project itself. When publishing clinical audit projects, it is essential to consider certain aspects:

1. Ethical Considerations: Ensure that patient confidentiality and privacy are protected when reporting the findings. Patient information should be de-identified and anonymized to prevent any potential breach of confidentiality.
2. Consent and Permissions: Obtain appropriate consent from patients, healthcare professionals, and other stakeholders involved in the audit project, particularly if their identifiable information or personal experiences are being shared in the publication.
3. Data Integrity and Validity: Verify that the data presented in the publication is accurate, reliable, and valid to ensure the integrity and credibility of the research. Clearly describe the methods used for data collection, analysis, and interpretation to provide transparency and enable replication by other researchers.
4. Authorship and Acknowledgments: Clearly define the criteria for authorship and acknowledge the contributions of all individuals involved in the audit project, including clinicians, researchers, and data collectors. Give credit to those who have made substantial contributions without meeting the criteria for authorship.
5. Conflict of Interest: Disclose any potential conflicts of interest that may exist among the authors or funding sources related to the audit project. Ensuring transparency in reporting potential conflicts of interest is important for maintaining the credibility and integrity of a publication.
6. Peer Review: Submit the publication to a peer-reviewed journal to undergo rigorous evaluation by experts in the field. Peer review helps ensure the quality, validity, and relevance of the published work.

7. Clear and Accessible Reporting: Present the findings in a clear and concise manner, using appropriate statistical analyses, charts, and tables. Write the publication in a way that is accessible to a wide range of readers, including clinicians, researchers, and policymakers.
8. Impact and Generalizability: Discuss the implications of the audit findings and their potential impact on clinical practice. Consider the generalizability of the results to other healthcare settings and populations and clearly communicate any limitations of the audit project.
9. Collaboration and Collaboration: Encourage collaboration and knowledge sharing by involving multiple stakeholders and institutions in the publication process. Collaborative efforts can enhance the credibility and impact of the publication.

Remember to comply with the ethical guidelines and policies of the journal in which you plan to publish your clinical audit project. Adhering to these considerations will help ensure the ethical and responsible dissemination of audit findings while contributing to the improvement of healthcare quality and patient care.

Success Stories 19

Contents

S. Sharma, *Quality Demystified*, https://doi.org/10.1007/978-981-95-0339-1_19

Sharing success stories plays an important role in promoting patient safety and enhancing quality in healthcare. These narratives highlight best practices, innovative solutions, and successful interventions, inspiring healthcare professionals to adopt proven approaches. By showcasing real-world examples of improvements, success stories create a culture of learning, encouraging teams to reflect on their practices and embrace positive changes. They also serve to recognize and motivate staff, reinforcing the importance of safety and quality in patient care. Ultimately, sharing these stories helps build trust among patients, staff, and stakeholders, fostering collaboration and a commitment to continuous improvement.

Accreditation of Primary Health Centres (PHCs) In India, the government has implemented a quality improvement initiative focused on accrediting Primary Health Centres (PHCs) to ensure the delivery of high-quality healthcare services at the grassroots level. The National Accreditation Board for Hospitals and Healthcare Providers (NABH) has been involved in the accreditation process. Through this initiative, PHCs are assessed against predefined quality standards and criteria, including infrastructure, human resources, patient care, infection control, and management processes. PHCs that meet the accreditation requirements are awarded certification, indicating their commitment to delivering quality healthcare.

One successful case study is the accreditation of a PHC in a rural area of Maharashtra. The facility underwent a comprehensive assessment and implemented various quality improvement measures, such as ensuring the availability of essential medicines, implementing infection control practices, and enhancing patient engagement through community outreach programs.

As a result of the accreditation process, the PHC witnessed improvements in several areas, including increased patient satisfaction, improved infection control practices, better management of medical records, and enhanced staff training and development. The accreditation also helped attract more patients and build trust within the community.

Another example is standardization of Surgical Practices in a Hospital. A hospital in Delhi implemented a quality improvement initiative to standardize surgical practices and enhance patient safety. The aim was to minimize variations in surgical procedures, improve outcomes, and reduce complications. The hospital established a multidisciplinary team comprising surgeons, anesthetists, nurses, and quality improvement experts. The team conducted a comprehensive analysis of surgical processes, identified areas of improvement, and developed standardized protocols and guidelines for preoperative, intraoperative, and postoperative care.

To implement the standardized practices, the hospital organized training sessions for the surgical team, emphasizing the importance of adhering to the protocols. Regular audits and performance reviews were conducted to monitor compliance and identify opportunities for further improvement.

The results of this quality improvement initiative were remarkable. The hospital experienced a significant reduction in surgical site infections, improved patient outcomes, and increased patient satisfaction. The standardized practices also facilitated seamless coordination among the surgical team and enhanced the overall efficiency of surgical services.

These case studies highlight the successful implementation of quality improvement initiatives in India's healthcare system. By focusing on accreditation, standardization, and continuous monitoring, healthcare facilities have been able to enhance the quality of care, improve patient outcomes, and build trust among the communities they serve.

I. A few success stories that demonstrate how clinical audits have contributed to improving patient outcomes and safety:

19.1 The Keystone ICU Project (United States)

The Keystone ICU Project is a landmark initiative aimed at reducing catheter-related bloodstream infections (CRBSIs) in intensive care units (ICUs) across Michigan. Through the use of a standardized checklist and improved adherence to infection prevention protocols, the project achieved remarkable results, reducing CLABSI rates by 66% and saving an estimated 1500 lives annually. The key points are:

Objective The project aimed to improve patient safety by reducing the incidence of CRBSIs through the implementation of evidence-based practices.

Interventions The project introduced several key interventions, including:

- **Use of Checklists**: Implementing a checklist to ensure adherence to infection control practices.
- **Safety Culture**: Promoting a culture of safety among healthcare staff.
- **Daily Goals Sheet**: Improving communication among staff by using a daily goals sheet.

Results The project achieved remarkable success, nearly eliminating CRBSIs in participating ICUs. The infection rates dropped significantly, demonstrating the effectiveness of the interventions.

Sustainability The reductions in infection rates were sustained over time, highlighting the long-term impact of the project on patient safety.

The Keystone ICU Project is a prime example of how targeted quality improvement initiatives can lead to substantial improvements in patient outcomes and safety.

For details read:

1. BMJ - Sustaining reductions in catheter related bloodstream infections
2. BMJ - Impact of a statewide intensive care unit quality improvement
3. Oxford Academic - Reducing Catheter-Related Bloodstream Infections
4. AHRQ - Impact of a statewide intensive care unit quality improvement initiative

19.2 The Safer Patients Initiative (United Kingdom)

The **Safer Patients Initiative (SPI)** was a pioneering program launched by the Health Foundation in the UK to enhance patient safety across hospitals. Some key points about the initiative:

Objective The primary goal of SPI was to reduce avoidable harm to patients by improving safety practices and creating a culture of safety within healthcare organizations.

Phases and Scope

- **Phase 1**: Launched in 2004, this phase involved four hospitals and focused on improving safety in specific clinical areas such as general ward care, critical care, perioperative care, and medicines management.
- **Phase 2**: Expanded in 2006 to include 20 additional hospitals, aiming for a 30% reduction in adverse events and a 15% reduction in mortality over 20 months.

Key Interventions

- **Use of Checklists**: Implementing checklists to ensure adherence to safety protocols.
- **Safety Culture**: Promoting a culture where staff feel empowered to identify and address safety issues.
- **Leadership Engagement**: Involving senior management to prioritize and support safety initiatives.

Outcomes and Impact The initiative led to significant improvements in patient safety at the micro-system level, such as specific wards or units. It raised awareness and galvanized action around patient safety, although organization-wide impact required more time and resources.

Lessons Learned The SPI highlighted the need for a comprehensive approach to patient safety, involving changes at every level of the healthcare system. It also underscored the importance of sustained effort and resource allocation to achieve lasting improvements.

The Safer Patients Initiative set the stage for future patient safety programs by demonstrating the potential for large-scale improvements through systematic and collaborative efforts.

For details read:

1. Health Foundation - Safer Patients Initiative
2. Health Foundation - Learning Report on Safer Patients Initiative

19.3 National Surgical Quality Improvement Program (NSQIP) (United States)

The NSQIP has been a significant success story in improving surgical outcomes in the United States. NSQIP is a risk-adjusted, outcomes-based program that uses clinical audit data to improve surgical care.

Origins and Development NSQIP was initially developed by the Department of Veterans Affairs (VA) in the late 1980s to address concerns about the quality of surgical care in VA hospitals. The program began as the National VA Surgical Risk Study (NVASRS) and evolved into NSQIP in 1994.

Data-Driven Approach NSQIP collects detailed data on surgical outcomes, including preoperative, intraoperative, and postoperative variables from participating hospitals. This data is risk-adjusted to account for patient differences, allowing for accurate comparisons and improvements.

Impressive Results The program has led to significant improvements in surgical care. For example, VA hospitals saw a 47% reduction in postoperative mortality and a 43% reduction in morbidity rates from 1991 to 2006. In the private sector, hospitals participating in NSQIP have reported reductions in wound complications, surgical site infections, and urinary tract infections.

Wider Adoption The success of NSQIP in VA hospitals caught the attention of the private sector, leading to its adoption by non-VA hospitals. This broader implementation has helped improve surgical outcomes across various healthcare settings.

Continuous Improvement NSQIP's focus on continuous quality improvement has made it a model for other quality improvement programs. Hospitals use the data to identify areas for improvement, develop protocols, and implement changes to enhance patient care.

Participating hospitals collect data on surgical procedures and outcomes, which are analyzed to identify areas for improvement. By implementing targeted interventions and sharing best practices, NSQIP has shown significant reductions in surgical complications, mortality rates, and hospital costs.

For details read:

1. History | ACS (facs.org)
2. ACS NSQIP | ACS (facs.org)
3. Velanovich V, Rubinfeld I, Patton JH, Ritz J, Jordan J, Dulchavsky S. Implementation of the National Surgical Quality Improvement Program: Critical Steps to Success for Surgeons and Hospitals. American Journal of Medical Quality. 2009;24(6):474-479. https://doi.org/10.1177/1062860609339937

19.4 Enhancing Recovery After Surgery (ERAS) Programs

ERAS is a multimodal, evidence-based approach to perioperative care aimed at optimizing patient outcomes and accelerating recovery. Clinical audits play a crucial role in monitoring adherence to ERAS protocols and identifying areas for improvement. Numerous studies have demonstrated the effectiveness of ERAS programs in reducing complications, shortening hospital stays, and improving patient satisfaction. Some key aspects of ERAS programs:

Multimodal Approach ERAS programs use a combination of techniques to reduce surgical stress and promote faster recovery. This includes optimizing nutrition, minimizing fasting, using multimodal pain management strategies, and encouraging early mobilization.

Preoperative Preparation Patients are educated about the surgery and recovery process. Preoperative optimization includes nutritional support, smoking cessation, and physical conditioning.

Intraoperative Care During surgery, techniques such as minimally invasive surgery, optimal fluid management, and maintaining normothermia are employed to reduce the body's stress response.

Postoperative Care Post-surgery, the focus is on early mobilization, resuming normal diet as soon as possible, and effective pain management with minimal use of opioids.

Improved Outcomes Studies have shown that ERAS programs lead to shorter hospital stays, reduced complications, and faster return to normal activities.

ERAS programs represent a significant shift in perioperative care, emphasizing a holistic and patient-centered approach to enhance recovery.

For details read:

1. AANA | Enhanced Recovery After Surgery
2. Enhanced Recovery After Surgery (ERAS) Program | Community Health Network (ecommunity.com)
3. Vikram Kate, Gurushankari Balakrishnan, S Sureshkumar. Enhanced Recovery After Surgery (ERAS) in Emergency Abdominal Surgery. http//Enhanced Recovery After Surgery (ERAS) in Emergency Abdominal Surgery: Background, Preoperative Components of ERAS, Intraoperative Components of ERAS (medscape.com)

19.5 Hand Hygiene Compliance Initiatives

Hand hygiene is a critical aspect of infection prevention in healthcare settings. Several hospitals and healthcare systems have implemented clinical audits to monitor hand hygiene compliance and identify opportunities for improvement. Through ongoing monitoring, feedback, and education, these initiatives have successfully increased hand hygiene compliance rates, leading to reduced healthcare-associated infections.

Hand hygiene initiatives aimed at improving public health and reducing the spread of infections:

1. **Hand Hygiene for All Global Initiative**:
   - **Overview**: Launched by the World Health Organization (WHO) and UNICEF, this initiative aims to implement global recommendations on hand hygiene to prevent and control the COVID-19 pandemic and ensure lasting infrastructure and behavior changes.
   - **Key Components**: It focuses on securing political leadership, strengthening institutional policies, ensuring the availability of hand hygiene stations, and promoting evidence-based behavior change.
2. **CDC's Global Water, Sanitation, and Hygiene (WASH) Program**:
   - **Overview**: The Centers for Disease Control and Prevention (CDC) partnered with various organizations to enhance WASH activities in healthcare, community, and household settings across multiple countries.
   - **Key Activities**: The program included assessing baseline WASH conditions, implementing new hygiene programs, and monitoring their impact to reduce the spread of COVID-19 and other diseases.
3. **SAVE LIVES: Clean Your Hands Campaign**:
   - **Overview**: This WHO campaign is part of a broader effort to improve hand hygiene in healthcare settings.
   - **Key Activities**: It promotes the use of alcohol-based hand rubs and proper handwashing techniques among healthcare workers to prevent healthcare-associated infections.
4. **National Hand Hygiene Initiative (Australia)**:
   - **Overview**: Implemented by the Australian Commission on Safety and Quality in Health Care, this initiative aims to improve hand hygiene compliance in healthcare settings.
   - **Key Components**: It includes educational programs, compliance monitoring, and the promotion of alcohol-based hand rubs.

These initiatives highlight the global effort to improve hand hygiene practices, which are crucial for preventing infections and promoting public health.

For details read:

1. WHO - Hand Hygiene for All Global Initiative
2. CDC - Global WASH Program
3. Australian Commission on Safety and Quality in Health Care - National Hand Hygiene Initiative

19.6 KayaKalp Initiative (India)

The **KayaKalp initiative** is a significant program launched by the Ministry of Health and Family Welfare, Government of India, aimed at rejuvenating public health facilities across the country. Some key aspects of the initiative are produced below:

Objective The primary objective of KayaKalp is to enhance cleanliness, hygiene, and infection control within public health facilities. It aims to enhance the overall quality of healthcare services by ensuring that hospitals and clinics maintain high standards of cleanliness.

Incentives and Recognition The initiative includes a system of awards and incentives for public health facilities that demonstrate excellence in maintaining cleanliness and hygiene. This encourages healthy competition between facilities to improve their standards.

Assessment and Monitoring Facilities are regularly assessed based on specific criteria, including waste management, sanitation, and infection control practices. The assessment process helps identify areas for improvement and ensures that standards are consistently met.

Impact Since its launch, KayaKalp has led to significant improvements in the cleanliness and hygiene of public health facilities. It has also raised awareness about the importance of maintaining a clean and safe environment for both patients and healthcare workers.

The KayaKalp initiative is part of a broader effort to improve public health infrastructure and ensure that healthcare services in India are safe, clean, and effective.

For details read:

1. Vikaspedia - Kayakalp: India's National Initiative for Rejuvenating Public Health Facilities
2. Godigit - Kayakalp Scheme: Objectives, Guidelines & Assessment Process
3. National Health Mission - Guidelines for Implementation of "Kayakalp"

II. Few reports that highlight the outcomes of audit and quality improvement initiatives in healthcare:

19.7 Improving Patient Outcomes Through Clinical Audit: A Review of Case Studies (National Institute for Health and Care Excellence, UK)

This report provides a comprehensive review of case studies from different healthcare settings that demonstrate the positive impact of clinical audit on patient outcomes in areas such as infection control, medication safety, surgical care, and chronic disease management. For example,

Stroke Care One notable example is the improvement in stroke care. Clinical audits have led to better adherence to guidelines, resulting in reduced mortality and improved recovery rates for stroke patients.

Lung Cancer Audits in lung cancer care have helped standardize treatment protocols, leading to earlier diagnosis and better survival rates.

Inpatient Falls The National Audit of Inpatient Falls (NAIF) has been instrumental in reducing the incidence of falls in hospitals by implementing targeted interventions based on audit findings

Implementation and Feedback Effective use of audit data involves continuous monitoring, feedback to healthcare providers, and engaging frontline staff in the improvement process. This approach ensures that changes are sustainable and lead to long-term improvements in patient care.

Challenges and Recommendations While clinical audits have shown significant benefits, challenges such as resource limitations, staff engagement, and the need for continuous education and training remain. NICE recommends integrating audits into the overall quality improvement strategies of healthcare organizations to maximize their impact

For details read

1. BMJ - Revitalising audit and feedback to improve patient care
2. BMC Health Services Research - How do healthcare providers use national audit data for improvement?

19.8 The Impact of Clinical Audit on Quality Improvement in Healthcare: A Systematic Review (Journal of Healthcare Quality Research)

This systematic review examines the impact of clinical audit on quality improvement across a range of healthcare domains. The report summarizes the findings of multiple studies and highlights the significant positive outcomes achieved through audit-driven interventions, including reduced mortality rates, improved patient safety, and enhanced clinical effectiveness. Some key findings from the review:

Effectiveness of Clinical Audits The review highlights that clinical audits are effective tools for improving healthcare quality. They help identify gaps in care, promote adherence to clinical guidelines, and facilitate the implementation of best practices.

Mechanisms of Improvement Clinical audits work through several mechanisms, including raising awareness among healthcare professionals, providing actionable feedback, and fostering a culture of continuous improvement.

Case Studies and Examples The review includes various case studies demonstrating the positive impact of clinical audits. For instance, audits in infection control have led to significant reductions in hospital-acquired infections.

Challenges and Recommendations Despite their benefits, clinical audits face challenges such as resource constraints, variability in implementation, and the need for sustained engagement from healthcare staff. The review recommends integrating audits into broader quality improvement frameworks and ensuring ongoing support and training for healthcare professionals.

For details read:

1. BMJ - Revitalising audit and feedback to improve patient care.
2. PLOS ONE - Understanding how and why audits work in improving the quality of hospital care
3. BMC Health Services Research - The impact of audit and feedback to support change behaviour in healthcare

19.9 Improving Quality and Safety Through Clinical Audit: A Comprehensive Report (Healthcare Quality Improvement Partnership, UK)

This comprehensive report presents the findings from a large-scale clinical audit program conducted in the UK. It showcases the outcomes achieved through audit cycles, including improvements in patient care, reduction in adverse events, and enhanced compliance with best practices. The report also emphasizes the importance of sustained audit activity in driving continuous quality improvement.

Definition and Purpose Clinical audits are systematic reviews of care against explicit criteria, followed by the implementation of change. They aim to improve patient outcomes by ensuring that clinical practice aligns with established standards.

Stages of Clinical Audit

Preparation and Planning: Identifying the audit topic, setting objectives, and planning the audit process.

Measuring Performance: Collecting and analyzing data to compare current practices against standards.

Implementing Change: Developing and applying strategies to address identified gaps.

Sustaining Improvement: Monitoring the impact of changes and ensuring continuous improvement.

Case Studies and Examples The report includes various case studies demonstrating the effectiveness of clinical audits. For instance, audits in infection control have led to significant reductions in hospital-acquired infections.

Challenges and Recommendations The report acknowledges challenges such as resource constraints and the need for staff engagement. It recommends integrating clinical audits into broader quality improvement frameworks and providing ongoing support and training for healthcare professionals.

Impact on Patient Safety Clinical audits have been shown to improve patient safety by identifying areas for improvement, promoting adherence to best practices, and reducing variability in care.

For details read

HQIP - Best Practice in Clinical Audit

19.10 Impact of Medication Safety Audits in Hospitals: A Multicentre Study (Journal of Patient Safety & Quality Improvement)

This study evaluates the impact of medication safety audits in hospitals and their effect on patient outcomes. Some key findings from the study:

Scope and Methodology The study involved 11 hospitals where staff tested an audit tool on 321 patients receiving a total of 17,361 doses of medication. The audit data was used to inform improvements in practice and for accreditation purposes.

Key Outcomes

- **Reduction in Medication Errors**: The audits helped identify and reduce medication errors, contributing to safer medication practices.
- **Improved Compliance**: Hospitals reported better compliance with medication safety protocols as a result of the audits.
- **Enhanced Patient Safety**: The study found that medication safety audits led to measurable improvements in patient safety, including a decrease in adverse drug events.

Feedback and Implementation Feedback from hospital staff indicated that the audit data was valuable for informing practice improvements. The study emphasized the importance of continuous monitoring and feedback to sustain these improvements.

Challenges and Recommendations The study also highlighted challenges such as the need for adequate resources and staff training. It recommends integrating medication safety audits into regular quality improvement processes to ensure ongoing benefits.

This study underscores the significant impact that medication safety audits can have on improving patient safety and reducing medication errors in hospital settings.

For details read:

Multicentre study to develop a medication safety package for decreasing medication errors

The findings demonstrate that medication safety audits led to significant reductions in medication errors, adverse drug events, and improved patient safety.

19.11 Clinical Audit and Its Role in Improving Patient Safety: A Case Study from a Tertiary Hospital (International Journal of Healthcare Management)

This case study explores the implementation of a clinical audit program in a tertiary hospital and its impact on patient safety. The report highlights how the audit process identified system failures, facilitated targeted interventions, and resulted in improved patient outcomes and reduced adverse events. Some key insights from the study:

Objective and Methodology The study aimed to evaluate the impact of clinical audits on patient safety by examining various departments within a tertiary hospital. The methodology included a combination of quantitative and qualitative data collection, such as patient record reviews, staff surveys, and direct observations.

Key Findings

- **Reduction in Adverse Events**: The clinical audits led to a noticeable reduction in adverse events, such as medication errors and surgical complications.
- **Improved Compliance**: There was an increase in compliance with clinical guidelines and protocols, which contributed to better patient outcomes.
- **Enhanced Safety Culture**: The audits fostered a culture of safety among healthcare staff, encouraging continuous improvement and vigilance.

Implementation and Feedback The study highlighted the importance of regular feedback and follow-up actions based on audit findings. Engaging healthcare professionals in the audit process was crucial for sustaining improvements.

Challenges and Recommendations The case study also identified challenges such as resource constraints and the need for ongoing training. It recommended integrating clinical audits into the hospital's overall quality improvement strategy to maximize their effectiveness.

This case study underscores the significant role that clinical audits can play in improving patient safety and quality of care in hospital settings.

For details read:

1. International Journal of Healthcare Management - Clinical Audit and Its Role in Improving Patient Safety

19.12 Improving Patient Outcomes Through Effective Hospital Administration

This article explores the critical role of hospital administration in enhancing patient outcomes. Key points from the review are:

Role of Hospital Administration The review emphasizes that effective hospital administration is crucial for improving patient safety, satisfaction, and overall well-being. Administrative decisions and practices significantly impact on the quality of care provided.

Key Components and Strategies

Patient-Centered Care: Focusing on patient-centered care and interdisciplinary collaboration is essential for better outcomes.

Data-Driven Measurement: Utilizing data-driven measurement and benchmarking helps in assessing hospital performance and fostering continuous improvement.

Future Trends and Challenges

Emerging Technologies: The review discusses the potential of emerging technologies and evolving healthcare policies to drive changes in hospital administration.

Healthcare Disparities: Addressing healthcare disparities and promoting value-based care are highlighted as ongoing challenges.

Recommendations: The authors call for a collective commitment from healthcare leaders and policymakers to prioritize the development of capable administrators, invest in technology, and ensure continuous improvement in patient care.

This comprehensive review underscores the importance of effective hospital administration in shaping the future of healthcare and improving patient outcomes.

For details read:

1. Bhati D, Deogade MS, Kanyal D. Improving Patient Outcomes Through Effective Hospital Administration: A Comprehensive Review. Cureus. 2023 Oct 26;15(10):e47731. https://doi.org/10.7759/cureus.47731. PMID: 38021686; PMCID: PMC10676194.

III. A few success stories from India that showcase the impact of clinical audits on improving patient outcomes and safety:

19.13 Surgical Site Infection (SSI) Reduction Program

In a tertiary care hospital in India, a clinical audit was conducted to assess and improve compliance with infection prevention protocols in surgical procedures. Through the implementation of evidence-based practices, including preoperative antibiotic prophylaxis, proper surgical hand hygiene, and wound care, the hospital achieved a significant reduction in surgical site infections. The audit-driven interventions resulted in improved patient outcomes and reduced healthcare costs.

For details read:

Seidelman JL, Mantyh CR, Anderson DJ. Surgical Site Infection Prevention: A Review. *JAMA*. 2023;329(3):244–252. https://doi.org/10.1001/jama.2022.24075.

19.14 Maternal and Neonatal Care Audit

In a government hospital in India, a clinical audit was conducted to evaluate the quality of maternal and neonatal care services. The audit focused on monitoring adherence to best practices in antenatal care, delivery, and postnatal care. Through the identification of gaps and implementation of evidence-based interventions, the hospital witnessed a reduction in maternal and neonatal complications, improved birth outcomes, and enhanced patient safety.

For details read:

1. FINAL LOGOS (qualityofcarenetwork.org)
2. Suman D Hegde, Nanjundappa S Harshavardhana, Mamatha M Hegde, Dinesh R Hegde. An Audit of Specialized Newborn Care Unit's (SNCU) Performance at a Taluk Hospital in India. Glob J of Ped & Neonatol Car. 3(3): 2021. GJPNC. MS.ID.000563.

19.15 Medication Safety Audit

A healthcare organization in India conducted a clinical audit to evaluate medication safety practices in their hospitals. The audit assessed various aspects, including medication prescribing, dispensing, administration, and monitoring. By identifying medication errors, developing targeted interventions, and improving medication management processes, the organization significantly reduced adverse drug events, medication-related harm, and enhanced patient safety.

19.16 Hospital-Acquired Infection (HAI) Control Initiative

In a multi-specialty hospital in India, a clinical audit was conducted to address hospital-acquired infections. The audit focused on monitoring hand hygiene compliance, disinfection practices, and adherence to infection control protocols. Through the implementation of robust infection prevention measures and regular audit-based feedback, the hospital achieved a substantial reduction in HAIs, leading to improved patient outcomes and reduced healthcare costs.

For details read:

1. Nair V, Sahni AK, Sharma D, Grover N, Shankar S, et al. Point prevalence & risk factor assessment for hospital-acquired infections in a tertiary care hospital in Pune, India," Indian Journal of Medical Research, 2017 Jun;145(6):824-832. https://doi.org/10.4103/ijmr.IJMR_1167_15. PMID: 29067985; PMCID: PMC5674553.

19.17 Emergency Department Wait Time Reduction

In a large hospital in India, a clinical audit was conducted to address long wait times in the emergency department (ED). The audit identified bottlenecks in the triage process, patient flow, and resource utilization. Key points are summarized below:

Reducing wait times in the ED is a critical challenge for large hospitals in India, as it directly impacts patient outcomes and satisfaction. Some effective strategies that have been implemented to address this issue are as under:

1. **Streamlined Triage Processes**: Implementing efficient triage systems helps prioritize patients based on the severity of their conditions. For instance, the All India Institute of Medical Sciences (AIIMS) in New Delhi successfully reduced their median waiting time to triage from 50 minutes to less than 30 minutes by enhancing staff training and reducing turnover.
2. **Optimized Patient Flow**: Optimizing patient flow within the ED and across the hospital is essential. This involves coordinating with other departments to ensure timely admissions and discharges, thus freeing up beds for new patients. Simulation modeling and optimization techniques have been used to allocate resources more effectively.
3. **Use of Technology**: Implementing EHR and real-time tracking systems can significantly reduce administrative delays. These technologies help with quick patient registration, tracking patient status, and ensuring timely interventions.
4. **Dedicated Fast-Track Units**: Creating fast-track units for less critical cases can help reduce congestion in the main emergency department (ED). These units handle minor injuries and illnesses, allowing the main ED to focus on more severe cases.
5. **Staff Training and Education**: Continuous training programs for ED staff on best practices and efficient patient management are crucial. This includes training on the use of new technologies, triage protocols, and patient communication skills.
6. **Leadership and Cultural Change**: Strong leadership and a culture that promotes teamwork and continuous improvement are vital. Hospital leadership must visibly support, and champion initiatives aimed at reducing waiting times. Staff must be encouraged to collaborate across departments and adopt new practices.
7. **Performance Monitoring**: Regular monitoring and analysis of performance metrics help identify bottlenecks and areas for improvement. Hospitals can use data analytics to track wait times, patient outcomes, and staff performance, ensuring that the implemented strategies are effective and sustainable.

   Through process optimization, staff training, and streamlining of workflows, the hospital achieved significant reductions in waiting times, improved patient satisfaction, and enhanced patient safety by ensuring timely access to care.

These success stories, reports, and studies demonstrate the effectiveness of clinical audits and quality improvement projects in driving quality improvement, enhancing patient outcomes, and ensuring patient safety across various healthcare settings. By identifying areas for improvement, implementing evidence-based interventions, and monitoring progress through audits, healthcare organizations in India are making strides in delivering safer and more effective care to their patients.

Specific Audit Worked Out Examples and Checklists

Annexure 1: Prescription Audit

Prescription audit is a type of clinical audit that focuses on evaluating the appropriateness, safety, and effectiveness of medication prescribing practices. It involves the systematic review of prescription records and aims to identify any deviations from established guidelines or best practices.

A prescription is a medico-legal document written by a prescriber to a dispenser for a patient. The purpose of a prescription is to authorize the dispensing and administration of specific medications to patients for the treatment of their medical conditions. Prescriptions serve several important purposes:

1. **Treatment:** The primary purpose of a prescription is to provide patients with appropriate medications to manage and treat their health conditions. It enables healthcare providers to recommend specific medicines, dosages, and instructions tailored to the individual patient's needs.
2. **Patient Safety:** Prescriptions help ensure patient safety by providing accurate and detailed information about the prescribed medication, including the name, strength, dosage form, dosage instructions, frequency of administration, and duration of treatment. This information helps to reduce the risk of medication errors and enhances proper medication management.
3. **Legal and Regulatory Compliance:** Prescriptions fulfil legal and regulatory requirements related to the dispensing and administration of medications. They establish a lawful basis for healthcare providers to prescribe and pharmacists to dispense medications, ensuring compliance with applicable laws, regulations, and professional standards.
4. **Communication:** Prescriptions facilitate effective communication between healthcare providers, pharmacists, and patients. They serve as a written instruction that allows pharmacists to accurately dispense the prescribed medication, and they provide patients with clear guidance on how to take their medications correctly.

S. Sharma, *Quality Demystified*, https://doi.org/10.1007/978-981-95-0339-1

5. **Documentation:** Prescriptions serve as important documentation in patients' medical records. They provide a record of the medications prescribed, enabling healthcare providers to track the treatment regimen and monitor patient progress over time. Prescriptions also support continuity of care when multiple healthcare providers are involved in a patient's treatment.
6. **Control of Medications:** Prescriptions help regulate the distribution and control of certain medications, particularly those classified as controlled substances or potentially harmful medicines. By requiring a prescription, regulatory authorities can monitor and control the access to these medications to prevent misuse, abuse, or unauthorized distribution.

The specific requirements of a prescription may vary depending on the country or jurisdiction, as regulations and practices can differ. However, in general, a prescription should include the following key elements:

1. **Prescriber Information:** This section includes the name, professional title, contact information (such as address and phone number), and sometimes the prescriber's registration or license number. This information is essential for identifying the healthcare provider who issued the prescription.
2. **Patient Information:** This section contains the patient's full name, date of birth, and sometimes additional identifying information, such as address or patient identification number. It ensures that the prescription is intended for the correct patient.
3. **Date:** The date on which the prescription was issued is mentioned to track its validity and ensure that it is used within an appropriate timeframe.
4. **Medication Details:** This information is essential for identifying the healthcare provider who issued the prescription. It also specifies the strength or dosage of the medication and the form in which it should be dispensed (e.g., tablets, capsules, liquid).
5. **Dosage Instructions:** This section provides clear and specific instructions on how the medication should be taken or administered. It includes the dosage amount, frequency of administration, and any additional instructions, such as taking with food or avoiding certain substances.
6. **Duration of Treatment:** The duration of treatment is mentioned to indicate the intended length of time the medication should be taken or used.
7. **Refills:** If authorized, the prescription may include the number of allowed refills or instructions on when refills can be obtained. This is relevant for medications that require ongoing use or management.
8. **Prescriber's Signature:** The prescription should be signed by the prescriber to authenticate and validate its issuance. The signature confirms the responsibility and accountability of the prescriber for the prescribed treatment.

Several common issues or problems can arise with prescription writing. Several studies have shown that prescription errors have been one of the leading causes of medication errors. Some common problems with prescription writing include:

1. **Illegible Handwriting:** One of the most significant issues is illegible handwriting, which can lead to errors in medication interpretation and dispensing. It can be challenging for pharmacists to accurately decipher the intended medication, dosage, and instructions. Pharmacies may struggle to accurately interpret the prescription, leading to medication errors. Illegible prescriptions can also hinder communication between healthcare providers.
2. **Incomplete or Missing Information:** Prescriptions may sometimes lack essential information, such as the patient's name, medication dosage, frequency, or duration of treatment. Incomplete prescriptions can cause confusion and potential medication errors.
3. **Unclear Dosage Instructions:** Prescribers may fail to provide clear and precise dosage instructions, leading to confusion for patients or caregivers. Vague instructions can result in incorrect dosing and compromised patient safety.
4. **Drug Name Confusion:** Some medications have similar-sounding names or spellings (LASA), which can lead to errors in prescribing or dispensing. Prescribers need to be cautious and use the correct drug names to avoid medication mix-ups.
5. **Drug-Drug Interactions:** Prescribers may overlook potential interactions between prescribed medications and a patient's existing medications, resulting in adverse drug reactions. It is essential for prescribers to consider a patient's complete medication profile and be aware of any potential drug-drug interactions.
6. **Overprescribing or Underprescribing:** Overprescribing refers to prescribing medications unnecessarily or in excessive quantities, which can contribute to medication misuse, abuse, or adverse effects. Underprescribing, on the other hand, occurs when patients do not receive adequate medications to manage their condition effectively.
7. **Lack of Patient Education:** Prescriptions should be accompanied by clear instructions and patient education regarding medication usage, potential side effects, and any necessary precautions. Insufficient patient education can lead to medication non-compliance or misunderstanding.

Annexure 1A: Prescription Audit Objectives

The objectives of prescription audit include:

- **Identifying and Preventing Medication Errors:** The primary objective of prescription audit is to identify and prevent medication errors in the prescribing process. By reviewing prescriptions, healthcare professionals can identify errors such as incorrect dosages, inappropriate drug selections, drug interactions, and prescribing discrepancies. The aim is to catch these errors before they reach the patient and take corrective actions to improve patient safety. Medication error related to prescription, administration, and dispensing can be detected while auditing the drug chart/prescription, interviewing the nursing in-charge, and counter-checking with the patient.
- **Evaluating Adherence to Guidelines:** Prescription audit helps evaluate the adherence of prescribers to evidence-based guidelines, protocols, and formularies. It assesses whether prescriptions align with recommended best practices and whether prescribers are following established guidelines for specific medical conditions or drug therapies. The objective is to ensure that prescribing practices are in line with current knowledge and standards of care.
- **Assessing Rational Drug Use:** Prescription audit aims to assess the rational use of medications by evaluating prescribing patterns and practices. It involves analyzing factors such as polypharmacy, inappropriate or unnecessary drug use, overprescribing or underprescribing, and the use of high-risk medications. The objective is to optimize medication use, reduce unnecessary drug costs, and minimize potential harm to patients.
- **Monitoring Medication Safety:** Prescription audit plays a role in monitoring medication safety by identifying adverse drug events, medication-related complications, and potential risks associated with specific medications or prescribing practices. It helps identify areas where improvements can be made to enhance patient safety and minimize the occurrence of adverse events. Prescription auditing is carried out to study a certain specific group of medicines, e.g., antibiotics, antihypertensives, analgesics, vitamins, anticancer, other groups, etc. Analysis is carried out to find out preferred medicines among these groups in a particular hospital.
- **Improving Prescribing Practices:** Prescription audit serves as a tool for quality improvement by identifying areas of concern and opportunities for enhancing prescribing practices. For example, prescription auditing can be used as an excellent tool to determine the extent of use of antibiotics in a hospital setting. It also helps to find out the variety and types of antimicrobials prescribed. This gives evidence to evaluate whether the antimicrobials are prescribed rationally or not. It provides feedback to prescribers, highlighting areas for improvement and promoting awareness about medication safety and best practices. The objective is to facilitate continuous learning and improvement among prescribers.
- **Enhancing Communication and Collaboration:** Prescription audit promotes communication and collaboration among healthcare professionals involved in

the prescribing process. It encourages discussions about medication choices, potential drug interactions, and patient-specific considerations. The objective is to foster interdisciplinary collaboration and shared decision-making for optimal patient care.
- **Educating and Training Healthcare Professionals:** Prescription audit provides an opportunity for education and training of healthcare professionals. It helps identify educational needs, gaps in knowledge, and areas where additional training or interventions are necessary to improve prescribing practices and enhance patient safety.

Completeness of Prescriptions

Ensuring the completeness of prescriptions is essential for patient safety and improving the quality of medication prescribing. The objectives of a prescription audit regarding completeness may include:

- Reviewing the prescription for essential elements: The audit aims to check if all the necessary components of a prescription are present, such as patient information (name, age, gender), date of prescription, prescriber's information, and signature.
- Verifying medication details: The audit assesses if the prescribed medications are clearly documented, including the drug name, strength, dosage form, route of administration, and dosage instructions. It ensures that the prescription provides clear instructions to the pharmacist and the patient regarding the medication.
- Assessing medication safety considerations: The audit looks for any potential medication safety issues, such as allergies or drug interactions, and evaluates if appropriate precautions and warnings are included in the prescription.

Detecting Adverse Drug Reactions (ADRs):

Certain medicines in prescription/drug chart can be used as a tracer drug or trigger tools to track or find out prevalence of ADRs, e.g., administration of anti-histaminic, corticosteroids, 50% dextrose, naloxone, flumazenil, etc. It may happen that patient develop allergies to a drug and to counteract that antihistaminic or corticosteroids are administered, or overdose has occurred, and antidotes (naloxone, flumazenil) are administered, or 50% dextrose administered to revert hypoglycemia caused by overdose of insulin or oral hypoglycemic agents. Many a times such cases are neither reported nor documented. In such cases an in-depth analysis helps to find out the ADR and drug causing it. Furthermore, different types of interactions can also be found out from it. Various ADRs can indeed result from interactions between medicines or between medicines and foods. The prescription auditing team can decipher any possible or probable interaction leading to an ADR by studying the drug chart/prescription in detail.

Checking legibility and clarity:

Prescription audit examines the legibility and clarity of the prescription to ensure that it can be easily understood by both the dispensing pharmacist and the patient. Illegible or unclear prescriptions can lead to medication errors and confusion.

Compliance with regulatory guidelines:

The audit assesses if the prescription adheres to regulatory guidelines, local laws, and professional standards regarding prescription writing. This includes checking the prescription/drug chart for drug name, strength, formulation, dose, rouéof administration, frequency, and duration of treatment. The prescription is complete if it contains all the above-mentioned components.

Details to Be Shown on Prescription Form

1. Prescriber's name, address, and telephone number(s). This will allow the patient or the dispenser/pharmacist to contact the prescriber for any clarification or potential problem with the prescription.
2. Date of the prescription.
3. Name, dosage form (syrups, tablets, injections, etc.), strength, frequency, and duration of the prescribed medicines. It is better if medicines are prescribed by generic name. Brand name can be written in bracket.
   Strength of the medicine should be stated in standard units using abbreviations that are consistent with System Internationale (SI). The terms "Micrograms" and "Nanograms" should be written out in full, rather than abbreviated, to avoid confusion. Abbreviating them as "g" can easily be misread as "mg", leading to a potentially dangerous 1000-fold overdose. If unavoidable, a zero should be written preceding decimal point, e.g., 0.1. Avoid writing trailing zero, for example, 1.0 mg.
   Give space between drug and strength as no space between drug and strength may be misread (e.g., Inderal 40 mg can be misread as Inderal 140 mg).
4. Directions
   Latin apothecary abbreviations/directions for use should be avoided.
   Instructions for medication should be clear and explicit, specifying the route, dose, and frequency. Avoid using vague phrases like "take as directed" or "take as before."
   For medications taken on an "as required" basis, it is important to state the minimum dose interval and, where relevant, the maximum daily dose. This ensures precise understanding and adherence, promoting patient safety and effective treatment.
   It is good practice to qualify such prescriptions with the purpose of medication (e.g., "every 6 h as required for pain" or "at night as required to sleep").
   It is good practice to explain the direction to the patients. These directions will then be reinforced by the label on the medicinal product and possibly by appropriate counselling of the patient by dispenser/pharmacist.
5. Quantity to Be Dispensed
   The quantity of the medicinal product to be supplied should be stated such that it is not confused with either the strength of the product or the dosage directions. Alternatively, the length of the treatment course may be stated (e.g., "for 5 days"). Whenever possible, the quantity should be adjusted to match the pack sizes available.

The quantity of the medicinal product to be supplied should be clearly stated to avoid confusion with the product's strength or dosage directions. Alternatively, the treatment duration can be specified (e.g., "for 5 days"). Whenever possible, adjust the quantity to match the available pack sizes. This practice ensures clarity and helps prevent medication errors.

For liquid preparations, the quantity should be stated in milliliters (abbreviated as "ml") or liters (abbreviated as "L," since the letter "l" could be confused with the "1").

Error-Prone Abbreviations to Avoid in Prescriptions

Unclear handwriting combined with abbreviated drug names can be lethal when medicines with similar name but with very different effects are available.

- "µg" for Microgram: Avoid using the abbreviation "µg" for micrograms, as it can be easily mistaken for "mg," which could result in a 1000-fold overdose. Instead, always write "micrograms" in full.
- U" for Units: Do not abbreviate "units" as "U" in handwritten prescriptions, since it can be misread as "0" or "4." Writing "units" in full helps prevent such confusion.
- Symbols ">" and "<": These symbols can be misinterpreted or confused with other characters.
- Abbreviations for medication names: Using abbreviations/acronyms for drug names, such as ASA (aspirin), PCM (paracetamol), and CPZ (chlorpromazine), can lead to confusion and errors. Write full medicine name.
- Stemmed drug names: Using shortened versions of drug names can lead to confusion, such as "Nitro" for nitroglycerine or "Norflox" for norfloxacin.
- Use of abbreviations "D/C" for discontinue, "TCA" for to come again, " CST" for continue same treatment, or discontinue 1, 2, 5, rest to continue, etc.
- Errors due to Mix-ups between numbers and alphabets: "l" & "1"; "O" & "0,"; "Z" & "2,"; "1" & "7." Q1d can easily be mistaken for QID leading to four times the dose.
- Use of error-prone abbreviations, symbols, and dose designations.

For details see chapter on Medication Safety.

Clinical Notes for IPD Patients

When writing prescriptions or clinical notes for inpatient (IPD) patients, it is essential to provide clear and comprehensive information to ensure safe and effective patient care. Some instructions for documenting prescriptions or clinical notes for IPD patients included:

1. **Patient Identification:** Begin by clearly stating the patient's full name, unique identifier (such as hospital ID number or medical record number), and relevant demographic information (e.g., age, sex).
2. **Date and Time:** Include the date and time when the prescription or clinical note is written to ensure accurate documentation and chronological order.

3. **Diagnosis and Reason for Treatment:** Clearly state the patient's diagnosis or condition for which treatment is being prescribed or documented. This helps provide context for the prescribed medications or interventions.
4. **Medication Orders:**
   - Use generic names: Preferably write prescriptions using generic names to ensure consistency and avoid confusion between different brand names.
   - Dosage and strength: Specify the exact dosage form (e.g., tablets, capsules, injections), strength (in standardized units), and quantity of medication to be administered.
   - Frequency and duration: Clearly state the frequency (e.g., once daily, every 6 h), the time of each medication to be taken, and duration of medication administration.
   - Route of administration: Indicate the appropriate route of administration, such as oral, intravenous (IV), intramuscular (IM), subcutaneous (SC), etc. Abbreviations for route of administration should be avoided, e.g., IJ, IN, IV, IM. In the case of injectable medicines, medium to be used for reconstitution (normal saline, dextrose, distilled water, etc.) and speed of injection such as infusion, bolus should be clearly written along with other related instructions, e.g., monitoring of vital signs.

     In case there is more than one injectable, clear instruction on mixing or not mixing of two injections should be written.

     In case medicine is known to cause hypersensitivity reaction, clear instructions on test dose should be written.
   - PRN medications: If prescribing medications on an as-needed basis, clearly state the indications and specify the minimum dose interval and maximum daily dose.
5. **Other Treatment Orders:**
   - Non-medication orders: Document any non-medication treatment orders, such as specific therapies, procedures, or interventions required.
   - Laboratory or diagnostic tests: Specify any specific tests or investigations to be conducted and the reason for ordering them.
6. **Allergies and Special Considerations:** Ensure any known allergies or adverse drug reactions the patient has experienced is documented. Also, include any special considerations or precautions related to the patient's condition or treatment.
7. **Signature and Credentials:** Sign the prescription or clinical note with your full name, professional designation, and contact information. This helps identify the responsible prescriber and enables communication if needed.

Remember, legibility, clarity, and accuracy are crucial when documenting prescriptions or clinical notes for IPD patients. Proper communication and collaboration among the healthcare team members are essential to ensure patient safety and continuity of care.

Annexure 1B: Methods of Prescription Auditing

Prescription auditing involves systematic review and analysis of prescription practices to identify any deviations from established standards and guidelines. There are several methods and approaches that can be used for prescription auditing. The specific method chosen for prescription auditing may vary based on the healthcare setting, objectives, available resources, and objectives of the audit. Some common methods of prescription auditing:

1. Retrospective Review: This method involves reviewing a sample of previously documented prescriptions to assess adherence to prescribing guidelines, accuracy, and completeness. The review can be conducted by manually examining paper-based prescription records or by analyzing electronic prescription data.
2. Prospective Audit: In this method, prescriptions are evaluated in real-time as they are written, before being dispensed or administered to patients. Pharmacists or other healthcare professionals review the prescriptions to ensure appropriateness, accuracy, and safety. Any necessary interventions or recommendations are made to the prescribers immediately.
3. Sampling: Rather than reviewing all prescriptions, a representative sample is selected for auditing purposes. The sample can be chosen randomly or based on specific criteria such as high-risk medications, specific patient populations, or a certain time period. Sampling helps streamline the auditing process while still providing a representative overview of prescription practices.
4. Benchmarking: Prescription auditing can involve comparing prescription practices within a healthcare facility or across multiple facilities to identify variations, outliers, and areas for improvement. Benchmarking allows for the identification of best practices and the establishment of performance targets.
5. Collaborative Auditing: This method involves collaborating with multiple stakeholders, including prescribers, pharmacists, nurses, and other healthcare professionals, to conduct prescription auditing collectively. By involving different perspectives and expertise, collaborative auditing can lead to more comprehensive assessments and facilitate shared learning and improvement.
6. External Auditing: External auditing involves engaging external experts or organizations to conduct prescription audits. These external auditors can provide an unbiased assessment of prescription practices and offer recommendations for improvement based on their expertise and experience.
7. Use of Clinical Decision Support Systems (CDSS): CDSS software integrated into electronic health records can help automate prescription auditing by providing real-time alerts and reminders to prescribers based on established guidelines and best practices. These systems can help identify potential medication errors, drug interactions, and dosage errors.

Prescription auditing involves the assessment of various indicators to evaluate the quality, safety, and appropriateness of prescribing practices. These indicators serve as measurable criteria for assessing prescription performance and identifying areas that require improvement. Some commonly used indicators for prescription auditing include:

1. **Compliance with Clinical Guidelines:** This indicator assesses the extent to which prescriptions align with established clinical guidelines or protocols. It involves evaluating whether the prescribed medications, dosages, and durations of therapy are consistent with recommended evidence-based practices.
2. **Adherence to Formulary:** This indicator examines whether prescriptions adhere to the facility's or healthcare system's approved formulary. It assesses whether the prescribed medications are on the formulary and whether there are any deviations or inappropriate use of non-formulary medicines.
3. **Drug-Drug Interactions:** This indicator focuses on identifying potential drug-drug interactions within prescriptions. It involves evaluating the prescribed medications to determine if there are any known interactions that could pose a risk to patient safety.
4. **Polypharmacy:** Polypharmacy refers to the use of multiple medications by a patient. While it can be necessary for treating complex or multiple health conditions, polypharmacy can also increase the risk of adverse drug reactions, interactions, and challenges in medication management. This indicator assesses the prevalence of polypharmacy and identifies cases where the number of medications exceeds recommended limits or where there may be inappropriate drug combinations.
5. **Inappropriate Medication Dosing:** This indicator evaluates whether prescribed medication dosages are appropriate for the patient's age, weight, renal or hepatic function, and other relevant factors. It helps identify cases of underdosing or overdosing that may compromise treatment effectiveness or patient safety.
6. **Medication Safety:** This indicator focuses on assessing prescribing practices to identify potential medication safety issues, such as prescribing medications with a high risk of adverse effects, prescribing medications with narrow therapeutic indices without appropriate monitoring, or prescribing medications with black box warnings.
7. **Legibility and Clarity:** This indicator examines the legibility and clarity of prescription writing. Illegible or unclear prescriptions can lead to medication errors and misunderstandings. This indicator assesses the readability of prescriptions to ensure that they can be accurately interpreted by pharmacists and other healthcare providers.
8. **Prescription Documentation:** This indicator evaluates the completeness and accuracy of prescription documentation, including the prescriber's name, patient information, medication names, dosages, frequencies, and duration of therapy. It ensures that all necessary information is included to facilitate safe and appropriate dispensing and administration of medications.

These indicators can be used individually or in combination to assess prescription practices and identify areas for improvement. The specific indicators chosen may vary based on the objectives of the audit, the healthcare setting, and the available data for analysis.

1. Core Drug Use Indicators are used as a first-line measures intended to measure specific aspects of behavior of the health providers in health facilities as described by WHO (Annexure 1D). Core drug use indicators are commonly used in medication management and prescription auditing to assess the appropriateness, safety, and effectiveness of drug use. These indicators provide insights into various aspects of drug utilization and help identify areas for improvement. Some examples of core drug use indicators are:

1. **Prescribing Indicators**:
   - Percentage of prescriptions with generic drug prescribing: This indicator measures the proportion of prescriptions that include generic drug names instead of brand names. It promotes cost-effective prescribing practices.
   - Percentage of prescriptions with antibiotics: This indicator assesses the appropriateness of antibiotic prescribing, as excessive or inappropriate antibiotic use can contribute to antimicrobial resistance.
   - Prescribing of high-alert medications: This indicator focuses on medications with a high potential for harm if used incorrectly. It evaluates the appropriate use and prescribing patterns of such medications.
2. **Patient Care Indicators**:

Patient care indicators are studied to measure key aspects of patients' experience at health facilities and how well they were prepared to deal with the medicines that were prescribed and dispensed.
   - The percent of prescribed medicines dispensed.
   - Dispensing accuracy rate: This indicator measures the accuracy of medication dispensing by comparing the dispensed medication against the prescribed medication.
   - Generic substitution rate: This indicator evaluates the proportion of prescriptions where generic substitution is implemented when available. It promotes cost-effective drug utilization.
3. **Patient Safety Indicators**:
   - Adverse drug events: This indicator tracks the occurrence of adverse drug events in patients, including medication errors, adverse drug reactions, and other drug-related incidents.
   - Drug-drug interaction rate: This indicator assesses the prevalence of potential drug-drug interactions in prescribed medications and helps identify risks to patient safety.
4. **Medication Adherence Indicators**:
   - Medication possession ratio (MPR): This indicator measures the proportion of time that a patient has access to their prescribed medication over a specified period. It reflects patient adherence to the prescribed medication regimen.
   - Proportion of medication refills on time: This indicator evaluates the percentage of medication refills that are filled on schedule. It indicates patient compliance with follow-up appointments and medication refills.

5. **Cost-Related Indicators**:
   - Average cost per prescription: This indicator calculates the average cost incurred for each prescription, including the cost of medications and associated healthcare services.
   - Prescription cost containment: This indicator assesses efforts to control prescription costs through measures such as promoting generic drug use, formulary management, and prescribing cost-effective alternatives.

These core drug use indicators provide valuable information for evaluating the quality, safety, and cost-effectiveness of medication use. They serve as benchmarks for assessing prescribing practices, identifying areas for improvement, and monitoring the impact of interventions to enhance drug utilization.

When a level of performance on any of the core drug use indicators is considered unacceptable or falls outside the desired range, it should trigger specific follow-up activities to address the identified issues. For example, average number of medicines prescribed per patient more than 3 for patients with no comorbidities, high antibiotic usage, high injection use, low percent medicines prescribed by generic name, high percentage of medicines prescribed outside the EML/Formulary or low percent of medicines dispensed, etc.

These follow-up activities could include:

1. **Individual or Group Interviews:**
   - Conducting interviews with prescribers, pharmacists, and other healthcare professionals involved in the medication management process to understand the reasons behind the poor performance on the indicator.
   - Asking targeted questions to gather insights into the factors contributing to the identified issues, such as excessive prescribing of medications, high antibiotic usage, low generic drug prescribing, etc.
   - Exploring potential barriers or challenges faced by healthcare professionals that may be influencing their prescribing or dispensing practices.
2. **Focus Group Discussions:**
   - Organizing focus group discussions with a diverse group of healthcare professionals to facilitate open and interactive conversations about the identified issues.
   - Using the listed questions as a guide to stimulate discussions and gather perspectives on the underlying causes and potential solutions for improving performance on the indicators.
   - Encouraging participants to share their experiences, insights, and suggestions for enhancing medication management practices and optimizing drug utilization.
3. **Root Cause Analysis:**
   - Conducting a detailed root cause analysis to identify the underlying factors contributing to the poor performance on the indicators.
   - Utilizing appropriate methodologies, such as the "5 Whys" technique or fishbone diagrams, to systematically identify the root causes and understand the interdependencies involved.

- Using the findings from the root cause analysis to develop targeted interventions and strategies to address the identified issues and improve performance on the indicators.

4. **Continuous Quality Improvement:**
   - Implementing a continuous quality improvement framework to monitor and evaluate the progress in addressing the identified issues.
   - Establishing quality improvement teams or committees to oversee the implementation of corrective actions, monitor the impact, and track the performance on the indicators over time.
   - Regularly reviewing and updating prescribing guidelines, formularies, and other relevant policies to align with evidence-based practices and promote optimal medication use.

By conducting individual or group interviews, focus group discussions, and utilizing methodologies like root cause analysis, healthcare organizations can gain valuable insights into the reasons behind poor performance on drug use indicators. This information can guide the development of targeted interventions and strategies to improve medication management practices, optimize drug utilization, and enhance patient safety and outcomes.

2. Complimentary indicators Complimentary indicators that can provide further insights into various aspects of medication use and prescribing practices include:

1. **% Patients Receiving Ten or More Medicines:**
   - This indicator assesses the proportion of patients who are prescribed a high number of medications, which can indicate polypharmacy.
   - It helps identify patients who may be at higher risk of medication-related issues or adverse drug events due to the complexity of their medication regimens.
2. **Most Frequently or Infrequently Used Medicines:**
   - This indicator identifies the medicines that are commonly prescribed or, conversely, rarely prescribed.
   - It helps in understanding prescribing patterns and the prevalence of certain conditions or diseases within the patient population.
3. **Most Expensive Medicines:**
   - This indicator highlights the medications that contribute significantly to healthcare costs.
   - It can assist in identifying opportunities for cost-saving measures, such as exploring generic alternatives or negotiating better pricing contracts.
4. **Cost of Individual Medicines and Classes of Medicines:**
   - This indicator quantifies the financial burden associated with specific medications or groups of medications.
   - It aids in resource allocation and budget planning by identifying areas where costs can be optimized or reduced.

5. **Percentage of Budget Spent on Specific Medicines or Classes of Medicines:**
   - This indicator provides an overview of the allocation of financial resources to specific medications or therapeutic classes.
   - It helps in monitoring expenditure patterns and evaluating the cost-effectiveness of different drug treatments.
6. **Comparing Two or More Medicines Used for the Same Condition or Indication:**
   - This indicator assesses the prescribing patterns for different medications used to treat the same medical condition.
   - It facilitates evidence-based decision-making by comparing the effectiveness, safety, and cost-effectiveness of alternative treatment options.
7. **% Patients Received Antibiotics as Empirical Therapy:**
   - This indicator measures the proportion of patients who receive antibiotics without a confirmed diagnosis.
   - It helps assess the appropriateness of antibiotic prescribing practices and supports antimicrobial stewardship efforts.
8. **% Patients Received Appropriate Perisurgical Antibiotic Prophylaxis:**
   - This indicator evaluates the adherence to guidelines for administering antibiotics before surgery to prevent surgical site infections.
   - It measures the appropriateness of antibiotic prophylaxis and identifies areas for improvement in surgical practices.
9. **Per Capita Consumption of Specific Products:**
   - This indicator calculates the average quantity or dosage of specific medications used per person within a given population.
   - It aids in assessing the utilization patterns of particular medicines and evaluating their potential impact on public health.
10. **% Prescription Complete in All Respects:**
    - This indicator assesses the completeness of prescriptions, ensuring that all necessary information is included, such as patient details, drug name, dosage, frequency, and duration.
    - It helps identify areas where prescribers may need additional education or support to improve prescription accuracy and patient safety.

By calculating and analyzing these complementary indicators alongside the core drug use indicators, healthcare organizations can gain a comprehensive understanding of medication use, prescribing practices, and resource utilization. This information can guide quality improvement initiatives, inform decision-making, and ultimately enhance patient care and outcomes.

Completeness of the prescription

The completeness of the prescription can be assessed, and scores are given for each component of the prescription and its correctness as below:

When assessing the completeness of a prescription, several criteria can be considered to ensure that all necessary information is included. Some common criteria for assessing the completeness of a prescription include:

1. **Patient Information:**
   - Full name of the patient
   - Age or date of birth
   - Gender
   - Patient identification number (if applicable)
   - Address/contact information (optional but useful)
2. **Prescriber Information:**
   - Prescriber's full name
   - Prescriber's professional designation (e.g., physician, nurse practitioner)
   - Prescriber's contact information (address, phone number)
3. **Date:**
   - The date when the prescription was written
4. **Medication Details:**
   - Drug name (preferably generic name)
   - Dosage form (e.g., tablet, capsule, syrup, injection)
   - Strength (specified in standard units, avoiding potentially confusing abbreviations)
   - Quantity (number of units or volume)
   - Directions for use (including route, frequency, and duration of administration)
   - Special instructions or precautions (if applicable)
5. **Signature:**
   - Prescriber's signature or electronic signature
   - Signature date (if separate from the prescription date)

Additional Considerations:

- Use of clear, legible handwriting or electronic prescription systems
- Avoidance of abbreviations that can be misinterpreted or are not universally understood
- Avoidance of ambiguous or vague instructions
- Allergies or intolerances (if applicable)
- Relevant diagnosis or indication for the medication
- Any necessary precautions or warnings
- Any required laboratory monitoring or tests
- Prescriber's contact information for clarification or follow-up
- Compliance with local regulations and guidelines regarding prescription format and content

By using this checklist, score can be calculated, and healthcare professionals can ensure that all necessary information is included in the prescription, reducing the risk of errors, misinterpretation, or incomplete treatment. It promotes patient safety, facilitates accurate dispensing, and supports effective medication management.

4. ***Legibility of the prescription***

   Assessing the legibility of a prescription is important to ensure that the information can be read and interpreted accurately. Some criteria for assessing the legibility of a prescription:

1. **Handwriting:**
   - Is the handwriting clear and easy to read?
   - Are the letters or numbers easily distinguishable from each other?
   - Are there any instances of overlapping or crowded writing?
2. **Clarity:**
   - Are the lines and strokes or cutting of the writing well-defined?
   - Is the spacing between words and characters consistent?
   - Are there any smudges, stains, or other factors that affect readability?
3. **Size and Style:**
   - Is the font size adequate to ensure readability?
   - Is the style of writing easy to comprehend (e.g., print or cursive)?
4. **Use of Abbreviations:**
   - Are abbreviations used appropriately and consistently?
   - Are commonly recognized and accepted abbreviations used?
5. **Overall Presentation:**
   - Is the prescription document neat and organized?
   - Are the sections and fields clearly labelled and separated?
6. **Feedback from Pharmacists:**
   - Has the prescription been successfully dispensed without the need for clarification or additional communication?
   - Have pharmacists or dispensing staff expressed concerns or difficulties in reading the prescription?

If any issues or concerns are identified, steps should be taken to improve the legibility to ensure accurate interpretation and safe dispensing of medications.

Rationale for prescribing should be ascertained as per STG guidelines. Assessing the rationale for prescribing is important to ensure that medications are being prescribed based on sound clinical judgment and evidence-based practices. Assessment criteria for evaluating the rationale for prescribing are as below:

1. **Diagnosis and Indication:**
   - Is there a clear indication or diagnosis documented for the prescribed medication?
   - Does the prescribed medication align with the diagnosed condition or indication?
2. **Evidence-Based Guidelines:**
   - Is the prescribed medication in line with established clinical guidelines or protocols?
   - Are there specific guidelines or recommendations supporting the use of the prescribed medication?

3. **Drug Selection:**
   - Has the prescriber considered the available therapeutic options and selected the most appropriate medication?
   - Are there specific reasons documented for choosing one medication over others in the same class?
4. **Safety and Efficacy:**
   - Has the prescriber considered the safety profile and efficacy of the prescribed medication?
   - Are there any contraindications or precautions documented that may affect the appropriateness of the medication?
5. **Patient-Specific Factors:**
   - Has the prescriber considered individual patient characteristics such as age, gender, comorbidities, allergies, and drug interactions?
   - Are there any documented justifications for tailoring the medication choice or dosage based on patient-specific factors?
6. **Duration and Follow-Up:**
   - Is there a documented plan for the duration of medication use?
   - Are there any instructions or follow-up recommendations provided for monitoring the patient's response or adjusting the treatment plan?

Sample Size Calculation

Depending on the total population size or the average number of prescriptions per day, sample size formula can be used:

$$\text{Sample Size} = \left(Z^2 * p * q\right) / E^2$$

where:

- Z represents the Z-score that corresponds to the desired level of confidence (e.g., 1.96 for a 95% confidence level).
- p is the estimated proportion of prescriptions with the characteristic of interest (if unknown, use 0.5 for a conservative estimate).
- q is $1 - p$.
- E represents the desired margin of error, expressed as a proportion.

Data Collection Methodology

The data collection methodology for a prescription audit can involve several steps and methods. General outline of the data collection process is as below:

1. **Define the objectives**: Clearly define the objectives and scope of the prescription audit. Determine what specific information to be collected and analyzed.
2. **Develop data collection tools such as prescription audit forms or checklists**. These tools should include the relevant indicators and criteria to be assessed.
3. **Select the sample**: Determine the sample size and sampling method based on your objectives and available resources. This could involve randomly selecting prescriptions from specific departments or specialties or using stratified sam-

pling to ensure representation across different categories. Usually for data collection in the OPD, collecting a minimum of 100 prescriptions per facility, with 30 prescriptions from each major specialty using a stratified sampling approach is sufficient. However, to determine the sample size for each specialty, we need to know the total number of prescriptions in each specialty or the population size. For inpatient prescription auditing, random selection of 10% of patients' prescriptions daily from the ICU and all major IPD departments over a period of one month is sufficient to give a snapshot of the prescribing patterns. One-time prescription auditing should be done. It is more practical to monitor one day's prescription rather than studying the prescription for the whole duration of the stay of the patient in the hospital.

4. **Train data collectors**: Provide training to the individuals who will be collecting the data. Ensure they understand the data collection tools, indicators, and criteria. Accuracy, consistency, and confidentiality are important in maintaining the integrity and trustworthiness of any work, especially in fields like healthcare, research, and data management.
5. **Collect the data**: Collect the prescription data according to the defined methodology. This may involve reviewing patient records, extracting information from electronic systems, or conducting direct observations. Exit interview is the best time to collect data.
6. **Ensure data quality**: Implement measures to ensure the quality and accuracy of the collected data. This can include periodic data verification, double-checking entries, and resolving any discrepancies or missing information.
7. **Data entry and management**: Enter the collected data into a database or software system for analysis. Ensure data security and confidentiality during this process.
8. **Data analysis**: To analyze the collected data effectively, choose the appropriate statistical methods or qualitative analysis techniques based on the nature of your data and research objectives. Calculate relevant indicators, identify patterns, and interpret the findings.
9. **Report and dissemination**: Prepare a comprehensive report presenting the results of the prescription audit. Include recommendations for improvement based on the identified issues or areas of concern. Disseminate the findings to relevant stakeholders, such as healthcare professionals, administrators, and policymakers.
10. **Follow-up actions**: Based on the audit findings, develop, and implement strategies to address identified issues or improve prescribing practices. Monitor the effectiveness of interventions and consider conducting regular follow-up audits to track progress over time.

Annexure 1C: Sample Prescription Audit Tool & Report

Evidence

High antibiotic use in the outdoor department and patient complaints regarding the availability of prescribed medicines in the hospital pharmacy have raised concerns. This prescription audit aims to study the hospital's prescribing pattern, specifically the extent of antibiotic use and the availability of essential medicines. The goal was to formulate strategies to improve prescribing patterns and enhance the availability and accessibility of essential medicines.

Aim

- To study the prescribing pattern of the hospital particularly extent of antibiotic use and availability of essential medicines in the hospital.
- To formulate strategies to improve prescribing pattern and availability and accessibility to essential medicines.

Audit Criteria

1. Conformity to Hospital Formulary or Essential Medicines List:
   - Target: 90% of prescribed medicines should conform to the hospital formulary or Essential Medicines List.
2. Antibiotic Prescribing for Outdoor Patients:
   - Target: Antibiotic prescribing rate should be below 40% for outdoor patients.
3. Prescription by Generic Name:
   - Target: All medicines should be prescribed using generic names.
4. Medicines Dispensed from the Hospital Pharmacy:
   - Target: At least 80% of prescribed medicines should be dispensed from the hospital pharmacy.

Method & Analysis

- Data is collected prospectively by conducting exit interview of the patient attending OPD and prescriptions details are recorded using the data indicator reporting form.
- At least 100 prescriptions were collected from the major clinical departments, namely medicine, surgery, obstetrics and gynecology, pediatrics, etc. Since the prescription audit was done hospital wide covering all departments, 30 prescriptions from each department were collected with a total of 600 prescriptions in any given survey.
- Prescriptions were collected on different OPD days to cover all weekdays, and all prescribers uniformly spread over the entire OPD timings over a period of two months. Patients were selected for interview randomly as they leave the hospital pharmacy after collecting their medicines. Care was taken not to take only samples in the early morning or in the afternoon only or covering some doctors and some days only.

- Data was collected by pharmacist trained to collect data.
- Data was analyzed using formula shown in the Drug indicator reporting form and percentage compliance was calculated and compared with the standard set.
- Areas/sections of non-compliance and find out reasons for non-compliance were noted.

Findings

1. Conformity to Hospital Formulary or Essential Medicines List:
   - Compliance Percentage: [Calculated Percentage]
   - Areas of Non-Compliance: [List areas/sections of non-compliance]
   - Reasons for Non-Compliance: [Specify reasons for non-compliance]
2. Antibiotic Prescribing for Outdoor Patients:
   - Compliance Percentage: [Calculated Percentage]
   - Areas of Non-Compliance: [List areas/sections of non-compliance]
   - Reasons for Non-Compliance: [Specify reasons for non-compliance]
3. Prescription by Generic Name:
   - Compliance Percentage: [Calculated Percentage]
   - Areas of Non-Compliance: [List areas/sections of non-compliance]
   - Reasons for Non-Compliance: [Specify reasons for non-compliance]
4. Medicines Dispensed from the Hospital Pharmacy:
   - Compliance Percentage: [Calculated Percentage]
   - Areas of Non-Compliance: [List areas/sections of non-compliance]
   - Reasons for Non-Compliance: [Specify reasons for non-compliance]

Recommendations: Based on the audit findings, the following recommendations are proposed to improve prescribing patterns and enhance the availability and accessibility of essential medicines:

1. [Provide specific recommendations for addressing non-compliance areas]
2. [Suggest strategies to reduce antibiotic prescribing]
3. [Propose initiatives to encourage prescription by generic name]
4. [Recommend measures to increase the proportion of medicines dispensed from the hospital pharmacy]

Drug Use Indicators, Their Purpose and Formula for Calculations

Prescribing Indicators

1. Prescribing indicators are designed to assess and evaluate the prescribing practices of healthcare providers.
2. These indicators are used to measure various aspects of the prescribing process, including the appropriateness, safety, and efficiency of medication prescriptions. The primary goals of prescribing indicators are to improve patient safety, optimize medication use, and promote evidence-based prescribing practices. Prescribing indicators help healthcare systems and providers monitor and evaluate key aspects of medication prescribing, such as Rational Drug Use: These indicators focus on assessing whether

(continued)

medications are prescribed based on evidence-based guidelines, appropriate indications, and the best available clinical evidence. They aim to minimize unnecessary or inappropriate drug prescribing.

3. Polypharmacy and Drug Interactions: These indicators assess the prevalence of polypharmacy (the use of multiple medications) and the potential interactions between medicines. They aim to identify situations where the number of prescribed medications may be excessive or where drug combinations may increase the risk of adverse reactions or drug interactions.
4. Compliance with Formularies and Essential Medicines Lists: These indicators evaluate the extent to which prescribers adhere to formularies or essential medicines lists, which are lists of approved and recommended medications for specific healthcare settings. They ensure that prescribing practices align with the available resources and prioritize essential and cost-effective medications.
5. Prescription Legibility and Clarity: These indicators focus on the legibility and clarity of prescription writing, including the use of standardized abbreviations, appropriate dosing instructions, and clear medication names. Improving prescription legibility reduces the risk of medication errors and enhances patient safety.
6. Prescribing of Antibiotics and Controlled Substances: These indicators assess the appropriate and judicious use of antibiotics and controlled substances to minimize antibiotic resistance and prevent drug misuse or abuse.

By monitoring these prescribing indicators, healthcare systems can identify areas for improvement, provide feedback to prescribers, implement interventions to enhance prescribing practices, and ultimately improve patient outcomes and medication safety.

5. Rational Drug Use:

These indicators focus on assessing whether medications are prescribed based on evidence-based guidelines, appropriate indications, and the best available clinical evidence. The

Indicator: Average number of medicines prescribed per encounter/prescription
Purpose: To measure the degree of polypharmacy.
Formula: Total number of medicines prescribed/number of prescriptions.
Follow-up questions:

- High number of medicines per encounter:
  - Is it due to patient demand?
  - Do prescribers lack training or appropriate diagnostic equipment?
  - Are there financial incentives to encourage polypharmacy?
- Low number of medicines per encounter:
  - Are there any constraints in the drug supply system, such as limited availability or user fees?
  - Are there administrative regulations that limit the number of medicines prescribed?

Indicator: Percentage of Encounters with Antibiotics

Purpose: To measure the overall level of antibiotic use.

Formula: Total number of cases where antibiotic was prescribed/number of cases X 100.

Follow-up questions:

- What type of antibiotics are commonly prescribed?
- Are they narrow or broad-spectrum antibiotics?
- What are the common ailments for which patients come to the OPD?
- Does the extent of antibiotics align with the morbidity pattern?
- What is the cost of antibiotics as a percentage of all prescribing?
- Are there particularly expensive forms of antibiotics prescribed?
- What proportion of antibiotics are systemic, dermatological, ophthalmological, etc.?
- Are there cultural beliefs in the community about antibiotics?
- Do patients have high expectations of receiving certain types of antibiotics?

Indicator: Percentage Encounters with Injections per Encounter

Purpose: To measure the overall level of injection use.

Formula: Total number of cases where injection was prescribed/number of cases X 100.

Follow-up questions:

- What specific health problems are injections given for?
- Are injections given more often to children or adults?
- Are there alternative modes of therapy, such as syrups or mixtures, available for small children?
- What are the beliefs of patients and health providers regarding the relative efficacy of injections vs. oral medication?
- Do prescribers report patient demand as a factor in determining injection use?
- Is there a financial incentive for health workers to administer injections rather than oral medication?

Indicator: Percentage of Medicines Prescribed by Generic Names

Purpose: To promote generic prescribing.

Formula: Total number of medicines prescribed by generic names / number of prescriptions * 100.

Follow-up questions:

- Are predominantly generic or branded forms of medicines available in the healthcare facility?
- Are medicines supplied by generic or brand names?
- Are branded products being prescribed that are not available in the healthcare facility?
- Do prescribers know the correct generic names for most medicines?
- How often are prescribers visited by pharmaceutical representatives, and do they leave promotional material behind?
- Does the training of prescribers affect their willingness to prescribe by generic name?

Indicator: Percentage of Medicines Prescribed from the Essential Medicines List (EML)

Purpose: To measure the degree of conformity to the State Drug Policy/EML/hospital.

Formula: Total number of medicines prescribed from EML / total number of medicines prescribed * 100.

Follow-up questions:

- What are the most common medicines being prescribed that are not on the EML/formulary?
- Which health problems are these medicines intended to treat?
- Is an equivalent of the prescribed medicines available in the healthcare facility?
- What is the value of non-EML medicines compared to EML medicines?
- Are prescribers consulted during the development/revision of the EML/formulary?
- Are prescribers aware of the existence of the EML/formulary and the availability of medicines?

Patient Care Indicators

The intent of patient care indicators is to assess and evaluate various aspects of patient care within healthcare facilities. These indicators provide insights into the quality and effectiveness of healthcare delivery, allowing healthcare providers and policymakers to identify areas for improvement and make informed decisions to enhance patient care. Patient care indicators aim to measure several key factors, including the average consultation time, average time taken to dispense medications, the percentage of prescribed medicines that are actually dispensed, patients' understanding of the correct dosage, the percentage of medicines with proper labeling, the availability of essential medications, and the duration of drug stock-outs. By monitoring these indicators, healthcare systems can assess the performance and efficiency of healthcare services, identify gaps or deficiencies in patient care, and implement strategies to enhance patient satisfaction, safety, and overall healthcare outcomes.

Average Consulting Time (P) This indicator measures the amount of time medical personnel spend with patients during consultation and prescribing. It is calculated by observing a minimum of 10 cases, adding up the total consulting times (O), and dividing it by the number of cases (N). The formula is P = O/N (in minutes). Follow-up questions for this indicator may include:

- What is the average workload of health staff?
- What is the volume of patients attending at different times of the day or week?
- Can chronic disease patients be scheduled at times when the workload is lower?
- Does the clinic layout provide privacy and confidentiality?
- What is the quality of interaction in terms of effective communication, illness understanding, treatment explanation, and expression of empathy?

Average Dispensing Time (S) This indicator measures the average time spent by personnel dispensing medicines to patients. Similar to the consulting time, a minimum of 10 cases is observed, and the total dispensing times (R) are divided by the number of cases (Q). The formula is S = R/Q (in seconds).

Some follow-up questions related to this indicator are:

- What is the average workload of dispensers? Do they have sufficient time to explain medications to patients?
- What is the volume of patients attending at different times of the day?
- Does the layout of the pharmacy allow for private interactions between pharmacist and patient?
- What is the quality of interaction in terms of effective communication, treatment explanation, and expression of empathy?
- Are dispensing supplies readily available?

Percentage of Medicines Actually Dispensed This indicator assesses the extent to which health facilities are able to provide the prescribed medicines. It is calculated by dividing the total number of medicines actually dispensed by the total number of medicines prescribed and then multiplying by 100. Follow-up questions for this indicator may include:

- Are there certain types of medicines that are routinely prescribed but not dispensed?
- Is this more common for a specific therapeutic class or medicines targeting a particular illness?
- Are medicines not being dispensed even when they are available in the health facility?

Patient's knowledge of correct dosage This indicator evaluates patient's knowledge regarding the correct dosage schedule for each prescribed drug. It is calculated by determining the total number of patients who have correct knowledge about the purpose, dose, frequency, and duration of each drug prescribed, divided by the total number of cases, and multiplied by 100. The scoring system assigns a value of 1 for correct knowledge of all medicines and parameters, and 0 if any medicine is not answered correctly. Follow-up questions related to this indicator include:

- What do patients actually understand about the medicines they have received, including their purpose, dosage instructions, and potential side effects?
- What are the reasons for expected or actual non-compliance with the recommended medicine therapies?

Percentage of Medicines Adequately Labelled This indicator assesses the degree to which essential information is recorded on medicine packages by dispensers. It is calculated by adding the number of medicines with adequate labels for each patient and dividing it by the total number of dispensed medicines, then multiplying by 100 to obtain a percentage. The minimum required information on packages includes the patient's name, medicine name, dose, and when the medicines should be taken. Follow-up questions for this indicator may include:

- Is labeling consistently done for the dispensed medicines?
- What elements of appropriate labeling are missing?
- Is the information written legibly?
- Is the information on the label written in a language that patients can understand?
- Are there any challenges related to packaging material availability?
- Do dispensers receive training on proper packaging and labeling of medicines?
- Do dispensers have enough time to package medicines effectively?

Facility Indicators

Facility indicators assess and evaluate various aspects of healthcare facilities, including their infrastructure, resources, and processes. These indicators provide insights into the quality, availability, and functionality of healthcare facilities, helping to identify areas of improvement and guide decision-making for healthcare planning and resource allocation.

Percent Availability of the Key Medicines This indicator assesses the availability of vital medicines that are required at all levels of health facilities. Since conducting a survey on the availability of all medicines is not feasible, the WHO recommends assessing availability based on a list of 18-20 key medicines that are considered essential for treating common health problems and should always be available. The calculation involves dividing the number of medicines in stock by the total number of medicines in the list of key medicines and multiplying by 100. This list may also include alternative medicines that are chemically or therapeutically equivalent to measure the availability of a more comprehensive range of essential medicines. The presence of any one of the alternatives is considered as being in stock. The quantity in stock is not considered. If a drug is not available, the duration of stock-out for each drug is recorded.

Follow-up questions for this indicator may include:

- Are there specific classes of medicines or particular dosage forms (such as Pediatric syrups) that are more prone to going out of stock?
- Is there any seasonal variation in stock-outs or in relation to the drug procurement cycle?
- What is the system in place for informing prescribers about medicine stock-outs?
- Are there therapeutic alternatives available for medicines that are found to be out of stock?

  Are there any mechanisms for therapeutic substitution by pharmacists?

Stock-Out Days at the Health Facility This indicator measures the duration of time that a drug is unavailable. The stock-out duration for each drug is recorded, and the durations are summed up.

Annexure 1D: WHO Core Drug Use Indicators

Indicator	Purpose	Formula	Follow-up questions
Prescribing indicators			
Average number of medicines prescribed per prescription/encounter	To measure the degree of polypharmacy	Total no. of medicines prescribed/no. of prescriptions. Combination medicines given as FDC are counted as one	**High no. of medicines per encounter** Is it due to patient demand? Do prescribers lack training or appropriate diagnostic equipment? Are there financial incentives to encourage polypharmacy? **Low no. of medicines per encounter** Are there any constraints in the drug supply system- only a few medicines are available/User fee? Are there administrative regulations that limit the no. of medicines prescribed?
Percentage encounters with antibiotics	To measure the overall level of use; commonly overused & costly	Total no. of cases where antibiotic was prescribed/ no. of cases multiplied by 100	What type of antibiotics are commonly prescribed? Relative narrow or broad spectrum What are the common ailments for which patients come to OPD? Does extent of use corroborate with the morbidity pattern? What is the cost of antibiotics as a proportion of total prescriptions? Particularly expensive forms of antibiotic Proportion of antibiotic represented by systemic, dermatological, ophthalmological products, etc. Cultural beliefs in the community about antibiotics Are patient's expectations of receiving certain type of antibiotics very high?

Percentage encounters with injections per encounter	To measure the overall level of use; commonly overused & costly	Total no. of cases where injection was prescribed/no. of cases multiplied by 100	What particular health issues require the administration of injections? Are injections given more often to children or adults What options are available for syrups and mixtures as alternative therapies for young children? What are the beliefs of patients and health providers about the relative efficacy of injections vs. Oral medication? Do prescribers report patient demand as a factor in determining injection use? Is there a financial incentive for a health worker to give an injection rather than an oral form of medication?
Percentage of medicines prescribed by generic names	To promote generic prescribing	Total no. of medicines prescribed by generic names/no. of prescriptions multiplied by 100	Are predominantly generic or branded forms of medicines available in the HF? Are medicines supplied by generic or brand names? Are branded products being prescribed which are not available in HF? Do prescribers know the correct generic name for most medicines? How often are prescribers visited by pharmaceutical representatives & promotional material left behind by them? Does the training of prescribers affect their willingness to prescribe by generic name?

(continued)

Indicator	Purpose	Formula	Follow-up questions
Percentage of medicines prescribed from the Essential Medicines List (EML)	To measure the degree to which practices conform to the State Drug Policy/State EML/hospital formulary	Total no. of medicines prescribed from EML out of total no. of medicines prescribed multiplied by 100	What are the most common medicines being prescribed which are not in EML/Formulary? Which health problems are these medicines intended to treat? Is the equivalent of the medicines prescribed available in HF? What is the value of non-EML medicine compared to EML medicine? Are prescribers consulted while developing/revising EML/ formulary? Are prescribers aware of the existence of the EML/Formulary and availability of medicines
Patient care indicators			
Average Consulting time (P)	To assess the duration medical staff dedicate to patient consultations and prescription processes	1. Number of cases observed (N) 2. Add all consulting times (O) and divide them by the number of cases (N) *P = O/N min* *Minimum number of cases to be observed is 10.*	What is the average workload of health staff? What is the volume of patients' attendances at different times, during the day/by week? Could chronic disease patients be scheduled at different time when the workload is less? Does the layout of the clinic allow for privacy and confidentiality? What is quality of interaction in terms of effective communication about illness, treatment, expression of empathy?

Average Dispensing time (S)	To measure the average time that personnel dispensing medicines spends with patients	1. Number of cases observed (Q) 2. Add all consulting times (R) and divide by the number of cases (Q) S = R/Q s *Minimum number of cases to be observed is 10*	What is the average workload of dispenser? Is there enough time for them to discuss medications with patients? What is the volume of patients' attendances at different times, during the day? Does the layout of the pharmacy allow for private-pharmacist-patient interactions? What is quality of interaction in terms of effective communication about illness, treatment, expression of empathy? Are dispensing supplies available?
Percentage of medicines actually dispensed	To assess how well health facilities can supply the prescribed medications	Total no. of medicines actually dispensed out of total no. of medicines prescribed multiplied by 100	Are there certain types of medicines that are routinely prescribed yet not dispensed? Is this more common for a specific therapeutic class or medicines for a particular illness/ Are medicines not being dispensed even when available in HF?
Patient's knowledge of correct dosage	Patients are questioned on their knowledge of when and in what quantity each drug given in their prescription should be taken to measure and evaluate whether appropriate information was given by the doctor, and the dispenser on the dosage schedule of the medicines they had received, and environment was conductive to patient communication	Total no. of patients having correct knowledge about purpose, dose, frequency, and duration of each medicines prescribed / total no. of cases multiplied by 100 *Score yes–1-* knowledge about all medicines with all parameters is correct *Score No- 0-* if any of the medicine is not replied correctly	What do patients actually understand about the medicines that they have received: what is the purpose of the medicine, how should they be taken, side effects and so forth? What are the reasons for expected or actual non-compliance with recommended medicine therapies?

(continued)

Indicator	Purpose	Formula	Follow-up questions
Percentage of medicines adequately labelled	To assess how well dispensers document crucial details on the medicine packages they distribute	Add the no. of medicines with adequate labels for each patient. Then divide by the total no. of medicines dispensed and multiply by 100 to make a percentage. *Minimum information required on packages is patient name, medicine name, dose and when the medicines should be taken*	Whether labeling is done? What element of appropriate labelling is missing? Is the information written legibly? Is information on label written in local language or which patients are likely to understand? Packaging material not available? Do dispensers have training in how medicines are to be packaged and labelled? Do dispensers have time to package medicines?
Facility indicators			
Availability of copy of the EML/Formulary/STGs at the pharmacy or in patient care areas	To know whether the prescribers understood the significance of EML/formulary and STGs	Yes – if could be produced physically at the time of survey	Are prescribers generally aware of which medicines are on the list? What efforts have been made to disseminate EML/Formulary to individual prescribers? Is the EML/Formulary clean with unbroken binding, or dirty indicating that it has been used?

Percent availability of the key medicines	To measure the availability of the vital medicines required at all levels of HF. Since a survey on availability of all medicines cannot be done, who recommends that availability be assessed based on key medicines, 18-20 key medicines considered as the most essential for the treatment of some common health problems and should be always available at all levels of health facilities	No. of medicines in stock / total no. of medicines in the list of key medicines multiplied by 100 This list may also include some alternatives (chemically or therapeutically equivalent form of the drug) to measure the availability of a more complete range of essential medicines. Presence of any one of the alternatives to be considered as being in stock. However, quantity in stock is not considered. If a drug is not in stock, the stock-out duration of each drug is recorded	Are there particular classes of medicines or a particular dose form (pediatric syrups) more likely to go out of stock? Is there any seasonal variation of the stock-outs or in relation to the drug procurement cycle? What is the system for informing the prescribers about medicine stock-outs? Are there therapeutic alternatives in stock for the medicines which are found to be out of stock? Are there any mechanisms for therapeutic substitution by pharmacists?
Stock-out days at the HF	To measure the length of time the drug is not available	Stock out duration for each drug then sum up	

Annexure 1E: Drug Indicator Reporting Form

Location

Data Collector

Prescription no.	Date	Diagnosis/ symptoms (optional)	Age	Sex	No. of medicines prescribed	Dosage form (optional)	# medicines prescribed by Generic name Yes (1) or No (0)	Injectable prescribed Yes (1) or No (0)	Antibiotic prescribed Yes (1) or No (0)	No. of medicines prescribed from EML	No. of medicines Dispensed Yes (1) or No (0)
		A.	**B.**	**C.**	**D.**	**E.**	**F.**	**G.**	**H.**	**I.**	**J.**
1.											
2.											
3.											
4.											
5.											
6.											
7.											
99											
100											
Total											
Average											
Percentage				**% male % female**			**% of total medicines (total of F/ total of D X100)**	**% of total cases (total of G/ total no. of patient X100)**	**% of total cases (total of H/total no. of patient X100)**	**% of total number of medicines prescribed (total of I / total of D X100)**	**% of medicines dispensed (total of J/ total of D X100)**

Prescription Audit Report

Prescription audit was carried out in the XYZ hospital and 631 prescriptions from outdoor department (OPD) of major clinical specialties namely medicine, pediatrics, obstetrics and gynecology, and surgery over a period of 2 months from May – July 2016 were collected and analyzed using the following indicators:

A. **Prescribing indicators** namely
   - Average number of medicines prescribed per encounter to measure the degree of polypharmacy.
   - Percentage encounters with antibiotics and injections per encounter to measure the overall level of use of these two important forms of drug therapy, but commonly overused and costly.
   - Percentage encounters with vitamins.
   - Percentage of medicines prescribed by generic names.
   - Percentage of medicines prescribed from the essential medicines list to measure the degree to which practices conform to the State Drug Policy.

B. **Patient care indicators** were studied to measure key aspects of patients' experience at health facilities and how well they were prepared to deal with the medicines that were prescribed and dispensed. Patient care indicators included:
   - Percentage of medicines actually dispensed to measure the degree to which health facilities were able to provide the medicines which were prescribed.

Results

A survey of drug use has brought out both favorable and less favorable features regarding availability, accessibility, and prescription pattern in the hospital.

Average No. of Medicines Prescribed per Prescription The average number of medicines prescribed per encounter was 2.6 (ranged from 1.95 to 2.89 across different departments in the hospital)l. More number of medicines per encounter were prescribed in the medicine department (2.89) and Obstetrics (2.56). The number of medicines prescribed ranged from 1 to 7. Though there is no objective norm for the average number of medicines prescribed per encounter, the accepted norm is that it should be between two to three.

Prescriptions with Antibiotics The mean percentage of prescriptions with antibiotics was 74%. Mean antibiotic use was highest (80.32%) and 76.89% at medicine and surgery department respectively.

Interpretation- The benchmark for optimum practice is 30% i.e., ~30% OPD patients may require antibiotics, however, in view of the current sanitation conditions and prevalence of infectious diseases prescription of antibiotics up to 40% of encounters can be viewed as adequate performance.

Top ten prescribed antibiotics are

1. Ofloxacin
2. Ciprofloxacin

3. Amoxycillin
4. Norfloxacin
5. Roxithromycin
6. Amoxycillin + Clavulanic acid
7. Gatifloxacin
8. Azithromycin
9. Erythromycin
10. Ceftriaxone

Prescription with Injections Generally, not more than 10% of the patients should be prescribed injections in the outpatient department. The number of injections per encounter was within acceptable limits as mean percent prescriptions with injections were only 5.54%.

Prescriptions Conforming to EML About 80% of medicines prescribed were from the Essential Medicines List (EML). The extent of prescriptions from EML was similar across different departments except medicine where only 68.51% of medicines only were from EML. Highest percentage of medicines prescribed from EDL (90%) was observed in the pediatrics department.

Prescribing by Generic Names Mean percent medicines prescribed by generic name was very low 24.76% and lowest was observed in medicine department (20%) followed by Surgery (8.17%) and obstetrics and gynecology (11.33%). However, in the pediatric department, medicines prescribed by generic names were highest (45%).

Percent of Medicines Dispensed Overall percent of medicines prescribed, which were actually dispensed to the patients was only 56.53%. 71% of medicines prescribed from the pediatrics department were dispensed. Minimum percent medicines (49.17%) were dispensed at surgery department followed by Obstetrics and gynae (43%).

Interpretation Average number of medicines prescribed per prescription was less than 3, and most medicines were prescribed from the Essential Medicines List, but most medicines were prescribed by brand names and were not dispensed from the hospital pharmacy. Though injection use was within acceptable limit but prescription for antibiotics was high. Only 60% of the medicines prescribed were actually dispensed, indicating poor availability and accessibility to essential medicines.

Annexure 1F and Figure 1A present the findings of the audit using core drug use indicators.

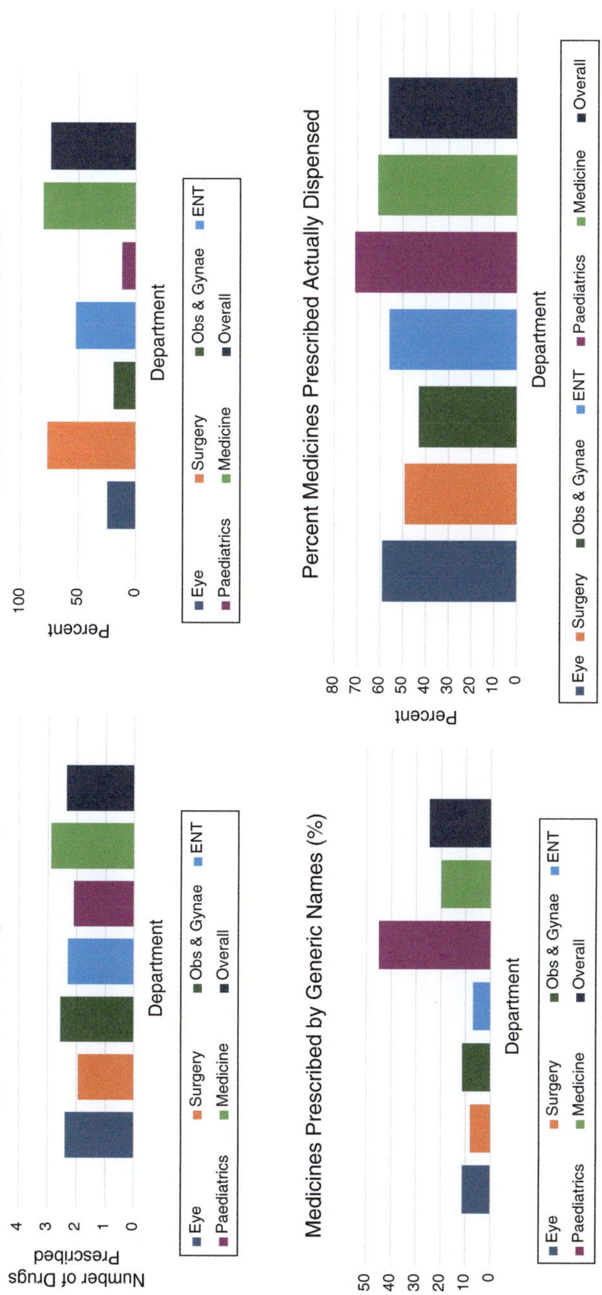

Fig. 1A Core drug use indicators

Annexure 1F: Sample Presentation of Results

Department	Average number of medicines prescribed	Percent medicines prescribed by generic name	Percent medicines prescribed from EML	Percent antibiotic prescribed	Percent injection prescribed	Percent medicines dispensed
Eye	2.40 (1 - 7)	11.33	76.83	25	2.77	59
Surgery	1.95 (1 – 3)	8.17	84.92	76.89	0.98	49.17
Obs & Gynae	2.56 (1 – 6)	11.33	68.51	19.46	7.96	43
ENT	2.30 (1 – 4)	7.00	89.89	52.32	6.97	56
Pediatrics	2.10 (1 – 4)	45.00	90.03	12.09	3.22	71
Medicine	2.89 (1 – 5)	20.00	68.51	80.32	12.24	61
Overall	**2.36**	**24.76**	**80.79**	**74.00**	**5.54**	**56.53**

Suggested Areas for Action and Improvement in Management

1. To ascertain reasons for high number of patients receiving antibiotics and medicines which are commonly prescribed but are not available in the hospital pharmacy in consultation with the concerned section.
2. To collect and whether extent of use of antibiotics corroborate with the recorded morbidity and mortality.
3. To organize a meeting with all the doctors to disseminate the findings of the prescription audit.
4. To discuss strategies for lowering prescription for antibiotics—to introduce Standard Treatment Guidelines and develop list of antibiotics to be available in the OPD and review and revise Essential Medicines List of the hospital accordingly.
5. Training program to be organized on rational use of medicines.
6. The prescription audit has generated some findings reflecting prevalent prescribing patterns which needs monitoring; therefore, prescription audit will be conducted every 3 months.

Acknowledgements Many thanks to all those who participated and .... for all their help and invaluable efforts in conducting the audit. Thanks also to .... for his/her advice and expertise especially in the design of the audit.

Annexure 2: Death Audit

Death audit, also known as mortality audit or mortality review, is a systematic process of analyzing and reviewing the circumstances surrounding a patient's death within a healthcare setting. The primary intent of death audit is to identify and analyze the factors contributing to a patient's death in order to improve the quality of healthcare and prevent similar incidents in the future.

Death audit serves as a logical, & structured approach to enhance the quality of healthcare by systematically evaluating clinical practices against established standards. While the long-standing Mortality and Morbidity (M&M) meetings have traditionally focused on reviewing in-hospital deaths and primarily involved surgeons for professional education, it is important to acknowledge that a majority of deaths occur outside of surgical settings.

These meetings offer a collaborative and transparent platform for healthcare professionals to conduct comprehensive reviews, encouraging accountability and facilitating the examination of medical practices. By analyzing deaths, identifying potential factors that could have been avoided or errors that may have occurred, valuable insights are gained to develop solutions aimed at reducing the risk of similar incidents in the future. This process leads to improvements in the quality of care and patient safety.

Regularly scheduled meetings also foster a culture of data collection, motivating healthcare providers to actively participate in the audit process. Moreover, studies have indicated that holding these meetings regularly is associated with a decline in mortality rates, demonstrating the positive impact they can have on patient outcomes.

Purpose

The purpose of a death audit is to systematically review and evaluate the circumstances surrounding a patient's death in order to identify potential areas of improvement in clinical practices and patient care. The primary goals of conducting a death audit are:

1. Quality Improvement: Death audits provide a structured approach to assess clinical practices against established standards and identify any shortcomings or errors that may have contributed to the patient's death. By understanding the factors involved, healthcare providers can implement changes and improvements in processes, protocols, and patient management to prevent similar incidents from occurring in the future.
2. Accountability and Learning: Death audits create a forum for healthcare professionals to discuss the case, encouraging accountability and fostering a culture of learning openly and transparently. It allows clinicians to reflect on their practices, identify areas for personal and professional growth, and develop strategies to enhance patient safety and quality of care.
3. Risk Reduction: By identifying avoidable factors or errors that may have contributed to the patient's death, death audits help healthcare providers develop targeted interventions and preventive measures to reduce the risk of similar inci-

dents in the future. This may include changes in protocols, training programs, guidelines, or resource allocation to ensure patient safety.

4. Continuous Quality Improvement: Death audits contribute to an ongoing cycle of quality improvement in healthcare. The findings and recommendations from audits are used to implement changes, monitor their effectiveness, and further refine clinical practices. This iterative process helps healthcare organizations maintain high standards of care and improve patient outcomes.
5. Professional Development: Through the process of death audit, healthcare professionals have the opportunity to enhance their knowledge, skills, and decision-making abilities. By critically reviewing cases and discussing them with colleagues, they gain valuable insights, learn from each other's experiences, and continuously improve their clinical practice.

Selected cases are presented with the purpose of a death audit to discuss management decisions, promote a culture of continuous learning, improvement, and accountability in healthcare, ultimately leading to better patient care, improved outcomes, and increased patient safety.

Clinical Departments

The primary objective is to determine the percentage of patients whose deaths were attributed to "problems in care." These problems may arise from healthcare processes, including both acts of omission (inactions), such as failure to diagnose and treat, and acts of commission (affirmative actions), such as incorrect treatment or management. The emphasis should be on evaluating the systems and processes of care rather than solely focusing on individual performance.

The recommendations derived from individual cases should primarily concentrate on implementing measures that can prevent similar outcomes or adverse incidents, as well as improving the overall care processes provided to hospital patients. It is crucial to foster an environment where blame or excessive focus on individuals is avoided, promoting a culture of continuous improvement and learning.

Managers

The main goal is to identify systemic issues, reduce organizational risk, and enhance organizational performance. By conducting audits and assessments, managers aim to uncover areas where the healthcare system can be optimized, mitigating risks, and improving the overall performance of the organization. This includes identifying and addressing systemic issues that may have contributed to adverse outcomes or incidents, promoting a proactive approach to risk management, and enhancing the quality of care provided by the organization as a whole.

Areas to Be Identified for Each Case

The selection of cases for presentation in a death audit should be done carefully to ensure a comprehensive review of the factors contributing to patient deaths. Considerations for selecting cases are as under:

1. Severity and Impact: Prioritize cases that involve significant patient harm or adverse outcomes. Focus on cases where the patient's death was unexpected, potentially preventable, or raises concerns about the quality of care provided.
2. Variety and Representativeness: Select cases that represent a range of medical conditions, specialties, and healthcare settings. This helps ensure a broader understanding of the challenges and issues across different clinical areas.
3. Frequency and Patterns: Identify cases that reveal patterns or recurring issues within the healthcare system. This can include cases with similar contributing factors or instances where common errors or problems have occurred.
4. Educational Value: Choose cases that provide valuable learning opportunities for healthcare professionals. This could involve complex diagnostic dilemmas, treatment challenges, or situations that highlight the importance of specific protocols or guidelines.
5. Systemic Factors: Consider cases that shed light on systemic issues or deficiencies in healthcare delivery. These cases help identify underlying problems in processes, communication, resource allocation, or organizational culture.
6. Stakeholder Input: Seek input from various stakeholders, such as clinicians, quality improvement teams, patient representatives, and healthcare administrators, to ensure a balanced and comprehensive selection of cases. Involving multiple perspectives can enrich the audit process.
7. Mix of Successes and Failures: Include cases where the care provided led to positive outcomes as well. This allows for recognition and sharing of best practices and successful interventions alongside cases with adverse events[1].

It is important to remember that the purpose of a death audit is to learn from adverse events and improve patient care. The selected cases should be presented in a manner that encourages open discussion, learning, and the identification of actionable recommendations for improvement.

Note: *An area of concern is where the clinician identifies that aspects of care could have been improved.*

Death or Mortality Audit Committee

The Death or Mortality Audit Committee is composed of individuals from the hospital who represent key departments, including management, treating doctors, and support departments. The committee is responsible for conducting mortality reviews and identifying areas for improvement in patient care and safety.

When it comes to attending the death review meeting, it is recommended that as many medical and nursing staff as possible participate. This ensures a comprehensive and multidisciplinary approach to understanding the factors contributing to patient deaths and finding solutions to address the identified problems.

[1] An Adverse Event is an unintended injury caused by medical management rather than by disease process, which is sufficiently serious to lead to prolonged hospitalization or to temporary or permanent impairment or disability of the patient at the time of discharge, or which contributes to or causes death.

By involving a wide range of healthcare professionals in the meeting, including those directly involved in patient care, the committee can benefit from different perspectives, experiences, and expertise. This collaborative approach fosters a sense of shared responsibility and ownership in improving patient outcomes and promotes a culture of continuous learning and quality improvement throughout the organization.

It is important to create an environment where staff feel encouraged and supported to contribute to the discussion without fear of blame or individual focus. The focus should be on system issues, identifying areas for improvement, and implementing necessary changes to enhance patient care and safety.

Sources of Data

Sources of data for death or mortality audits can include:

1. Monthly statistical reports: Each department or hospital typically maintains monthly statistical process reports that capture relevant data on patient outcomes, including deaths. These reports provide an overview of key indicators and trends within the department or hospital, serving as a starting point for identifying cases that warrant further investigation.
2. Summaries of routine data collection: Feedback from routine data collection processes, such as hospital data and death certificates, can provide valuable information about each hospital death. Summaries of this data can highlight important details, such as the cause of death, contributing factors, and any potential issues in care delivery.
3. Routine hospital data: Information captured in the routine hospital data systems can serve as a valuable resource for identifying patterns or trends related to patient deaths. This data may include patient demographics, medical diagnoses, treatment interventions, and outcomes. Analyzing this data can help identify specific cases or areas that require a more in-depth investigation through case-note audits.
4. Case-note audit: In-depth investigations can be conducted by reviewing the case notes of individual patients who experienced adverse outcomes, including deaths. Case-note audits involve a detailed examination of the patient's medical history, diagnostic tests, treatment plans, and any documented actions or decisions taken by healthcare providers. This process helps identify specific areas where problems in care may have occurred and informs recommendations for improvement.

By utilizing these various sources of data, death or mortality audit committees can gain a comprehensive understanding of patient outcomes, identify areas for improvement, and develop targeted interventions to enhance the quality and safety of patient care.

Cases presented at department death audit meetings can be identified from various sources, including:

1. All deaths occurring after 48 h of admission: This criterion ensures that deaths that occur within the hospital setting are reviewed and analyzed to identify any potential issues in care delivery.
2. Interesting and unique cases: Cases that present unique circumstances or provide opportunities for new learning and inquiry are also considered for review. These cases can offer insights into uncommon situations or challenges that may require further investigation and improvement.
3. Referrals via clinical risk management strategy: Cases referred through the organization's clinical risk management strategy are important considerations for death audit meetings. This includes sentinel events, critical incidents, near misses, and complaints, which indicate potential areas of concern and the need for analysis.

During the death audit process, the death case sheets are examined for both qualitative and quantitative adequacy. Several parameters are assessed, including:

- Diagnosis, investigation, and treatment: The appropriateness and adherence to standard practices in terms of diagnosis, investigations, and treatment are evaluated to determine if there were any deviations or potential areas for improvement.
- Delay in examination, investigation, or initial treatment: Any delays in crucial aspects of patient care, such as examination, investigation, or initial treatment, are identified and assessed to understand their impact on patient outcomes.
- Types of consultations obtained and recorded: The involvement of various healthcare professionals and the documentation of their consultations are reviewed to assess the multidisciplinary approach to patient care and the quality of communication among healthcare providers.
- Daily monitoring of progress: The regular monitoring of patient progress, including vital signs, response to treatment, and other relevant parameters, is evaluated to determine if appropriate actions were taken based on the patient's condition.

To facilitate the death audit process, a standardized audit tool can be developed. This tool focuses on reviewing specific areas of care, such as the quality of clinical observations, prescribing and medicines management, infection control and treatment, thromboprophylaxis, appropriateness of location of care, and communication between health professionals. The audit tool ensures a structured and consistent evaluation of these key aspects, enabling targeted improvements and promoting patient safety and quality of care.

Number of Cases to Be Reviewed in Each Meeting

The number of cases to be reviewed in each death audit meeting should be determined based on the capacity and resources of the committee, as well as the frequency of deaths in the hospital or clinic. It is important to strike a balance between ensuring a thorough discussion of each case and avoiding overwhelming the meeting with a large number of cases. Some considerations are given as below:

1. Agreeing on the number of cases: The committee should agree on the appropriate number of deaths to be presented and discussed in each meeting. This number should allow for a comprehensive analysis of each case while also considering time constraints.
2. Focus on deaths with avoidable factors: Cases where there are obvious avoidable factors contributing to the death should always be included in the review. These cases provide valuable opportunities for identifying areas of improvement and preventing similar incidents in the future.
3. Include cases with important lessons: Deaths that offer significant learning opportunities should also be included in the review, even if there are no obvious avoidable factors. These cases may involve complex medical conditions, unique circumstances, or challenges that can provide valuable insights for enhancing care delivery.
4. Consider the frequency of meetings: In larger hospitals with frequent deaths, it may be practical to hold mortality review meetings on a weekly basis. This frequency allows the staff to recall details of the patient's condition and management more easily. In smaller hospitals and clinics with fewer deaths, meetings can be scheduled on a monthly basis.

By considering the number of cases to be reviewed and tailoring the frequency of meetings accordingly, the death audit committee can ensure that a meaningful and comprehensive review of each case is conducted, promoting continuous quality improvement and patient safety.

Death Review Presentation

During the death review presentation in a meeting, the following key points should be included:

1. Number of deaths and mortality rates: The total number of deaths that occurred within a specific period should be presented, along with the corresponding mortality rates. This provides an overview of the scale and impact of mortality within the hospital or clinic.
2. Summarized clinical records: A carefully prepared summary of the clinical records for each death case should be presented. This summary should highlight relevant details such as the patient's medical history, diagnoses, treatments, and significant events leading to the outcome. The purpose of the summary is to provide a clear understanding of the case to facilitate discussion.

Steps in Managing a Death Review Meeting

The following are the steps involved in managing a death review meeting:

1. Agree on time and venue: Determine a suitable date, time, and venue for the meeting. Ensure that it is convenient for all participants and book the venue accordingly.
2. Invite relevant staff: Send out invitations to all relevant staff members who should attend the meeting. Provide them with sufficient advance notice and information about the cases that will be reviewed.
3. Choose a chairperson: Select a chairperson who will lead and facilitate the meeting. This individual should be impartial and capable of maintaining a productive and respectful discussion.
4. Identify all deaths: Compile a list of all the deaths that occurred within the defined period under review. Ensure that the list is comprehensive and includes relevant details such as patient demographics and dates of death.
5. Assign case preparation and presentation: Assign someone the responsibility of preparing and presenting the cases for review. This person should carefully read and summarize the clinical records of the deceased patients, focusing on key information and pertinent details.
6. Prepare case summaries: Prepare concise summaries of each case to be presented at the meeting. These summaries should highlight important aspects of the patient's medical history, diagnoses, treatments, and any factors that may have contributed to the death.
7. Employ a standardized review process: Establish a short, standardized review process that emphasizes identifying avoidable deaths and contributory factors. This process should allow for staff involvement and accommodate specialty-specific questions when necessary.
8. Keep a summary of discussions: Maintain a record or summary of the cases discussed during the meeting. Include findings, identified avoidable and unavoidable factors related to the deaths, and the management plan agreed upon to address those factors.
9. Allocate sufficient time for case discussions: Allow ample time for in-depth discussions of each case, utilizing review questions or prompts to guide the conversation. Encourage a focus on system and process variations rather than individual competence.
10. Conduct an open review: Foster an open and transparent environment for the review, ensuring that all participants feel comfortable sharing their insights and perspectives. Summarize the actions agreed upon at the end of the meeting and track their progress in subsequent meetings.
11. Record meeting minutes: Keep detailed minutes of the meeting, documenting the key points discussed, decisions made, and any action items assigned. These minutes will serve as a valuable resource for future audits and performance analysis.

By following these steps, a death review meeting can be effectively managed to facilitate a thorough examination of cases, identify areas for improvement, and implement measures to prevent similar deaths in the future.

3. Identification of primary and final causes of death: The primary cause and final cause of death should be identified and discussed. This involves determining the underlying medical conditions or events that directly contributed to the patient's death. Understanding these causes helps in identifying potential areas for improvement.
4. Identification of avoidable factors: The meeting participants should actively identify any avoidable factors, missed opportunities, or instances of substandard care that may have played a role in the patient's death. This includes examining whether there were any preventable errors, delays in diagnosis or treatment, or failures in following established protocols. Discussing these factors allows for a critical analysis of the care provided.
5. Discussion of preventive measures: Based on the identified avoidable factors, the participants should engage in a discussion to develop a management plan aimed at preventing similar deaths in the future. This may involve implementing changes in protocols, enhancing staff training, improving communication channels, or addressing any system-level issues that contributed to the adverse outcome.

By actively engaging in the discussion and focusing on identifying causes, avoidable factors, and preventive measures, the death review meeting aims to promote a collaborative and proactive approach to improving patient care and reducing mortality rates.

Problems That May Occur with Mortality Review Meetings

Mortality review meetings, like any organizational process, can encounter several challenges. Some common problems that may arise with mortality review meetings include:

1. Attendance: One of the main challenges is ensuring the attendance of relevant staff members. It can be difficult to coordinate schedules and ensure that all necessary participants, including medical and nursing staff, are present at the meeting.
2. Time constraints: If there are a large number of deaths or cases to review, there may not be enough time to discuss each case in detail. Time constraints can limit the thoroughness of the review and hinder in-depth analysis of individual cases.
3. Lack of preparation: Inadequate preparation, such as incomplete case summaries or insufficient data analysis, can hinder the effectiveness of the meeting. Without proper preparation, the discussions may lack focus and fail to identify meaningful insights or improvement opportunities.
4. Blame culture: If the review process becomes focused on assigning blame or individual performance, it can create a defensive atmosphere among staff members. This blame culture can hinder open and constructive discussions, preventing the identification of system-level issues and improvement opportunities.
5. Inadequate follow-up: Without a structured mechanism for tracking and implementing the identified actions or recommendations from the meeting, the review process may lack accountability and fail to drive meaningful change. Proper

follow-up is essential to ensure that improvements are made, and lessons are learned from the review process.

6. Data limitations: The quality and availability of data can impact the effectiveness of mortality review meetings. Incomplete or inaccurate data can lead to incomplete understanding of the cases and hinder the ability to draw meaningful conclusions or identify patterns or trends.
7. Resistance to change: Implementing changes or improvements based on the findings of the mortality review meetings may face resistance from staff members or the broader organizational culture. Overcoming resistance to change and fostering a culture of continuous improvement can be a significant challenge.

Addressing these problems requires a proactive approach, clear communication, and a commitment to learning and improvement. It is essential to establish a supportive and blame-free environment that encourages open discussion, provides adequate resources for preparation and follow-up, and emphasizes the importance of collective learning and system-level improvement.

Patients Who Pass Away at Home After Being Discharged Are Not Included in the Death Audit

Including patients who die at home after discharge in the death audit can provide valuable insights into the quality of care and potential areas for improvement. While these patients may not have died within the hospital setting, their deaths are still relevant to the overall assessment of patient outcomes and healthcare processes.

- By including patients who die at home after discharge, the death audit can capture the continuum of care beyond the hospital stay and identify any potential issues or gaps in the transition from hospital to home care. It allows for a more comprehensive evaluation of the effectiveness of healthcare interventions, adherence to treatment plans, and the impact of post-discharge support and follow-up.
- Analyzing the cases of patients who die at home after discharge can help identify factors such as delayed or inadequate follow-up care, insufficient patient education, medication errors, or other issues that may have contributed to adverse outcomes. It provides an opportunity to review the effectiveness of discharge planning, communication between healthcare providers, and the provision of appropriate resources and support for patients after leaving the hospital.
- Including these cases in the death audit can help identify system-level improvements to enhance patient safety, optimize care transitions, and ultimately improve patient outcomes. It allows for a more comprehensive assessment of the healthcare system's performance and supports a patient-centered approach to quality improvement.

Some potential issues include:

- Staff involved in specific cases may be unable to attend.
- Individual staff members may feel threatened when management problems are discussed.
- Confidentiality issues may arise.

- It can lead to a blame game to identify the staff who made a mistake.
- Patient records may be missing or incomplete.
- Cases and data may not be adequately prepared.
- Lessons learned are not utilized to improve care.

Should Patients Who Are Brought Dead or Die Within a few Hours Before Assessment Is Done Be Included in Death Audit?

In general, patients who are brought dead or die within a few hours before assessment may not be included in the death audit. The purpose of a death audit is to review and analyze cases in which patient care was provided and to identify areas for improvement in healthcare processes and practices.

Patients who are brought dead or die shortly before assessment typically do not receive substantial medical interventions or care within the healthcare facility, which limits the opportunity for meaningful analysis and improvement. Including such cases in the death audit may not yield significant insights into the quality of care provided or contribute to identifying preventable factors or areas for improvement.

However, it is important to note that the specific criteria for inclusion in a death audit may vary depending on the institution or healthcare setting. Some organizations may have their own guidelines and policies regarding which cases should be included in the audit process. It is advisable to follow the established guidelines and criteria set by the institution or regulatory bodies to ensure consistency and relevance in the audit process.

It is worth considering alternative mechanisms to capture information and learn from cases where patients are brought dead or die shortly before assessment. For example, conducting separate reviews or investigations to understand the circumstances surrounding such cases and identify any system-level issues that may have contributed to the outcome. These findings can then be used to inform process improvements and enhance patient care in similar situations.

Annexure 2A: Sample Death Audit Form

General Information

Hospital details

Hospital Name:		
Department:	Unit Name/Number:	
Patient Registration No.	Ward:	Bed:
Name & Designation of Treating Doctor		
Name, & Name, Designation & Signature of Reporting Person		

Patient's Particular

Patient Name:	Age:	Sex:
Registration no	Occupation:	
Full Address:		
Diagnosis:	Disease Code (ICD-10):	

Dates & Times of Admission & Death

Date of Admission:	Time of Admission:
Date of Death:	Time of Death:
Type of admission (elective, emergency) and comorbidity	

Cause of Death

Underlying Cause of Death:
Contributory Cause of Death:
Immediate Cause of Death:
Note: Cardio-respiratory failure or cardiac arrest should not be listed as the cause of death since it merely describes the mode of death.

Legal Categorization of Death

Natural	Unnatural (Brought dead /Accidental / Suicidal / Homicidal)

Case summary (in the form of narrative including complaints at the time of admission, chronology of events up to death of the patient)

Treatment given: Medical/Surgery/Procedure/ Radiotherapy/ Chemotherapy/ Others (specify)

Investigations Done and Their Reports

SL No	Investigation	Report	Remarks
1			

SL No	Investigation	Report	Remarks
2			
3			
4			
5			
6			
7			

Medications Received

S. No	Medications	Date started	Date stopped	Remarks
1				
2				
3				
4				
5				
6				
7				

AUTOPSY/ POSTMORTEM

Performed:	Yes	No	Unknown

If performed, please report the gross findings

Case Assessment

If there any areas of CONCERN or ADVERSE EVENTS in the management of this patient (Comments on potential avoidable factors, missed opportunities, and sub-standard care)

Describe the key occurrences throughout the hospital treatment period.

Yes	No

Note any areas of:

Concern

Adverse Event

Note if these areas caused any of the following:

- Made no difference to the outcome
- May have contributed to death
- Caused death of patient who would otherwise be expected to survive

(a) If surgery was performed were there any Areas of Concern, or Adverse Events in any of the following areas:

Discussion points	*Yes*	*No*	*NA*
Pre-anesthetic check-up/fitness for surgery/ treatment			
Decision to operate			
Choice of operation			
Timing of operation (too late, too soon, wrong time of the day			
Intra-operative process			
Problems in functioning of OT			
Grade/experience of surgeon deciding			
Grade/experience of surgeon operating			
Postoperative period			
Others, please specify			

Was the patient admitted to an ICU or HDU for critical care during this stay?

Yes	No

If not, was it necessary for this patient to receive critical care in the ICU/HDU?

(b) In case of medical management what delays in the facility contributed to the death

Yes	No

What delays in the facility contributed to the death	*Yes*	*No*	*NA*
Delay in seeking help by patient			
Refusal of treatment or admission by patient			
Lack of blood products			
Delay in receiving treatment on admission			
Inability to treat problem at this facility			
Inadequate in-house clinical expertise			
Delay in getting specialist referral			
Diagnostic problems			
Lack of essential medicines			
Inadequate monitoring			
Inadequate equipment			

Audit committee's assessment of the general mortality risk:
Was the death preventable?

Definitely	☐
Probably	☐
Probably not	☐
Definitely not	☐
Don't know	☐

Quality of medical records: Comment on the key data elements missing from the patient's file

Legibility:	1. Good	2. Poor

Final recommendations and follow-up actions to address audit findings of the mortality audit committee

1.
2.
3.

Minimal	☐
Mild	☐
Moderate	☐
Severe	☐

Attestation by the mortality audit committee members

	Name	Designation	Signature
1			
2			
3			
4			
5			
6			

Date:

Annexure 2B: Sample Death Audit Report

[Hospital Name] Death Audit Report

Date: [Date of Death Audit] Venue: [Meeting Venue]

1. Introduction: The death audit meeting was held on [Date of Meeting] at [Meeting Venue]. The purpose of the meeting was to review and analyze recent patient deaths within the hospital to identify any areas of improvement in patient care processes and practices. The meeting was attended by key representatives from various departments, including management, treating doctors, and support departments.
2. Attendees: The following individuals were present at the meeting:
   - [Name, Department/Role]
   - [Name, Department/Role]
   - [Name, Department/Role]
   - [Name, Department/Role]
   - [Name, Department/Role]
   - [Name, Department/Role]
   - [Name, Department/Role]
   - [Name, Department/Role]
   - [Name, Department/Role]
3. Cases Reviewed: During the meeting, the following cases were reviewed and discussed:

Case 1:

- Patient Name: [Name]
- Date of Death: [Date of Death]
- Primary Cause of Death: [Primary Cause]
- Summary of Clinical Record: [Brief summary of the patient's medical history, diagnosis, and treatment]
- Discussion Points: [Key points discussed during the review]
- Identified Avoidable Factors: [Any factors identified that could have been prevented]
- Management Plan: [Actions agreed upon to prevent similar deaths in the future]

Case 2:

- Patient Name: [Name]
- Date of Death: [Date of Death]
- Primary Cause of Death: [Primary Cause]
- Summary of Clinical Record: [Brief summary of the patient's medical history, diagnosis, and treatment]
- Discussion Points: [Key points discussed during the review]
- Identified Avoidable Factors: [Any factors identified that could have been prevented]
- Management Plan: [Actions agreed upon to prevent similar deaths in the future]

4. Findings and Recommendations: Based on the discussions held during the meeting, the following findings and recommendations were made:

- Overall, the death audit highlighted the need for improved communication between healthcare professionals involved in patient care.
- It was noted that delays in diagnosis and treatment were contributing factors in several cases reviewed.
- There is a need to enhance staff awareness and adherence to infection control protocols.
- Further training and education programs should be implemented to improve the quality of clinical observations and prescribing practices.
- The importance of timely and accurate documentation was emphasized, along with the need for regular monitoring of patient progress.
- A comprehensive review of thromboprophylaxis protocols and practices should be conducted to minimize the risk of thrombotic events.

5. Action Plan: The following action plan was agreed upon to address the identified findings and recommendations:

- Develop and implement a communication protocol to improve interdisciplinary communication and collaboration.
- Conduct regular training sessions on early diagnosis and timely treatment interventions.
- Enhance infection control measures and provide ongoing education to healthcare staff.
- Establish guidelines and protocols for appropriate clinical observations and prescribing practices.
- Implement a system for regular monitoring and review of patient progress.
- Form a committee to review and update thromboprophylaxis protocols and guidelines.

6. Next Meeting: The next death audit meeting is scheduled to take place on [Date of Next Meeting]. Participants were requested to ensure their availability and to prepare for the discussion of additional cases.
7. Conclusion: The death audit meeting provided valuable insights into the areas of improvement in patient care processes and practices within the hospital. The identified findings and recommendations will serve as a basis for implementing necessary changes and enhancing the quality of care provided to patients. The minutes of the meeting will be documented for audit and performance evaluation purposes.

[Signature] [Name of Chairperson] [Position/Department] [Date]

An Example of a Change in Practice with Death Audit Recommendations with Improved Governance

During the monthly departmental Mortality and Morbidity (M&M) meeting, all 18 deaths that occurred in the previous month were reviewed. The cases were initially assessed by two junior doctors in consultation with the M&M meeting chair. A comprehensive PowerPoint presentation was prepared, summarizing each case, and providing details about the clinical aspects and causes of death.

Out of the 18 cases, four cases were identified where the patients' deaths were unexpected. These cases were subjected to further examination and discussion led by the meeting chair. One particular case involved an acutely ill patient admitted to a medical ward who did not receive timely specialist input, despite a referral being made. The M&M meeting concluded that this lack of appropriate clinical care was a contributing factor to the patient's death and identified "Process issues and communication problems" as factors leading to the inadequate response to the referral. Additionally, the case was reviewed in the corresponding specialist M&M meeting.

Actions Taken

1. Specialty Consultant Involvement: The M&M meeting decided to invite a specialty consultant to the next meeting for an open discussion regarding the identified issues and to explore possible solutions.
2. Joint Referral Guidelines: It was agreed upon that both departments would collaborate in creating referral guidelines to ensure clear and effective communication between specialties, thereby improving the timeliness of specialist input.

Problem Recurrence

1. Failure to Obtain Prompt Specialist Input: Despite the initial actions taken, there was a similar occurrence where prompt specialist input was not obtained.

Actions Taken:

1. Escalation to Divisional M&M Meeting: The meeting chair raised the issue at the quarterly divisional M&M meeting, highlighting the persistent problem and the need for a broader intervention.
2. Organizational Guidance: The head of the organization emphasized the importance of compliance with the new referral guidance, advising all divisions about the need to follow the established protocols for obtaining timely specialist input.

By implementing these actions, the M&M meeting aimed to address the identified gaps in clinical care, improve communication and processes, and ensure that prompt specialist input is obtained when needed. This example showcases the significance of death audit recommendations and the role of improved governance in driving positive changes and enhancing patient outcomes.[Signature] [Name of Chairperson] [Position/Department] [Date]

Annexure 3: Medical Audit

Annexure 3A: Improving Medication Safety

Objective: The objective of this medical audit is to assess and improve medication safety practices within the hospital.

Methodology

1. Audit Team: A multidisciplinary team consisting of a pharmacist, nurse, and a quality improvement officer was formed to conduct the audit.
2. Sample Selection: A random sample of 100 patient records was selected from the past six months. These records included both inpatient and outpatient cases.
3. Audit Criteria: The audit criteria were developed based on established medication safety guidelines, hospital policies, and best practices. The criteria focused on the following areas: a. Prescription accuracy and completeness. b. Medication reconciliation. c. Proper documentation of medications. d. Adherence to medication administration protocols. e. Patient education on medication usage and potential side effects. f. Reporting and management of medication errors.
4. Data Collection: The audit team reviewed the selected patient records and collected data based on the predetermined audit criteria. They documented any deviations or non-compliance with the criteria.
5. Data Analysis: The collected data was analyzed to identify patterns, trends, and areas of concern. The team looked for common issues and recurring errors related to medication safety.

Findings: The audit revealed several areas of improvement in medication safety practices:

1. Incomplete documentation of medication allergies in patient records.
2. Lack of medication reconciliation during transitions of care.
3. Inconsistent documentation of medication administration times.
4. Insufficient patient education on medication usage and potential side effects.
5. Inadequate reporting and follow-up on medication errors.

Recommendations and Actions: Based on the audit findings, the following recommendations were made to improve medication safety practices:

1. Enhance documentation processes to ensure accurate recording of medication allergies.
2. Implement a standardized medication reconciliation process during transitions of care to ensure continuity and accuracy in patients' medication regimens.
3. Develop guidelines for proper documentation of medication administration times.
4. Enhance patient education programs on medication usage and potential side effects.
5. Establish a robust system for reporting and investigating medication errors, with a focus on learning and prevention.

The hospital's quality improvement committee reviewed the audit findings and recommendations. They prioritized the implementation of the recommended actions and assigned responsible individuals or teams to oversee each improvement initiative. Regular progress reviews and follow-ups were scheduled to monitor the effectiveness of the interventions.

Conclusion: This medical audit on medication safety provided valuable insights into areas requiring improvement within the hospital's medication management practices. By implementing the recommended actions, the hospital aimed to enhance patient safety, reduce medication errors, and improve overall medication management processes.

[Signature]

[Namc of Audit Tcam Lcad]

[Position/Department]

[Date]

Annexure 3B: Dengue Management Clinical Audit Example I

Evidence:

During monsoon months an outbreak of dengue cases occurs and 10% patients presenting with dengue fever and dengue hemorrhagic fever die because of lack of appropriate and timely fluid management and inappropriate use of blood and blood products deaths.

Objective: The objective of this dengue management clinical audit is to assess the effectiveness of fluid management and appropriate use of blood and blood products in reducing mortality due to dengue fever (DF) and dengue hemorrhagic fever (DHF) and improving patient outcomes.

Methodology

1. Audit Scope: The audit focused on dengue cases leading to death within the hospital setting during the monsoon months.
2. Data Collection: A review of individual patient records of dengue cases resulting in death was conducted. Data on fluid management, use of blood and blood products, and clinical outcomes were collected.
3. Audit Indicators: The following clinical indicators were used for analysis:

a. Case fatality rate (DF): Number of deaths due to dengue fever divided by the total number of clinically diagnosed dengue fever cases.
b. DHF fatality rate: Number of deaths due to dengue hemorrhagic fever (DHF) or dengue shock syndrome (DSS) divided by the total number of clinically diagnosed DHF/DSS cases.
c. Appropriateness of blood and blood component usage: Number of cases with significant bleeding that received platelet concentrates, fresh frozen plasma, whole blood, or packed cells divided by the total number of DF/DHF cases with significant bleeding.

Findings

1. Case fatality rate (DF): The fatality rate for dengue fever was above the target of <5%.
2. DHF fatality rate: The DHF fatality rate was also higher than the desired target of zero deaths due to DHF/DSS.
3. Appropriateness of blood and blood component usage: The audit revealed variations in the usage of blood and blood products in cases with significant bleeding, indicating a need for improvement in adherence to guidelines.

Recommendations

1. Enhance clinical protocols and guidelines for fluid management in DHF/DSS cases, emphasizing appropriate and timely administration.
2. Strengthen clinical training and awareness among healthcare professionals on the management of dengue cases, with a focus on early recognition of DHF/DSS and appropriate intervention.
3. Develop standardized guidelines for the use of blood and blood products, ensuring that they are utilized appropriately and in line with clinical indications.
4. Implement regular audits and monitoring of dengue management practices, including fluid management and blood product usage, to track progress and identify areas for improvement.

Action Plan: The audit findings and recommendations were presented to the hospital's management and relevant departments. An action plan was developed, which included the following actions:

1. Revision and dissemination of updated clinical guidelines on dengue management, specifically addressing fluid management, blood product usage, and DHF/DSS recognition and intervention.
2. Conducting training sessions and workshops for healthcare professionals to enhance their knowledge and skills in dengue management.
3. Establishing a regular audit cycle to monitor adherence to guidelines and assess the impact of interventions on reducing mortality rates.
4. Collaborating with the blood bank and transfusion services to ensure appropriate availability and utilization of blood and blood products for dengue patients.

Conclusion: The dengue management clinical audit highlighted the need for improvement in fluid management, appropriate blood product usage, and overall management of dengue cases. By implementing the recommended actions and monitoring progress through regular audits, the hospital aims to reduce mortality rates, improve patient outcomes, and enhance the quality of care provided to dengue patients.

[Signature]

[Name of Audit Team Lead]

[Position/Department]

[Date]

Annexure 3C: Dengue Management Clinical Audit Example II

Objective: The objective of this dengue management audit is to assess the adherence to established guidelines and protocols in the management of dengue cases within a healthcare facility.

Methodology

1. Audit Team: A multidisciplinary team consisting of infectious disease specialists, nurses, laboratory technicians, and quality improvement officers was formed to conduct the audit.
2. Sample Selection: A random sample of 50 patient records of confirmed dengue cases from the past year was selected. The sample included both pediatric and adult patients.
3. Audit Criteria: The audit criteria were developed based on national and international guidelines for dengue management. The criteria focused on the following areas:
   a. Timely and accurate diagnosis of dengue
   b. Adequate monitoring of vital signs and laboratory parameters
   c. Appropriate fluid management
   d. Proper use of antipyretics and analgesics
   e. Identification and management of warning signs and complications
   f. Patient education on self-care and follow-up instructions
4. **Data Collection**: The audit team reviewed the selected patient records and collected data based on the predetermined audit criteria. They documented any deviations or non-compliance with the criteria.
5. **Data Analysis**: The collected data was analyzed to identify areas of non-compliance, patterns, and potential gaps in dengue management. The team looked for common issues and areas requiring improvement.

Findings: The audit revealed several areas of improvement in dengue management:

1. Delayed diagnosis due to inadequate testing or misinterpretation of test results.
2. Inconsistent monitoring of vital signs and laboratory parameters.
3. Variations in fluid management approaches and inadequate documentation of fluid balance.
4. Inappropriate use of antipyretics and analgesics without considering individual patient needs.
5. Inadequate identification and management of warning signs and complications.
6. Limited patient education on self-care measures and follow-up instructions.

Recommendations and Actions: Based on the audit findings, the following recommendations were made to improve dengue management practices:

1. Enhance training and awareness among healthcare professionals regarding dengue diagnosis and interpretation of test results.
2. Implement standardized protocols for the monitoring of vital signs and laboratory parameters.
3. Develop clear guidelines for fluid management and ensure proper documentation of fluid balance.
4. Promote evidence-based use of antipyretics and analgesics, considering patient factors and symptom severity.
5. Strengthen protocols for the identification and management of warning signs and complications.
6. Enhance patient education materials and ensure comprehensive instructions on self-care and follow-up.

The healthcare facility's quality improvement committee reviewed the audit findings and recommendations. They prioritized the implementation of the recommended actions and assigned responsible individuals or teams to oversee each improvement initiative. Regular progress reviews and follow-ups were scheduled to monitor the effectiveness of the interventions.

Conclusion: This dengue management audit provided valuable insights into areas requiring improvement within the healthcare facility's dengue management practices. By implementing the recommended actions, the facility aimed to enhance patient outcomes, improve adherence to guidelines, and optimize dengue management protocols.

[Signature]

[Name of Audit Team Lead]

[Position/Department]

[Date]

Annexure 3D: Adult Community-Acquired Pneumonia (CAP) Management Audit

Evidence

The number of deaths in adult patients with community-acquired pneumonia (CAP) has increased during the last 3 months reportedly because of delay in investigations and initiation of treatment.

1. **Objective:** The objective of this audit is to assess the management of adult patients with community-acquired pneumonia (CAP) in order to establish a baseline, improve outcomes, increase adherence to local CAP guidelines, and reduce inappropriate antibiotic use particularly:
   - Reduce the proportion receiving antibiotics before the diagnosis is confirmed.
   - Reduce unnecessary use of parenteral (IV) and combination antibiotics for low-severity CAP.

Criteria

1. 80% of the prescription should be as per guidelines.
2. In 80% of the cases choice of antibiotic should be as per local CAP treatment guidelines.
3. Antibiotics should be administered only after confirmation of the diagnosis.

Methodology:
1. Study Design: Retrospective random study design.
2. Sample Size: Randomly selected sample of 30 case notes from the past 6 months out of the total number of cases admitted with a diagnosis of CAP.
3. Audit Parameters: Severity of CAP (using CURB65 score), referral to critical care, average length of stay, investigations performed, treatment given, and outcomes were reviewed from the case note files.
4. Data Analysis: Assessment of adherence to guidelines, appropriateness of antibiotic choice, timing of antibiotic administration, and identification of areas for improvement.

Results
Patient Profile and Outcome:
- The mean age of cases was 71 years, with over half (52%) being female.
- Comorbid illnesses included chronic heart disease (23%), hypertension, and COPD (22% each).
- Based on the CURB65 score, 40% had low severity CAP (Score o-1), 30% had moderate severity CAP (Score 3 to 5), and 30% had high severity CAP (Score 3-5).
- The average length of stay was 5 days, with critical care admission required in 7% of cases.
- Death occurred in 21.3% of patients, and the readmission rate within 30 days of discharge was 8.3%.

Processes of Care

- 52% of patients received the first dose of antibiotics 8 h after admission, while 20% received antibiotics ≥ 4 h after admission.
- The time interval from admission to a chest x-ray was generally long, with 41% getting it <4 h, 43% between 4 and 6 h, and 11% beyond 6 h.
- In 74% of cases, the chest x-ray was reviewed before antibiotics were given, but in 21% of cases, the interval between the x-ray and the first dose of antibiotics exceeded 4 h.

Use of Antibiotics

Adherence to local CAP guidelines for antibiotic use was only 54%.

- Initial empirical antibiotics were given intravenously (IV) in 77% of cases.
- There was an overuse of combination antibiotics in patients with low-severity CAP (53% received beta-lactam + macrolide combination).
- In patients with moderate to high severity CAP, there was an underuse of combination antibiotics.

Areas for Improvement in Management

1. Increase adherence to local CAP guidelines.
2. Reduce delays in obtaining chest x-rays and administering the first dose of antibiotics.
3. Increase the proportion of patients with high-severity CAP receiving combination antibiotics.
4. Reduce the proportion of patients receiving antibiotics before the diagnosis is confirmed.
5. Reduce unnecessary use of parenteral (IV) and combination antibiotics for low-severity CAP.

Conclusion: The audit revealed areas of improvement in the management of adult patients with CAP, including adherence to guidelines, timing of interventions, and appropriate antibiotic use. By implementing the suggested improvements, such as increasing guideline adherence, reducing delays, and optimizing antibiotic selection, healthcare facilities aim to improve outcomes, reduce inappropriate antibiotic use, and enhance patient care.

[Signature]

[Name of Audit Team Lead]

[Position/Department]

[Date]

Annexure 4: Surgical Audit

A surgical audit is a systematic and critical analysis of medical care in surgical practice, encompassing various aspects such as diagnosis, treatment procedures, resource utilization, and patient outcomes. It involves both qualitative and quantitative measurements of activity and aims to improve the quality of care provided to patients.
The scope of a surgical audit can cover several areas, including:

1. 30-day mortality and significant morbidity: Assessing the rates of death and major complications within 30 days of surgery.
2. Length of hospital stay: Evaluating the duration of hospitalization for surgical patients.
3. Unplanned readmission or re-operation rates: Examining the rates of unexpected readmissions or the need for repeat surgeries.
4. Positive and negative outcomes: Analyzing both favorable and unfavorable outcomes resulting from surgical interventions.
5. Operation-specific complications: Focusing on complications that are specific to particular surgical procedures.
6. Process of care, such as preoperative care: Assessing the quality of care provided before surgery, including preoperative evaluations and preparations.
7. Time on the waiting list: Evaluating the duration patients have to wait for their surgical procedures.
8. Numbers waiting for outpatient appointment: Monitoring the number of patients waiting for appointments in outpatient surgical clinics.
9. Use of investigations: Examining the appropriateness and frequency of diagnostic investigations performed before and after surgery.
10. Justification of management: Assessing the decision-making process and appropriateness of the management plans for surgical patients.
11. Patient satisfaction: Measuring patient satisfaction with the care they received during their surgical experience.
12. Adverse event incident reports: Reviewing and analyzing reports of adverse events or incidents related to surgical procedures.

Peer review plays a crucial role in surgical audits, ensuring the credibility and validity of the data collected. By comparing audit data to historical records and established standards, healthcare professionals can identify areas for improvement, implement changes, and ultimately enhance patient care and outcomes.

Overall, surgical audits are an essential component of quality improvement in surgical practice, facilitating the identification of areas of excellence and areas requiring improvement to enhance patient safety and care delivery.

Adverse events are unintended harms that occur during healthcare interventions and are not a direct result of the underlying disease process. These events can range from major complications to minor incidents, and monitoring and reporting them is crucial for quality improvement in surgical practice. Some of the adverse events include:

1. Death in a surgical patient: This refers to any patient who dies as a result of surgical intervention.
2. Unplanned readmission within 28 days: When a patient requires readmission to the hospital within 28 days of their initial surgery.
3. Unplanned readmission to ICU from ward: The need for an unplanned transfer of a patient from a regular hospital ward to the Intensive Care Unit (ICU).
4. Unplanned reoperation: Any surgical procedure that is performed on a patient as a result of a complication or unexpected outcome from the initial surgery.
5. Unplanned blood transfusion: The administration of blood or blood products to a patient due to unforeseen circumstances or complications.
6. Transfer for more complex care: When a patient needs to be transferred to a higher level of care or a different healthcare facility for specialized treatment.
7. Complication (Grade 3): Refers to a significant adverse event or complication that occurs during or after surgery, graded according to severity.
8. Complication prolonging anticipated hospital stay by more than 7 days: When a complication arises that extends the patient's hospital stay beyond the initially anticipated duration by more than 7 days.
9. Inadvertent perforation of a viscus: Accidental perforation of a hollow organ or structure during a surgical procedure.
10. Serious drug reaction or interaction: An adverse reaction or interaction between medications that results in significant harm or complications for the patient.
11. Medication error: Any mistake or error related to medication administration, prescribing, or dispensing that can potentially harm the patient.
12. Cardiac or respiratory arrest: The cessation of normal heart or respiratory function, requiring immediate intervention and resuscitation.
13. Medical Emergency Team call: Activation of the medical emergency team due to a critical deterioration or sudden change in the patient's condition.
14. Fall: Any incident where a patient falls and sustains an injury during their hospital stay.
15. Pressure sore: The development of a pressure ulcer or bed sore during the patient's hospitalization.
16. Reportable infection: Infections that meet specific criteria for reporting, such as surgical site infections or healthcare-associated infections.
17. Booked for theatre and cancelled: When a patient's scheduled surgery is canceled or postponed for various reasons.

It is important to monitor and report these adverse events to identify areas for improvement, implement preventive measures, and enhance patient safety and outcomes.

The process for evaluating surgical audits typically involves several key steps. General outline of the process is as under:

1. Data Collection: The first step is to collect relevant data related to the surgical procedures or interventions being audited. This may include patient demographics, procedure details, outcomes, complications, length of hospital stay, readmission rates, and other relevant metrics.

2. Data Analysis: Once the data is collected, it needs to be analyzed to identify patterns, trends, and areas for improvement. Statistical analysis may be performed to determine the significance of findings and to compare the results with established benchmarks or standards.
3. Comparison with Standards: The audit findings are compared with established standards or guidelines for surgical care. These standards can be based on national or international guidelines, best practices, or previous audit results. By comparing the audit data with standards, areas of non-compliance or opportunities for improvement can be identified.
4. Peer Review: The audit findings are subjected to peer review, where they are critically assessed by a panel of experts or peers in the field. This helps ensure the accuracy and validity of the audit results and provides additional insights and recommendations.
5. Identification of Areas for Improvement: Based on the audit findings and peer review, specific areas for improvement are identified. These can include changes to surgical protocols, modifications to preoperative or postoperative care, improvements in patient selection or risk assessment, or enhancements to surgical techniques or processes.
6. Development of Action Plan: An action plan is created to address the identified areas for improvement. This plan outlines specific steps, interventions, and strategies to be implemented to enhance the quality and safety of surgical care. It may involve changes in protocols, staff training, resource allocation, or other interventions as deemed necessary.
7. Implementation of Interventions: The action plan is put into action, with the necessary changes, interventions, and improvements being implemented in the surgical practice. This may involve changes in surgical protocols, staff training, implementation of new technologies or equipment, or other measures to enhance patient safety and improve outcomes.
8. Post-Intervention Evaluation: After the interventions have been implemented, a follow-up evaluation is conducted to assess their effectiveness and impact. This may involve a reassessment of the audit metrics, comparing post-intervention data with pre-intervention data, and assessing whether the desired improvements have been achieved.
9. Continuous Monitoring and Feedback: Surgical audits are an ongoing process, and continuous monitoring is essential to ensure sustained improvement. Regular data collection, analysis, and feedback are important to track progress, identify new areas for improvement, and make further adjustments as needed.

By following this evaluation process, surgical audits can provide valuable insights and drive continuous quality improvement in surgical care, leading to better outcomes and enhanced patient safety.

In addition, other important components for comprehensive evaluation include:

1. Meetings should be held regularly: Regular meetings provide a platform for discussing audit findings, sharing insights, and fostering collaboration among team members. These meetings serve as a forum for reviewing data, identifying areas for improvement, and developing action plans.
2. All teams in a particular specialty should be involved: Involving all teams within a specific specialty ensures that a comprehensive evaluation of surgical care is conducted. This allows for a broader perspective and insights from different practitioners involved in the care pathway.
3. All members of the team from Intern to Consultant should participate: Including all team members regardless of their role promotes a multidisciplinary approach to evaluation. Each team member can contribute unique insights based on their expertise and experience, leading to a more comprehensive understanding of the audit findings.
4. All patients admitted under each service should be included in the review process: The inclusion of all patients admitted under a particular service ensures a representative sample for evaluation. This helps in capturing a comprehensive picture of the surgical care provided within that service and identifying trends or patterns in outcomes.
5. All morbidity and mortality occurring in the unit should be discussed: Reviewing all morbidity (complications) and mortality cases is crucial to identify adverse events and understand their causes. Analyzing these cases can help identify areas for improvement in patient care, surgical techniques, or perioperative processes.
6. There should be a mechanism for altering practice based on the results of evaluation: Evaluation alone is not sufficient; it should be followed by actionable steps for improvement. Establishing a mechanism for implementing changes based on the audit findings ensures that the evaluation process leads to meaningful improvements in surgical care.
7. The final stage of the "audit loop" should include re-evaluation of any alterations made in practice: It is essential to assess the impact of the implemented changes. Re-evaluation helps determine whether the alterations have resulted in the desired outcomes and identifies the need for further adjustments or improvements.
8. A written record should be kept of the audit process: Maintaining a written record of the audit process ensures documentation of the findings, actions taken, and outcomes. This record serves as a valuable resource for future reference, tracking progress over time, and sharing information with relevant stakeholders.

By following these steps, the evaluation of surgical audits can be systematic, collaborative, and effective in driving continuous improvement in surgical care.

Annexure 4A: Sample Safe Surgery Audit Report

The sample safe surgery audit plan provides an outline of a comprehensive approach to evaluating compliance with the Surgical Safety Checklist and identifying incidents related to non-compliance with the Hospital Policy and Procedure for Safe Surgery.

Audit Plan

Objectives of the Audit

1. To ensure the Surgical Safety Checklist is completed appropriately.
2. To identify if there were any incidents related to non-compliance with the Hospital Policy and Procedure for Safe Surgery Methodology.

Audit 1: Retrospective Healthcare Records Audit

1. Random selection of patient charts: A sample of patient charts should be randomly selected from the theatre register.
2. Sample size: The sample should be large enough to be representative but small enough for efficient data collection.
3. Data collection form: A separate form should be completed for each patient, ensuring data protection by using respondent numbers instead of patient healthcare record numbers.
4. Analysis: Calculate the percentage compliance for each criterion identified in the data collection form.

Audit 2: Observational Audit of the Checklist

1. Observation of checklist utilization: The audit tool should include questions 8-36, focusing on observing the checklist being used by the operating team.
2. Assigned observer: A member of the operating team should be assigned to complete the observational audit.

Audit 3: Review of Incident Report Forms

1. Review of incidents: All incidents related to non-compliance with the Hospital Policy and Procedure for Safe Surgery should be reviewed.
2. Results sharing: Share the audit results with the Operating Theatre Manager and Staff, Clinical Lead for Surgery/Obstetrics/Dental, and the hospital's Clinical Governance/Audit Committee.
3. Action plan: Develop an action plan that outlines the steps to address areas of non-compliance. Assign responsibilities and set completion dates for each action.
4. Annual report: Write an annual report on compliance with the Hospital Policy and Procedure for Safe Surgery. Include evidence from the audits and adverse incident reporting.

By implementing this audit plan, one can systematically evaluate compliance with the Surgical Safety Checklist, identify areas of non-compliance, and take appropriate actions to improve safe surgery practices.

Safe Surgery Checklist Audit Tool

Number of Observations: A minimum of 10 procedures per quarter.

Area of Observation: Observations should be made in both the operating room and other procedure areas.

Training of Observers: Observers should be trained on the checklist elements and the form
to be used.

Annexure 4B: Audit Tool Safe Surgery Checklist

Respondent number: _______

	Consent		
1.	Is there a written and signed consent form available in the patient's clinical Record?	**Yes**	**No**
2.	Is the consent form legible?	**Yes**	**No**
3.	Is the consent form signed by a doctor (name, designation, signature, and date) who was present during the surgery?	**Yes**	**No**
4.	No abbreviations used on the consent form	**Yes**	**No**
	Surgical Safety Checklist		
5.	Is there an addressograph on the checklist?	**Yes**	**No**
6.	Is the operation date documented on the checklist?	**Yes**	**No**
7.	Was the checklist filed with the theatre documentation in the HCR?	**Yes**	**No**
	"SIGN IN" Were each of the following checks completed?	**Yes**	**No**
8.	Patient confirmed identity, site, procedure, and consent	**Yes**	**No**
9.	Surgical site marked/not applicable	**Yes**	**No**
10.	Anesthetic checklist completed	**Yes**	**No**
11.	Known allergies checked	**Yes**	**No**
12.	Blood loss risk documented	**Yes**	**No**
13.	Vein thrombosis (VTE) Prophylaxis check	**Yes**	**No**
14.	American Society of Anaesthesiologists' (ASA) grade checked	**Yes**	**No**
15.	Sign In section signed	**Yes**	**No**
16.	Sign In section timed	**Yes**	**No**
	"TIME OUT" Were each of the following checks completed?		
17.	All team members introduced themselves	**Yes**	**No**
18.	Verbal confirmation of patient's name, procedure, and incision site	**Yes**	**No**
19.	Verification that patient positioned correctly	**Yes**	**No**
20.	Essential imaging displayed/not applicable	**Yes**	**No**
21.	Antibiotic prophylaxis/not applicable	**Yes**	**No**
22.	Patient-specific concerns: Surgeon	**Yes**	**No**
23.	Patient-specific concerns: Anesthetist	**Yes**	**No**
24.	Patient-specific concerns: Nursing Team	**Yes**	**No**
25.	Equipment issues: Surgeon	**Yes**	**No**
26.	Equipment issues: Nursing	**Yes**	**No**
27.	Time Out section signed	**Yes**	**No**
28.	Time Out section timed	**Yes**	**No**
	"SIGN OUT" Were each of the following checks completed?		
29.	Name of procedure confirmed	**Yes**	**No**
30.	Completion of instrument, sponge, and needle count	**Yes**	**No**
31.	Specimen labelling	**Yes**	**No**
32.	Patient-specific post-op concerns: Surgeon	**Yes**	**No**
33.	Patient-specific post-op concerns: Anesthetist	**Yes**	**No**
34.	Patient-specific post-op concerns: nursing	**Yes**	**No**
35.	Sign Out section signed	**Yes**	**No**
36.	Sign Out section timed	**Yes**	**No**

Intra-operative Record Checklist

	Intra-operative record	Yes	No
1	Is there an intra-operative record in patient's clinical record		
2	Is there documented evidence on the intra-operative record of		
	Surgeon's name and designation?		
	Assistant surgeon(s) and attending staff names?		
	Operation/procedure performed?		
	Date of operation/procedure?		
	Time in and time out of theatre/procedure room?		
	Indications/reason for operation/procedure?		
3	Results/outcomes of the operation/procedure?		
4	Follow-up arrangements?		

Anesthetic Record Checklist

	Anesthetic record	Yes	No
1	Is there an anesthetic record present in the patient's clinical record?		
2	If yes: are the following recorded		
	Date on the anesthetic chart?		
	Evidence on the anesthetic chart that the Anesthetist conducted a preoperative interview with the patients?		
	Evidence of assessment and recording of patient allergies/ previous reactions?		
	Evidence of medications, doses, route of administration given during anesthesia		
	Evidence of monitoring of patient's vitals and IV fluids		
	ASA score documented		
	Anesthetic type (e.g., general anesthesia, regional, sedation, local)		
	Anesthetist's name and signature present		
	Surgery Start and End times		
1	Was the surgery initiated within 5 min of the planned start time?		
2	Did the surgery finish within 30 min of the intended finish time?		

Criteria

- Surgeries should start within 5 min of their intended start time
- Surgeries should end within 30 min of the intended finish time

Method & Analysis

- Data collection should span a 2-week period (10 working days).
- Data should be gathered either by the doctor or by another staff member.
- Data analysis involves dividing the number of surgeries that start on time (Y) by the total number of surgeries audited. For example, over a 2-week period, a doctor starts 25 surgeries within 5 min of the intended start time out of a total of 30 surgeries. Therefore, the compliance percentage is calculated as 25/30 x 100 = 83%, which should be compared to the set standard.

Annexure 5: Significant Event Analysis

What Happened?

A patient attended the outpatient department to pick up a prescription for phenytoin from a tertiary care hospital. He was given the prescription for phenytoin, but in addition there was also a prescription for phenytoin from another hospital which had been lying from the month before and he was also given this prescription. The patient therefore had a large amount of phenytoin at home and over the following few days an overdose was taken, with hospital admission and monitoring required.

- The patient received two prescriptions for phenytoin, one from the current visit and another from a previous hospital visit.
- The patient had a large quantity of phenytoin at home, leading to an overdose and the need for hospital admission and monitoring.

Why Did It Happen? Possible Issues for Discussion

1. System to identify uncollected prescriptions: The hospital may lack a robust system to identify prescriptions that have not been collected by patients, leading to unintentional duplication.
2. Minimizing quantity of dangerous medicines: The incident highlights the need to assess and minimize the quantity of potentially dangerous medicines available to patients, especially when there is a risk of an overdose.
3. Follow-up of patients on dangerous medicines: The incident raises questions about the follow-up process for patients with significant illnesses who are prescribed potentially dangerous medicines.
4. Practice of multiple prescriptions: The practice of multiple prescriptions being reviewed in tertiary care settings may contribute to the risk of duplication.

What Would You Change? Potential Solutions to Consider

1. Reviewing prescriptions and medications during patient visits: Healthcare professionals should thoroughly review all prescriptions, including old prescriptions for duplicate medicines, during a patient's visit to the hospital.
2. Identification and explanation of duplicate prescriptions: Duplicate prescriptions should be identified and destroyed, and patients should be informed about the reason for their removal.
3. Tagging repeat prescriptions: Repeat prescriptions for potentially dangerous medicines can be "tagged" in the system, allowing healthcare professionals to check if the previous prescription has been collected before issuing a new one.
4. Weekly dispense marking: Prescription labelling can include a "weekly dispense" marking to minimize the volume of medication available to patients at a given time.
5. Increased review of patients on specific medications: Patients on specific medications that pose a higher risk, like phenytoin, may benefit from increased monitoring and review to assess their risk profile.

What Are the Main Learning Points?

- There is no foolproof system to completely prevent patients from hoarding medication and potentially overdose. However, the risk can be significantly reduced through system improvements.
- Implementing changes to regularly check for "old" prescriptions and taking appropriate actions when necessary is crucial in preventing incidents like this from occurring.

By analyzing this significant event and implementing the proposed changes, healthcare providers can enhance patient safety, minimize the risk of medication errors, and improve the overall quality of care.

Sample Report Form: Significant Event Analysis

Date of significant event:
Date of significant event meeting:
Date report compiled:

What Happened?

(Provide a detailed account of what transpired. Include specifics such as how the incident occurred, where it took place, who was involved, and the impact or potential impact on the patient, the team, the organization, and others).

Why Did It Happen?

Provide a comprehensive analysis of the main and underlying reasons—both positive as well as negative—that contributed to the event. Consider factors such as the professionalism and conduct of the team, the presence or absence of an effective system, any system failures, gaps in knowledge, and the complexity and uncertainty surrounding the event.

What Have You Learned?

(Demonstrate that reflection and learning have occurred on an individual or team level, involving relevant team members in the event analysis. Consider factors such as a lack of education and training, the necessity to follow systems or procedures, and the critical importance of teamwork and effective communication).

What Have You Changed?

After reviewing the event, actions were agreed upon and implemented where feasible. For instance, a protocol was amended, updated, or introduced to address the issues identified. This process involved key team members who contributed to the changes. Additionally, it is good practice to attach any documentary evidence of the changes made, such as a letter of apology to the patient or the new protocol document.

Annexure 6: Medical Records Completion and Legibility Audit

An audit of patient records completion and legibility focuses on ensuring that medical records are completed accurately, thoroughly, and in a legible manner. The audit aims to identify any deficiencies or areas of improvement in documentation practices.

Objectives

i. To evaluate the completeness and legibility of patient case records.
ii. To assess adherence to documentation standards and guidelines.
iii. To identify areas for improvement in documentation practices.

Criteria

i. The patient's record should contain an up-to-date summary card.
ii. All documents should be filed in chronological order within the patient's case record.
iii. All sections in the patients' record should be complete.

Method and Analysis

Determine the Audit Sample

- Make a list of the number of patients discharge during a period.
- Out of the total number of discharged patients select 20% of the patients' files for audit randomly. If a total of 100 patients were discharged within one month, then select 20 files for audit, e.g., 2nd, 7th, 14th, 20th, 32nd file from a list of all patients discharged, etc.

Develop an Audit Tool

- Create a standardized checklist or form to assess various aspects of patient record completion and legibility.
- Include criteria such as patient identification, medical history, physical examination findings, diagnostic test results, treatment plans, and progress notes.
- Consider incorporating specific guidelines or requirements from regulatory bodies or professional organizations.
- Microsoft Excel sheet for the checklist can automate these calculations by feeding formula as shown in the checklist.

Conduct the Audit

- Review each patient record in the selected sample using the audit tool.
- Assess the completeness and accuracy of documented information
- Evaluate the legibility of handwritten entries or readability of electronic records.
- Note any missing or incomplete sections, illegible handwriting, or unclear documentation.
- Review each file using the Medical Record Completion, Legibility & Timeliness Audit Master Sheet

Analyze the Audit Findings

- Calculate the percentage of compliance for each criterion assessed. Calculate number of files where criteria are fulfilled and then calculate percent criteria fulfilled as shown in the check list for each component of the medical record.
- Calculate overall score for compliance by adding up score for fulfilling for each section.
- Identify areas of non-compliance or deficiencies in documentation practices.
- Compare with the set standard.
- Analyze common patterns or recurring issues.

Generate an Audit Report

- Summarize the findings of the audit, including overall compliance rates and specific areas of concern.
- Provide recommendations for improvement based on identified deficiencies.
- Include a comparison with relevant standards or guidelines, if available.
- Emphasize the critical role of accurate and legible documentation in ensuring effective patient care, facilitating clear communication among healthcare providers, and fulfilling legal requirements.

Implement Improvement Measures

- Develop action plans to address the identified deficiencies.
- Assign responsibilities for implementing recommended changes.
- Provide training or education on documentation standards and best practices.
- Monitor the progress of improvement initiatives.

Conduct Follow-up Audits

- Repeat the audit process at regular intervals to assess the effectiveness of implemented changes.
- Track improvements in documentation completeness and legibility over time.
- Adjust strategies or interventions as needed based on ongoing audit results.

By conducting regular audits of patient records completion and legibility, healthcare organizations can ensure that accurate and comprehensive information is documented, promoting patient safety, continuity of care, and effective communication among healthcare professionals.

Annexure 6A: Medical Record Completion & Legibility Audit Checklist

Medical record completion, legibility & timeliness audit master sheet					Page 1 of ….		
Sr. No. of Audit	Date of Audit: Number of Cases Audited	Chairperson for Audit: Names of Auditors:		Signatures of Auditors:			
S. No.	Criteria	(1) number Fulfilled	(2) No. Not Fulfilled	(3) No. Not Applicable	n = (1)+(2)	% Fulfilled= (1)X100 / n	Comments
Admission Record Form – Demographic Data							
1.	Case records file registration No. (CRF No.) present						
2.	Registration No. on file present						
3.	Age of patient mentioned						
4.	Sex (M/F) mentioned						
5.	Ward and Unit No. mentioned						
6.	Patient's name mentioned and legible						
7.	Monthly income mentioned						
8.	Full patient's address mentioned						
Admission Record Form – Clinical Details							
9.	Date of Admission mentioned						
10.	Time of Admission mentioned						
11.	Type of Admission mentioned (only psychiatric patient's) see box under section						
12.	Admission diagnosis (provisional) mentioned						
13.	Signatures of admitting SR present						
14.	Name of admitting SR present						
15.	Signatures of admitting consultant present						
16.	Signed by designated Medical Officer (in psychiatry patient's only)						

In MLC cases only							
17.	Brought by name present						
18.	Name of police station mentioned						
19.	Name of court issuing reception order mentioned (in psychiatry patient's only)						
20.	Name of corresponding court mentioned (in psychiatry patient's only)						
Discharge Details							
21.	Duration of stay mentioned						
22.	Date of discharge mentioned						
23.	Time of discharge mentioned						
24.	Final diagnosis mentioned						
25.	ICD code mentioned						
26.	Signatures of discharging SR with name mentioned						
27.	Signatures of discharging consultant with name mentioned						
28.	Referred by details mentioned						
29.	Type of discharge mentioned						
30.	Status at discharge mentioned						
31	Cause of death (in case of death of patient only)						
Income declaration form							
32.	Form filled properly						
33.	Signed by patient/guardian						
34.	Date mentioned						

(continued)

Generic consent form						
35	Form filled properly					
36.	Signed by patient/guardian					
37.	Relationship of signatory with patient mentioned					
38.	Complete address of signatory mentioned					
39.	Date mentioned					
40.	Placed mentioned					
Leaving Against Medical Advise						
41.	Filled properly					
42.	Signed by patient/guardian					
43.	Name of Signatory Mentioned					
44.	Relationship of signatory with patient mentioned					
45.	Date mentioned					
46.	Time mentioned					
47.	Place mentioned					
48.	Signature of Staff Nurse In charge on the statement that "No valuable things are left with the patient at the time of admission into the ward"					
Patient's Discharge Declaration						
49.	Filled properly					
50.	Signed by patient/guardian					
51.	Address of Signatory mentioned					

Application form for admission as applicable							
52.	Filled properly						
53.	Date mentioned						
54.	Place mentioned						
55.	Name of Signatory mentioned and legible						
56.	Father's name of signatory present						
57.	Address of signatory present						
58.	Approval of admission by the concerned SR present						
59.	Approval of admission by the concerned consultant present						
60.	Approval of admission by the concerned designated MO in-charge present						
61.	Computerized Demographic Sheet of patient's record present						

(continued)

Main Patient's Record							
62.	Name of the informant with relationship to patient, written						
63.	Presenting complaints written						
64.	Present History written						
65.	Physical examination written						
66.	Specific examination written						
67.	Signatures of Dr. with name written						
68.	Date written						
69.	Follow up sheets numbering done						
70.	Doctor's orders dated						
71.	Doctor's orders timed						
72.	Doctor's orders signed						
73.	Doctor's orders follow his/her full name after signatures						
74.	Telephonic/Verbal orders by the admitting physician counter signed						
75.	All entries in medical record are made in ink and no pencil used						
76.	All orders are written in designated places marked for the purpose						
77.	All records arranged in chronological order						
78.	All cancelled or corrected orders are cut with a single strike through with a pen and not obliterated or made illegible						
79.	All procedures performed on the patient are recorded as separate notes						
80.	These notes are signed by the concerned medical officer						
81.	Informed consent forms for specific procedures are attached						
82.	These forms are legible and complete						
83.	All consent forms were signed by the appropriate person						

Nursing Records							
84.	T.P.R., B.P. Charting complete						
85.	Nursing notes sheet complete						
86.	Ward observation chart complete						
87.	Nursing notes legible						
88.	Nursing notes signed by the nursing in-charge						
89.	All investigation performed are attached in original in the case file						
90.	These are attached in chronological order						
91.	Referral/Transfer forms are used for all referrals						
92.	These forms are present in the case file						
93.	Referral/Transfer forms are complete and legible						
Discharge Summary/Slip							
94.	Copy of discharge summary attached in case file						
95.	Discharge summary is complete with all boxes filled up						
96.	Discharge summary/ slip mentions about whom and where to contact in case of emergency						
97.	The patient's file is tagged/bound properly						

Annexure 7: Dialysis Safety Audit

A hemodialysis safety audit is conducted to assess and improve the safety measures and practices related to hemodialysis procedures. It aims to identify potential risks, deficiencies, and areas for improvement in order to enhance patient safety and the quality of care provided with the following objectives:

i. Identifying potential risks and hazards associated with hemodialysis procedures.
ii. Evaluating compliance with safety protocols, guidelines, and best practices.
iii. Assessing the adequacy of staff training and competency in hemodialysis safety.
iv. Identifying areas for improvement to enhance patient safety.

Annexure 7A: Hemodialysis Injectable Medication Preparation

Evidence:
High infection rate observed in patients undergoing hemodialysis. There could be various reasons for high infection rates which may include as staff does not follow hand hygiene or do not adhere to safety precautions for preparation of medications.

Criterion

1. 80% of instances of injectable medication preparation all steps to check infections are followed
2. 90% Hand hygiene is observed

Method & Analysis

1. Determine the audit sample:
   - Select a representative sample of hemodialysis sessions or patients for review.
   - Consider including different shifts, days of the week, and patient populations to capture a comprehensive picture.
2. Develop an audit tool:
3. Conduct the audit:
   - Review the selected hemodialysis sessions or patient records using the audit tool.
   - Evaluate adherence to safety protocols and optimal practices.
   - Identify any deviations from established standards or potential safety risks.
   - Observe staff adherence to infection control practices and proper use of personal protective equipment (PPE).
   - Evaluate the functionality and maintenance of hemodialysis equipment.
4. Analyze the audit findings:
   - Calculate the percentage of compliance for each criterion assessed.
   - Identify areas of non-compliance/deviations, potential risks, and improvement areas.

- Analyze common patterns or recurring issues in hemodialysis safety practices.

5. Generate an audit report:
   - Summarize the findings of the audit, including overall compliance rates and specific areas of concern.
   - Provide recommendations for improving hemodialysis safety based on identified deficiencies and risks.
   - Include a comparison with relevant guidelines, regulations, and best practices.
   - Emphasize the importance of patient safety and the need for ongoing monitoring and improvement efforts.
6. Implement improvement measures:
   - Create a strategy to tackle the recognized shortcomings and potential threats.
   - Assign responsibilities for implementing recommended changes.
   - Provide training or educational programs on hemodialysis safety.
   - Enhance infection control measures and reinforce the proper use of PPE.
   - Improve equipment maintenance and ensure regular checks and calibrations.
7. Conduct follow-up audits:
   - Repeat the audit process periodically to assess the effectiveness of implemented changes.
   - Monitor compliance with hemodialysis safety protocols and track improvements over time.
   - Adjust strategies or interventions as needed based on ongoing audit results.

By conducting regular hemodialysis safety audits, healthcare organizations can proactively identify and address potential risks, enhance patient safety, and ensure the delivery of high-quality care during hemodialysis procedures.

Hemodialysis Injectable Medication Preparation Audit Tool

Facility Name:	Observer:
Date(s):	Location of Medication Preparation:

1. Observe at least 20% of the medication preparation sessions using following checklist.
2. Use a "√" to indicate if an action is performed correctly, and an "X" if it is not performed or performed incorrectly. If the action is not observed, leave it blank. To count the procedure as successful, all applicable actions within a row must have a "√".
3. Calculate % occasions where medication preparation sessions were adhered using formula= Number of sessions performed correctly divided by total number of sessions observed.

Annexure 7B: Hemodialysis Injectable Medication Preparation Observation Checklist

Day (i.e., M, Tu, W)	Shift (i.e., 1 - 4)	Discipline	Med prep done in designated area	Med prep area is clean	[a]All vial(s) are inspected	[b]Hand hygiene performed	Septum of all vial(s) disinfected	All vials entered with new needle and new syringe	Med prep done aseptically	All single dose vial(s) discarded	All multi dose vial(s) discarded or stored properly

Discipline: *P* physician, *N* nurse, *T* technician, *S* student, *O* other

[a]Injectable medications must be prepared in a designated clean area that is free from any obvious contamination sources, such as blood, body fluids, contaminated equipment, or tap water.
[b]If the sterility of a vial is questionable, or if the expiration date or beyond use date has been exceeded, the vial should be discarded. If a multidose vial will not be immediately discarded after use, the vial should be labelled upon opening to indicate the beyond use date.

Additional comments/observations, if any:

Annexure 7C: Hemodialysis Hand Hygiene Observations

For each observed "hand hygiene opportunity," mark with a "√". For "opportunity successful," indicate success with a "√" and leave it blank if not successful.

Discipline	Hand hygiene		Provide details of any overlooked attempts, such as those occurring during medication preparation, between patient interactions, or following blood contamination.
	Hand hygiene opportunity	Opportunity successful*	

Discipline: *P* physician, *N* nurse, *T* technician, *S* student, *D* dietitian, *W* social worker, *O* other

Duration of observation period = minutes

- Number of successful hand hygiene opportunities observed =
- Total number of patients observed during the audit =

Total number of hand hygiene opportunities observed during audit =

Kindly pay attention to the details outlined in this session.			
	Yes	No	Comments
Supply of alcohol-based hand sanitizer is adequate			
Supply of soap at hand washing stations is adequate			
Supply of paper towels at hand washing stations is adequate			
There is visible and easy access to hand washing sinks or hand sanitizer			

*** See hand hygiene opportunities on back page

Hand Hygiene Opportunities in Hemodialysis

Hand hygiene opportunity category	Specific examples
1. Prior to touching a patient	Before entering the station to attend to the patient Before touching the vascular access site Before adjusting or removing the cannulation needles
2. Prior to aseptic procedures	Before cannulation or accessing catheter Before performing catheter site care Before parenteral medication preparation Before administering IV infusions or medications
3. After body fluid exposure risk	Following exposure to blood or bodily fluids Following contact with other contaminated substances (e.g., used dialysate) Following the handling of used dialyzers, blood tubing, or prime buckets. After performing wound care or dressing changes
4. After touching a patient	After completing patient care and departing the station. After removing gloves
5. After touching patient surroundings	After touching dialysis machine After handling different objects at the dialysis station After using chair-side computers for charting When leaving station After removing gloves

Annexure 8: Adequacy of Referral Letters

Good patient care requires comprehensive services, especially when patients have multiple co-morbid conditions or complex medical issues. Effective care depends on a robust referral policy that ensures collaboration between departments for specialized consultations, diagnostics, and treatment. Inter-departmental and inter-hospital referrals facilitate interdisciplinary cooperation and allow for optimal care outcomes by leveraging the expertise of various specialties and availability of the services.

However, challenges such as delays, communication breakdowns, and inconsistent referral practices can hinder the process. To address these, hospitals must standardize referral protocols, streamline workflows, and provide proper training for healthcare providers. Monitoring and quality assurance are crucial to evaluating the referral process and ensuring it complies with the best practices.

Currently, referral mechanisms are often inconsistent, with each department following its own practices, leading to delays and conflicts. Junior residents handle most referrals, even when higher expertise is needed.

A referral policy should be developed to improve the referral process. Key principles include involving consultants in referral decisions, ensuring residents discuss referrals with senior staff, and maintaining clear communication between departments. Hospitals should also maintain updated rosters of available staff and provide patients and families with clear explanations of the referral process and outcomes. Proper communication and coordination between teams will enhance patient care and learning among healthcare providers.

Do's and Don'ts for Inter-departmental and Inter-Hospital Referrals at Hospitals

Do's

1. Initiate referrals promptly when specialized care or diagnostics beyond the admitting department's/hospital's scope is needed.
2. Accurately document referrals, including clinical details, expected outcomes, patient preferences, and specific instructions.
3. Ensure clear communication with receiving departments, sharing necessary clinical context for proper evaluation and management.
4. Acknowledge receipt of referrals quickly and update referring providers on the referral's status and patient appointments.
5. Collaborate with receiving departments or specialists or hospital for continuity of care, shared decision-making, and optimal treatment outcomes.
6. Follow up on referrals to coordinate care transitions and ensure timely follow-up appointments and treatments.
7. Maintain patient confidentiality when sharing information across departments or hospitals.
8. Seek feedback from referring and receiving departments/hospital to improve the referral process and its efficiency.

9. Adhere to hospital policies and regulations governing referrals to ensure patient safety and compliance.
10. Prioritize patient-centered care by considering patient preferences and goals in the referral and treatment planning process.

Don'ts

1. Don't delay referrals, as this may jeopardize patient care and lead to adverse outcomes.
2. Don't omit essential clinical information when initiating referrals, as this can hinder appropriate care by the receiving department.
3. Don't assume all referrals are non-urgent; assess each patient's condition to determine the appropriate level of urgency.

For further reading refer to the https://valueadded.in/wp-content/uploads/2024/06/Referral-guidelines.pdf

In special situations during the referral process, specific strategies must be employed to address challenges effectively:

1. **Referral Denial/Rejection**: If a referral is denied by the specialist or department or hospital, the reasons should be communicated to the referring team along with alternative care options. Additional information may be provided to support the request.
2. **Patient Non-Attendance**: For patients missing their referral appointments, follow up to understand the reasons and reschedule as needed, providing reminders and support.
3. **Urgent/Emergent Referrals**: In urgent cases, expedite the process by contacting the specialist directly, prioritizing appointments and coordinating diagnostic services.
4. **Complex Care Coordination**: For patients with complex needs, a multidisciplinary team approach should be used to ensure a comprehensive, coordinated care plan involving various specialists and services.
5. **Discharge Planning Referrals**: Ensure smooth transitions during hospital discharge by coordinating follow-up appointments, medications, and other support services for post-discharge care.
6. **Inter-Departmental Transfers**: When transferring patients between departments, clearly document the reason for the transfer, maintain a detailed patient progress report, and ensure continued follow-up after the transfer. Departments should collaborate seamlessly during transfers and discharge.
7. **Inter-Hospital Referrals**: When patients are referred to other hospitals, communication between both institutions should ensure seamless care transfer, timely follow-up, and proper documentation.
8. **Referral in Outpatient Department (OPD)**: Set up an "EXIT" counter for patients referred between OPDs to avoid delays. This helps patients avoid unnecessary waiting times and minimizes transport and wage loss for additional visits.
9. **Patient Death**: If a referred patient dies, the department where the patient was at the time of death must complete all documentation before handing over the body.

Referral Audit

An audit of the adequacy of referral letters aims to assess the quality and completeness of referral letters received from healthcare providers when referring patients to a specialist or a specialized healthcare service. The audit helps identify areas for improvement in the content, clarity, and information provided in referral letters.

Define the Objectives

- Determine the specific aspects of referral letters to be assessed, such as content, comprehensiveness, clinical information, and referral criteria.
- Identify the standards or guidelines to be used as a benchmark for evaluating the adequacy of referral letters.

Criteria

Referral letters must include comprehensive information such as the history of the presenting complaint, the reason for the referral, current or recent medications, and any relevant clinical warnings.

Method & Analysis

1. Develop an audit tool:
    - Create a checklist or scoring system that covers key elements of a referral letter, including patient identification, referring healthcare provider details, reason for referral, medical history, clinical findings, diagnostic results, and relevant medications.
    - Consider including additional criteria specific to the specialty or healthcare service being audited.
2. Determine the audit sample:
    - Select a sample of referral letters received within a specific timeframe or from a particular healthcare provider or department.
    - Ensure the sample is representative and includes a range of medical conditions and specialties. For example, review referral letters issued within 4-week period (24 working days) if many referrals are issued or review all referral letters in a given period.
3. Conduct the audit:
    - Review the selected referral letters using the audit tool.
    - Assess each criterion to determine if it is adequately addressed or included in the referral letter.
    - Evaluate the clarity, completeness, and relevance of the information provided.
    - Identify any missing or insufficient information that may affect the quality of patient care or cause delays in the referral process.

4. Analyze the audit findings:
   - Calculate the percentage of compliance for each criterion assessed.
   - Identify common deficiencies or patterns of inadequacy in referral letters.
   - Analyze the impact of inadequate referral letters on patient care, patient safety, and healthcare service delivery.
5. Generate an audit report:
   - Summarize the findings of the audit, including overall compliance rates and specific areas of concern.
   - Provide recommendations for improving the adequacy of referral letters based on identified deficiencies.
   - Include a comparison with relevant standards, guidelines, or best practices.
   - Emphasize the importance of clear and comprehensive referral letters for effective communication and continuity of care.
6. Implement improvement measures:
   - Develop educational initiatives or training programs for healthcare providers involved in writing referral letters.
   - Provide guidelines or templates to assist in the creation of comprehensive and informative referral letters.
   - Establish feedback mechanisms or communication channels to facilitate dialogue between referring healthcare providers and specialists.
   - Promote awareness of the importance of high-quality referral letters and their impact on patient care outcomes.
7. Conduct follow-up audits:
   - Repeat the audit process periodically to assess the effectiveness of implemented changes.
   - Monitor the quality of referral letters over time.
   - Measure the impact of improvement measures on the adequacy of referral letters.

- Regular audits of the adequacy of referral letters help improve communication between healthcare providers, enhance the quality of patient care, and facilitate efficient and timely referrals to specialized services.

Referral Checklist

A referral checklist helps ensure that referral letters contain all the necessary information for effective communication between healthcare providers and for proper patient management. While the specific components of a referral checklist may vary depending on the specialty or healthcare service, A sample included components:

1. Patient information:
   - Full name
   - Date of birth
   - Contact information (address, phone number)
2. Referring healthcare provider information:
   - Name
   - Specialty
   - Contact information (clinic address, phone number, email)
3. Reason for referral:
   - Clear and concise description of the patient's symptoms, condition, or diagnostic findings necessitating the referral.
   - Relevant clinical history
4. Relevant medical history:
   - Pre-existing medical conditions
   - Previous surgeries or treatments
   - Allergies or adverse drug reactions
   - Current medications
5. Clinical findings:
   - Physical examination findings
   - Results of relevant investigations (e.g., laboratory tests, imaging studies)
   - Diagnostic criteria, if applicable
6. Treatment provided or attempted:
   - Medications prescribed and their effectiveness
   - Procedures performed
   - Therapies or interventions already initiated
7. Referral request:
   - Specific services or specialty requested
   - Urgency or priority level of the referral
   - Any additional specific requirements or preferences for the referral
8. Supporting documentation:
   - Attachments of relevant medical reports, test results, imaging films, or other supporting documentation
   - Ensure proper labelling and organization of attachments
9. Contact information for further communication:
   - Preferred method of communication (phone, email, fax)
   - Availability for consultation or discussion

10. Referring healthcare provider's signature and date:
    - Confirmation of the referring healthcare provider's authorization and responsibility for the referral.

Post-referral advice to include

Reference Seen by: Name of the doctor/unit who saw the referral, Date and time of the consultation

History and Examination: Detailed patient history and findings from the examination

Diagnosis: Primary diagnosis based on the referral assessment

Advice: Treatment plan or further recommendations provided; Instructions for the referring department and patient

Signature: Doctor's signature, Name and stamp of the attending doctor/unit

Follow-up: Date for the next follow-up appointment

Name and unit responsible for the follow-up

- **Repeat Referrals, if any**: Details of the unit to which any repeat referrals should be sent

Annexure 8A: Sample Referral Audit Checklist

Referral audit checklist	Yes	No	NA
Patient information: Full name, Date of birth in newborns & infants, Contact information (address, phone number)			
Referring Physician's details (name, Address phone number)			
Date and time of referral (name of the referred facility/specialist)			
Date of admission			
Type of Reference: Immediate/Urgent/Routine/Bedside/Medico-legal			
Whether facility where patient is being referred contacted before referring			
Details of the facility where referred is provided			
Details of the patients (Name, age, sex)			
Details of presenting complaints			
Patient's Medical History and Records (Medical History, Surgeries/Procedures, etc.)			
Diagnosis mentioned			
Reason for referral mentioned			
Details of the treatment given before referral			
Details of the laboratory tests done			
Details of X-ray or any other diagnostic procedure done along with their report			
Whether facility to which patient is being referred contacted			
Patient transferred in an ambulance			
Caregiver explained about red flag signs and treatment to be given during transfer			
Any personnel accompanying the patient			
Total score			

NA not applicable

Annexure 9: Response Time in the Emergency Department

An audit of response time in the Emergency Department (ED) focuses on assessing the timeliness of patient care and the efficiency of the ED staff in responding to patients' needs. Some components that can be included in an audit of response time in the ED:

1. **Define the metrics**: Determine the specific response time metrics that will be measured. This could include:
   - Door-to-Triage time: The time from when a patient arrives at the ED to when they are assessed by a triage nurse.
   - Triage-to-Provider time: The time from when a patient is triaged to when they are seen by a healthcare provider.
   - Door-to-Provider time: The total time from when a patient arrives at the ED to when they are seen by a healthcare provider.
   - Length of Stay: The total time a patient spends in the ED until disposition (admission, discharge, or transfer).
2. **Data collection**: Collect data on response times for a defined period. This can involve reviewing electronic medical records, ED logs, or other documentation systems to gather the necessary information. Make certain that the data is accurate, complete, and anonymized.
3. **Analysis and comparison**: Analyze the collected data and compare it against established benchmarks or targets. This could include comparing response times to national or local guidelines, best practices, or historical data from previous audits. Identify any areas where the response times fall below the expected standards.
4. **Identify contributing factors**: Evaluate the factors that may contribute to delays in response time. This could include staffing levels, patient volume, triage processes, availability of resources (e.g., examination rooms, diagnostic equipment), and any identified bottlenecks or inefficiencies in the workflow.
5. **Root cause analysis**: Perform a root cause analysis for any identified delays or prolonged response times. Determine the underlying reasons for the delays, such as system issues, communication breakdowns, or resource constraints. This analysis will help identify areas for improvement and guide the development of targeted interventions.
6. **Recommendations and interventions**: Based on the findings of the audit and the root cause analysis, develop recommendations and interventions to address the identified issues. These could include process improvements, workflow modifications, staffing adjustments, enhanced communication protocols, or resource allocation changes.
7. **Implementation and monitoring**: Implement the recommended interventions and monitor their impact on response times. Regularly assess and track response time metrics to evaluate the effectiveness of the interventions. Adjustments may be required based on continuous monitoring and feedback.

8. **Reporting and communication**: Prepare a report summarizing the audit findings, including response time data, contributing factors, root cause analysis, and recommended interventions. Share the report with relevant stakeholders, such as ED staff, hospital administration, and quality improvement committees. Effective communication of the audit results is essential for creating awareness, obtaining buy-in, and fostering a culture of continuous improvement in the ED.

By conducting regular audits of response time in the ED and implementing appropriate interventions, healthcare organizations can strive to optimize patient flow, enhance patient satisfaction, and improve overall ED efficiency.

Response Time in the Emergency Department Audit

- Reducing waiting times is a key issue in the provision of quality services for patients and for the improvement of the working environment of doctors. Patients reported long response time of around 30 min.
- Define the audit objective: The objective is to evaluate the proportion of patients who are seen within 15 min of their appointment time.
- Determine the data collection period: Set a 2-week period (10 working days) for data collection.
- Data collection method: Use a paper-based system to record the following information for each patient:

  Name of the doctor.

  Patient's reporting time (the time when the patient arrives for the appointment).

  Whether the patient was seen within 15 min of their appointment time (Yes or No).

- Data analysis: Calculate the percentage of patients who were seen within 15 min of their appointment time. Use the following formula: Number of patients seen within 15 min / Total number of patients * 100.
- Perform the analysis: Analyze the data collected over the 2-week period to determine the percentage of patients seen within 15 minutes.
- Compare with the criterion: Compare the calculated percentage with the criterion that states patients should be seen within 15 min of their appointment time. In this case, if the percentage is below 100%, it indicates that improvements are needed.
- Identify areas for improvement: Review the findings to identify factors contributing to delays in meeting the criterion. These may include staffing issues, scheduling inefficiencies, patient flow management, or other operational challenges.
- Develop recommendations: Based on the analysis and identified areas for improvement, develop recommendations to address the issues. These may include process modifications, scheduling adjustments, resource allocation, or staff training.

- Implementation and monitoring: Implement the recommended interventions and monitor their effectiveness. Continuously track response times and assess the impact of the implemented changes.
- Reporting: Prepare a report summarizing the audit findings, including the calculated percentage of patients seen within 15 min, areas for improvement, and recommended interventions. Share the report with relevant stakeholders, such as healthcare providers, administrators, and quality improvement teams.

  By conducting this audit, healthcare facilities can identify areas for improvement in response times and implement strategies to enhance patient satisfaction and optimize the efficiency of their services.

Annexure 10: Audit of Discharge Summaries

Objective: The objective of the audit is to assess the quality and completeness of discharge summaries for patients being discharged from the healthcare facility.

Audit Components

1. **Sample Selection**: Randomly select a sample of discharge summaries from a specific time period (e.g., one month). The sample size should be large enough to provide a representative overview but manageable for data collection.
2. **Audit Criteria**: Define the criteria for evaluating the discharge summaries. Some potential criteria include:
   - Mention of patient demographics (name, age, date of birth (in case of infants), gender, address)
   - Date of admission and discharge
   - Discharge diagnosis and reason for admission
   - Summary of hospital course and key findings
   - Details of procedures, surgeries, or interventions performed
   - List of medications prescribed at discharge
   - Follow-up instructions and recommendations
   - Contact information of the healthcare provider responsible for the patient's care
   - Any necessary referral information
3. **Data Collection**: Review each discharge summary in the sample and assess its compliance with the defined criteria. Use a standardized data collection form to record the findings for each discharge summary. This can be done manually or through electronic means, depending on the available systems.
4. **Data Analysis**: Calculate the percentage of discharge summaries that meet each criterion. Identify areas of strengths and weaknesses based on the compliance rates. This analysis provides an overview of the quality and completeness of the discharge summaries.
5. Review the audit results to pinpoint areas requiring enhancement. This may include incomplete or missing information, inconsistent documentation practices, or other deficiencies. Determine the root causes contributing to these issues.
6. **Develop Recommendations**: Based on the identified areas for improvement, develop recommendations to enhance the quality and completeness of discharge summaries. Recommendations may include staff education and training, process improvements, documentation templates, or the use of electronic health record systems.
7. **Implementation and Monitoring**: Implement the recommended interventions and monitor their effectiveness in improving the quality of discharge summaries. Monitor adherence levels and evaluate the effects of the changes made over a period.
8. **Reporting**: Prepare a comprehensive report summarizing the audit findings, areas for improvement, and the recommended interventions. Include any relevant data analysis, compliance rates, and specific examples or case studies. Share the report with the healthcare providers involved in the discharge process, quality improvement teams, and relevant stakeholders.

Regularly conducting audits of discharge summaries helps ensure that vital information is accurately communicated to patients and receiving healthcare providers, promoting continuity of care and patient safety.

Annexure 10A: Discharge Summary Audit Checklist

		Yes	No
1	Is there documented evidence that a discharge summary been completed by the medical officer?		
2	If yes: Does the discharge summary contain:		
a.	Patient details – 3 unique identifiers		
b.	Referring doctor- name and contact details?		
c.	Consultant's name		
d.	Hospital unit discharge?		
e.	Discharge date?		
f.	Principal diagnosis?		
g.	Provisional diagnosis?		
h.	Other current complications/comorbidities?		
i.	Details of any procedure performed, if any?		
j.	Significant investigations/results documented?		
k.	Recommendations for follow-up?		
l.	Discharge unit and contact number?		
m.	Medications on discharge, including any medications ceased during admission?		
n.	Discharge summary contain emergency contact numbers		
o.	Signature of the authorized medical officer?		
p.	Printed name and designation of the authorized medical officer		
q.	Date of completion by the medical officer		

Annexure 11: Medication Safety Audit

Objective: The objective of the medication safety audit is to assess the adherence to medication safety practices and identify areas for improvement in medication management processes within the healthcare facility.

Audit Components:

1. **Sample Selection**: Randomly select a sample of patient medication records or medication orders from a specific time period (e.g., one week or one month). The sample size should be large enough to provide a representative overview but manageable for data collection.
2. **Audit Criteria**: Define the criteria for evaluating medication safety practices. Some potential criteria include:
   - Accuracy and completeness of medication orders
   - Proper documentation of medication administration
   - Identification of allergies and adverse drug reactions
   - Appropriate prescribing practices
   - Medication reconciliation upon admission, transfer, and discharge
   - Availability and accuracy of drug information resources
   - Compliance with high-alert medication protocols
   - Use of barcode scanning or other medication verification systems
   - Adequate storage and handling of medications
   - Staff education and training on medication safety
3. **Data Collection**: Review the medication records or medication orders in the sample and assess their compliance with the defined criteria. Use a standardized data collection form to record the findings for each medication record or order. This can be done manually or through electronic means, depending on the available systems.
4. **Data Analysis**: Calculate the percentage of medication records or orders that meet each criterion. Identify areas of strengths and weaknesses based on the compliance rates. This analysis provides an overview of medication safety practices within the healthcare facility.
5. **Identify Areas for Improvement**: Review the audit findings and identify areas where improvement is needed. This may include medication errors, incomplete or incorrect documentation, non-compliance with medication safety protocols, or other deficiencies. Determine the root causes contributing to these issues.
6. **Develop Recommendations**: Based on the identified areas for improvement, develop recommendations to enhance medication safety practices. Recommendations may include staff education and training, process improvements, standardization of medication order sets, implementation of technology solutions (e.g., barcode scanning), or improvements in medication storage and handling procedures.
7. **Implementation and Monitoring**: Implement the recommended interventions and monitor their effectiveness in improving medication safety practices. Track compliance rates and assess the impact of the implemented changes over time.

Continuously monitor and evaluate medication safety practices to identify any emerging issues or trends.

8. **Reporting**: Prepare a comprehensive report summarizing the audit findings, areas for improvement, and the recommended interventions. Include any relevant data analysis, compliance rates, and specific examples or case studies. Share the report with the healthcare providers involved in medication management, quality improvement teams, and relevant stakeholders.

Regularly conducting medication safety audits is crucial to identify and address potential risks and improve patient safety related to medication use. It helps ensure that medication processes are safe, accurate, and consistent throughout the healthcare facility.

Annexure 11A: Medication Safety Record Completion Audit Checklist

		Yes	No	NA
1.	Is there evidence that the medication history is documented			
2.	If yes: where is the medication history documented? Select all that apply Medication chart Medication action plan			
3.	Is there documented evidence of medication allergies and adverse drug reactions (ADR) status (including nil known and unknown) at a uniform place on head case note sheet in the medication chart			
1.1.4.	If no, where in the clinical record is medication allergy or ADR history documented			
	Patient history section			
	Drug treatment chart Nursing treatment chart			
5.	Is there proof of medication reconciliation documented on the Medication Action Plan (MAP), such as in the reconcile column, the discharge medication record, or the interim medication administration record, indicating that the change column is filled out?			
6.	Are discrepancies between prescribed and actual medications identified and resolved?			
	Medication Administration			
7.	Are there clear protocols for medication administration?			
8.	Are medication administration records (MARs) accurately maintained?			
9.	Is there a system in place for double-checking medications before administration?			
	Documentation			
10.	Are all medications documented, including over-the-counter and PRN (as needed) medications?			
11.	Are any medication errors or adverse reactions documented and reported?			
12.	Is there documented evidence of a Venous-thrombosis (VTE) risk assessment in the medication chart			
13.	Is there documented evidence of a medication management plan in the patients' clinical record (start/stop/review date)			
14.	Is the patient identification (minimum of name, Registration No) complete on all pages of the medication chart			
15.	Does the patient have documented evidence of a medication order in the medication chart			
16.	If yes; does each medication order have: Generic name of medication Route Dose Frequency Start date Authorized prescriber's signature with date & time Authorized prescriber's printed name Authorized prescriber's designation			

(continued)

		Yes	No	NA
17.	Any prohibited abbreviations used in the case sheet			
18.	If yes, fill details as below			
19.	Does the facility allow nurse-initiated medications			
20.	If yes: is there documented evidence for the nurse-initiated medication(s) have been prescribed/administered according to facility policy			
	Telephonic and Verbal Order			
21.	Does the facility allow for verbal medications orders			
22.	If yes: is there documented evidence that any verbal medication orders have been countersigned by the prescriber within 24 h of the order being received			
23.	Does the facility allow for telephone medication orders?			
24.	If yes: is there documented evidence that any telephone orders have been countersigned by the prescriber within 24 h of the order being received			
	Medication Storage			
25.	Are the medications stored in their original packaging with the labels preserved?			
26.	Are medications stored in a secure, locked area?			
27.	Are medications stored at the correct temperature (e.g., refrigerated if necessary)?			

Annexure 11C. Prohibited drug name abbreviations:

The following medical abbreviations must not be used in clinical documentation in the form of free text (either handwritten or electronic) or any pre-printed forms

Abbreviation used	Yes	No	NA
Drug name abbreviation (PCM, CPM, etc.) or stemmed drug name (Nitro, cipro, etc.)			
Any Latin apothecary abbreviations (OD, BD, TDS, QIH, SOS, qhs, etc.)			
Symbols (, >, <, +, ^{0}, X)			
Decimal (naked decimal or trailing 0) .5, 1.0, 1.25 mg, etc.)			
Not using a thousand separator (100000 for 1,00,000)			
Abbreviations for IJ, IU, U hand written			
Standard drug dosage micrograms (mcg) for μ)			
Non standard abbreviations (TCA, CST, D/C, etc.			
Discontinue 1, 2, 6 rest to continue, etc.			

Annexure 12: Blood or Blood Product Audit

Objective: The objective of the blood or blood product audit is to assess the appropriate and safe use of blood and blood products within the healthcare facility. The audit aims to ensure adherence to established guidelines, policies, and protocols for transfusion practices.

Audit Components

- **Sample Selection**: Randomly select a sample of patient records involving blood transfusions or blood product administration from a specific time period (e.g., one month). The sample size should be sufficient to provide a representative overview of transfusion practices within the facility.
- **Audit Criteria**: Define the criteria for evaluating the appropriateness and safety of blood transfusions. Some potential criteria include:
  - Documentation of indications for blood transfusion
  - Verification of the patient's blood type and crossmatch compatibility
  - Compliance with pre-transfusion testing (e.g., blood grouping, antibody screening)
  - Verification of informed consent for transfusion
  - Proper administration of blood or blood products
  - Monitoring and documentation of transfusion reactions or adverse events
  - Timely completion of transfusion-related documentation
  - Adherence to transfusion trigger thresholds (e.g., hemoglobin level)
  - Availability and use of transfusion guidelines or protocols
  - Adherence to the guidelines for temperature monitoring and storage of blood products.
- **Data Collection**: Review the selected patient records and assess their compliance with the defined criteria. Collect data on each criterion using a standardized data collection form. Record any deviations or non-compliance with the established guidelines.
- **Data Analysis**: Analyze the collected data to determine the compliance rates for each criterion. Identify areas of strengths and weaknesses in blood transfusion practices. Calculate the percentage of records that meet each criterion and identify any trends or recurring issues.
- **Pinpoint Opportunities for Enhancement**: Examine the audit results to determine where enhancements are necessary. This may include issues such as incomplete documentation, deviations from transfusion guidelines, inadequate monitoring of transfusion reactions, or non-compliance with pre-transfusion testing protocols. Identify the root causes contributing to these issues.
- **Develop Recommendations**: Based on the identified areas for improvement, develop recommendations to enhance blood transfusion practices. Recommendations may include staff education and training, reinforcement of transfusion protocols, improved documentation practices, strengthening monitoring and reporting of transfusion reactions, and ensuring availability of transfusion-related guidelines and protocols.

- **Implementation and Monitoring**: Implement the recommended interventions and monitor their effectiveness in improving blood transfusion practices. Track compliance rates and assess the impact of the implemented changes over time. Continuously monitor and evaluate transfusion practices to identify any emerging issues or trends.
- **Reporting**: Prepare a comprehensive report summarizing the audit findings, areas for improvement, and the recommended interventions. Include any relevant data analysis, compliance rates, and specific examples or case studies. Share the report with the healthcare providers involved in transfusion practices, quality improvement teams, and relevant stakeholders.

Regularly conducting blood or blood product audits is essential to ensure the appropriate and safe use of these resources, minimize the risks associated with transfusions, and optimize patient outcomes. It promotes adherence to transfusion guidelines, identifies areas for improvement, and supports the implementation of evidence-based transfusion practices within the healthcare facility.

Annexure 12A: Postpartum Hemorrhage Audit

Evidence

The number of deaths in postpartum hemorrhage cases has increased during the last 6 months reportedly because of delay in investigations and initiation of treatment.

Objective: The objective of this postpartum hemorrhage clinical audit is to assess the quality of care provided to patients experiencing severe hemorrhage during the postpartum period, identify areas of improvement, and reduce maternal mortality rates.

Methodology

Step 1 – Establish Criteria for Severe Hemorrhage

A hemorrhage occurring at the time of pregnancy outcome (birth, abortion, caesarean section, ectopic pregnancy) was defined as severe if blood loss: equaled 1500 ml or more (if measured) or the use of plasma expanders was necessary, amounting to 2500 ml or more within a 24-h period, or an equivalent volume in packed cells, leading to transfusion, hysterectomy, or maternal death.

Step 2 – Structure for Case Review

1. Expert Committee: An expert committee comprising obstetricians, anesthetists, and epidemiologists was formed to review and assess the cases qualitatively.

Step 3 – Establish Best Practices

Evaluation of Quality of Care: Specific criteria were identified to evaluate the quality of care provided in managing severe hemorrhage. These criteria included time intervals for diagnosis, interventions, administration of medications, and consideration of surgical or radiological interventions as below:

Selected Criteria for Evaluation of Quality of Care

The duration from delivery to the diagnosis of the hemorrhage was under 45 min.

Was the manual removal of the placenta or uterine examination performed within 15 min following the diagnosis of the hemorrhage?

Time elapsed in minutes between diagnosis of the hemorrhage and anesthetic administration (in the absence of previous epidural).

In the case of manual placenta delivery, manual uterine exploration, or caesarean delivery, were prophylactic antibiotics administered?

Was oxytocin administered within 15 min of observation of uterine atony?

Were prostaglandins administered within 20 min of determining the failure of oxytocin?

If hemoglobin fell below 7 g/dL, did colloid replacement take place or were blood products administered?

If no treatment succeeded within 60 min of diagnosis of the hemorrhage or 120 min after delivery, was surgical or radiological intervention considered?

The quality of care was classified into three categories:

- Appropriate (all the criteria were met)
- Insufficient (only one or two criteria were not met, or the judgment of the experts was not clear-cut).
- Totally inadequate (the majority of the criteria were not met).

Step 4 – Identify Cases and Collect Information

Trained investigators collected information retrospectively for each case using a questionnaire. The questionnaire covered demographic details, medical history, antenatal care, delivery, postpartum period, anesthesia, treatment received, transfusion, and organizational aspects of the obstetric facility.

Information was also collected on the organizational aspects of the obstetric facility relating to each patient: volume of annual deliveries; staffing levels, including 24-h on-site presence of anesthetist and obstetrician consultants; on-site blood bank/storage unit; on-site intensive care unit; guidelines for treating hemorrhage.

Step 5 – Review of the Life-Threatening Morbidity Cases

Cases were randomly assigned to teams consisting of an obstetrician and an anesthetist for classification. Discrepancies were discussed by the entire group of experts to reach a consensus.

Findings

1. Case Distribution: Among the 165 cases examined, 51% resulted in vaginal deliveries, 19% were operative vaginal deliveries, and 30% were caesarean deliveries.
2. Uterine atony was the leading cause of hemorrhage.
3. Quality of Care: Overall, 62% of cases received appropriate care, 14% received insufficient care, and 24% received totally inadequate care.
4. Areas for Improvement: The audit identified several areas requiring improvement, including delays in diagnosis, delays in performing manual placental delivery or uterine exploration, and underutilization of prostaglandins in cases where oxytocin was ineffective.
5. Contributing Factors to Substandard Care: Lack of a 24-h on-site anesthetist and low volume of deliveries in the facility were associated with substandard care.

Recommendations

1. Improve Diagnosis and Reaction Time: Implement measures to reduce the time between delivery and diagnosis of hemorrhage, diagnosis and manual placental delivery or uterine exploration, and consider alternative interventions when oxytocin is ineffective.
2. Enhance Staffing Levels: Ensure the presence of a 24-h on-site anesthetist and obstetrician consultants to facilitate prompt and appropriate management of severe hemorrhage.

3. Guidelines and Training: Develop and disseminate guidelines for the management of postpartum hemorrhage, emphasizing best practices and evidence-based interventions. Organize frequent training programs for healthcare professionals to improve their expertise and abilities.
4. Collaboration and Resource Allocation: Strengthen collaboration between obstetric facilities, blood banks, and storage units to ensure the availability of blood products when needed. Assess and allocate resources based on the volume of deliveries to optimize patient care.

Conclusion: The postpartum hemorrhage clinical audit revealed variations in the quality of care provided, with a significant number of cases receiving insufficient or totally inadequate care. By implementing the recommended interventions and addressing the identified areas for improvement, healthcare facilities aim to enhance the management of severe hemorrhage, reduce maternal mortality rates, and improve patient outcomes.

[Signature]

[Name of Audit Team Lead]

[Position/Department]

[Date].

Annexure 12B: Blood or Blood Product Clinical Patient Record Checklist

A Blood or Blood Product Clinical Patient Record Checklist is essential for ensuring the safe and effective administration of blood transfusions.

If the patient received a blood or blood product in the current admission complete the form below:

1.	Transfusion of blood or blood products during this admission.	Yes	No	NA
2.	Patient identification (using at least two identifiers, e.g., name and date of birth).			
3.	What was the recorded reason for the transfusion, if applicable?	Hb Clinical indication Hb and clinical indication No indication		
	Pre-transfusion checks			
4.	Is there evidence of crossmatch report			
5.	If yes: is there evidence that the			
6.	Product type is complete			
7.	Product number is complete			
8.	Group is complete			
9.	Expiry date			
10.	Verification of patient, product, and label has been completed and endorsed by TWO clinical staff members.			
11.	Is there any evidence of a prior adverse reaction to a blood or blood product if the patient has received one? Select "NA" if there is no history of blood or blood product transfusion.			
12.	Any premedication prescribed to prevent transfusion reaction			
	Blood or blood product transfusion consent			
13.	Is there proof of consent for a blood products transfusion?			
14.	If yes: is there evidence that the blood products accepted have been documented on the first page			
15.	Is there evidence that the patient's name, signature, date are complete OR legally authorized person's name, signatures, relationship, date are complete			
16.	Is there evidence that the doctor's name, designation, signatures, and date are complete			
	Blood prescriptions			
17.	Is there proof of a blood prescription order			

(continued)

18.	If yes: is there evidence on the order form of Date of transfusion Type of blood product Volume/quantity/number to be given Special requirements listed. If yes to special requirements listed: what are they Frusemide Irradiated Warmer Pre-medications Rate of transfusion Doctor's signature Doctor's printed name Fluid order form signed by TWO nurses			
19.	Is there evidence that the patient's full name, date of birth, registration number were confirmed against the transfusion departure report and patient's arm band/ registration number			
20.	Is there documented proof that the product type was checked against the fluid order, transfusion department report and compatibility label			
21.	Is there documented proof that the product bag is intact, no signs of deterioration, contamination, clots, or discoloration			
22.	Is there documented proof of confirmation of blood product will not expire before transfusion is complete			
23.	Is there documented proof that the following were completed: Transfusion start time Transfusion stop time Volume infused Non-urgent blood given out of hours (20.00 to 07.00)			
24.	Is there documented proof that the vitals were completed before starting transfusion: Pulse Temperature Respiratory rate Blood pressure Oxygen saturation			
25.	Is there proof of commencement observation completed (within 15 min of commencement of transfusion			
26.	If yes: is there documented proof that the following vitals are completed: Pulse Temperature Respiratory rate Blood pressure Oxygen saturation			
27.	Is there evidence of hourly observations completed			

(continued)

28.	If yes: is there documented evidence that the following vitals are completed: Pulse Temperature Respiratory rate Blood pressure Oxygen saturation			
29.	Is there evidence of observation on completion of transfusion (within 2 h of completion of transfusion			
30.	If yes: is there documented evidence that the following vitals are completed: Pulse Temperature Respiratory rate Blood pressure Oxygen saturation			
31.	Was the blood infused within four hours			
32.	If no: please state, why			
33.	Is there evidence that the patient has an adverse reaction to blood transfusion (e.g., symptoms include fever, >1^{0}C above baseline, rigors, chest or abdominal pain, hypotension, tachycardia, rashes/itching, etc.)			
34.	If yes is there evidence the medical officer was notified			
35.	Is there evidence the adverse reaction was recorded in the incident reporting system			
36.	Is there evidence of the transfusion outcome in the chart			
37.	**Post-transfusion care documentation**			
38.	**Post-transfusion reaction**			
	Surgical procedure in the current admission			
37.	If patient has undergone a surgical procedure AND there is evidence of informed consent: is there evidence that the: The patient has been informed that the procedure could involve a blood transfusion as part of the consent process.			
	The patient has received the information sheet regarding blood and blood products transfusion.			
38.	Patient refused the administration of certain blood or blood products during this admission.			
39.	Has the patient declined blood or blood product transfusion Or declined a specified blood or blood product to be transfused in the current admission			
40.	If yes: is there evidence of a refusal or refusal form filled			
41.	If yes: is there evidence that the blood product accepted has been documented on the form			
42.	Patient's/Legally authorized person's name, signature and date are complete			
43.	Doctor's name, designation, signature and date are complete			

Annexure 13: Checklist for Monitoring Injection Safety in a Hospital

Injection safety is a fundamental aspect of patient care within healthcare settings, encompassing the administration of medications, vaccines, and other therapeutic fluids directly into the body. Ensuring the highest standards of injection safety is crucial not only for preventing healthcare-associated infections (HAIs) and minimizing medication errors but also for safeguarding the well-being of both patients and healthcare professionals. Unsafe injection practices can lead to severe consequences, including the transmission of infectious diseases, adverse drug reactions, and compromised patient trust in the healthcare system.

This Checklist for Monitoring Injection Safety provides healthcare facilities with a structured and comprehensive tool for evaluating and maintaining safe injection practices. By systematically addressing each checkpoint, hospitals and clinics can ensure adherence to best practices, comply with regulatory requirements, and foster a culture of safety and accountability.

Check point	Frequency of observation	Responsible person	Status (Yes/No)	Remarks
Staff training and competency	Annually / upon hire	Nursing supervisor / HR		Ensure all staff complete updated training
Hand hygiene compliance	Each injection	Infection control officer		Verify hand hygiene before and after injections
Use of single-use needles and syringes	Ongoing	Supply chain manager		Confirm single-use items are not reused
Proper disposal of sharps	After each use	All healthcare staff		Check sharps containers are available and used correctly
Sterile technique	Each procedure	Supervising nurse		Monitor adherence to aseptic techniques
Patient identification	Each injection	Administering nurse		Use at least two patient identifiers
Medication verification	Each injection	Administering nurse		Double-check medication, dosage, and expiry
Proper storage of medications	Daily	Pharmacy manager		Ensure medications are stored correctly
Documentation and record-keeping	Each injection	Administering nurse		Accurately document all injection details
Equipment maintenance	Weekly	Biomedical engineer		Inspect and maintain injection equipment
Environmental cleanliness	Daily	Housekeeping staff		Maintain cleanliness in injection areas
Patient education	Each injection	Administering nurse		Inform patients about the injection procedure
Incident reporting	As occurs	All healthcare staff		Report and investigate any injection-related incidents

(continued)

Check point	Frequency of observation	Responsible person	Status (Yes/No)	Remarks
Audit and compliance checks	Monthly / Quarterly	Quality assurance team		Conduct regular audits of injection practices
Vaccination programs monitoring	Each vaccination session	Vaccination coordinator		Monitor safety protocols during vaccinations
Use of alcohol swabs	Each injection	Administering nurse		Ensure alcohol swabs are used for site disinfection
Backup supplies availability	Weekly	Supply chain manager		Confirm the availability of backup injection supplies
Emergency protocols	Annually	Clinical lead		Review and update emergency response protocols
Patient allergies and history	Each injection	Administering nurse		Check and document patient allergies and history
Labeling of medications	Each injection	Pharmacy staff		Ensure all medications are clearly labeled
Temperature monitoring of refrigerated supplies	Daily	Pharmacy technician		Record temperatures of refrigerated medications
Multidose vial handling	Each use	Pharmacy staff		Multidose vials are dedicated to patients preferably; if used for multiple purposes, it should be stored in central medication area Inspect multidose vials for signs of contamination before use Disinfect the rubber stopper of the vial with alcohol swab in a circular motion from the center outward. Allow it to air dry completely to ensure proper disinfection Do not touch the vial stopper or dip needles into the vial after it has been opened Do not leave needle in septum Label the vial with the date and time of first use and discard after the recommended period (Beyond use date) Use a new sterile needle to reduce the risk of contamination and ensure patient safety

(continued)

Check point	Frequency of observation	Responsible person	Status (Yes/No)	Remarks
Bags and IV line handling	Each use	IV nurse		Ensure IV bags are properly labeled and stored. Check IV lines for kinks, leaks, or contamination before use. Use aseptic techniques when connecting IV lines to prevent infections Replace IV tubing according to institutional protocols or manufacturer recommendations, typically every 72-96 hours, to minimize infection risks associated with prolonged use
Medication administration tubing and connector	Each injection	Administering nurse		Inspect tubing and connectors for integrity and proper connection before administration. Replace any damaged or worn tubing immediately. Avoid using expired or compromised connectors
Pre-drawn medications in syringe handling	Each use	Pharmacy Staff		Prepare pre-drawn syringes in a sterile environment. Label syringes immediately with patient information, medication name, dosage, and expiration date Do not store pre-drawn syringes for extended periods; use them within the recommended timeframe to maintain sterility and medication efficacy

Annexure 14: Delivery Case Audit

Objective: The objective of the delivery case audit is to assess the quality and safety of care provided during childbirth. The audit aims to evaluate adherence to established protocols and guidelines, identify areas for improvement, and enhance the overall quality of maternal and neonatal care.

Audit Components:

1. **Sample Selection**: Randomly select a sample of delivery cases from a specific time period (e.g., one month) for review. The sample size should be sufficient to provide a representative overview of the delivery care provided within the facility.
2. **Audit Criteria**: Define the criteria for evaluating the quality and safety of care during childbirth. Some potential criteria include:
   - Documentation of antenatal care and risk assessment
   - Adherence to labor management protocols and guidelines
   - Monitoring and management of fetal well-being during labor
   - Use of evidence-based practices for pain relief during labor
   - Adherence to infection control measures
   - Availability and utilization of skilled birth attendants
   - Documentation of maternal and neonatal vital signs
   - Identifying and addressing issues that arise during childbirth.
   - Adherence to protocols for active management of the third stage of labor
   - Timely initiation of breastfeeding and support for neonatal care
   - Documentation of postpartum care and follow-up plans
3. **Data Collection**: Review the selected delivery case records and assess their compliance with the defined criteria. Collect data on each criterion using a standardized data collection form. Record any deviations or non-compliance with the established protocols and guidelines.
4. **Data Analysis**: Analyze the collected data to determine the compliance rates for each criterion. Identify areas of strengths and weaknesses in the delivery care practices. Calculate the percentage of records that meet each criterion and identify any trends or recurring issues.
5. **Identify Areas for Improvement**: Examine the audit results and pinpoint areas requiring enhancement. This may include issues such as incomplete documentation, deviations from established protocols, inadequate monitoring of maternal and fetal well-being, lack of adherence to infection control measures, or gaps in postpartum care. Identify the root causes contributing to these issues.
6. **Develop Recommendations**: Based on the identified areas for improvement, develop recommendations to enhance the quality and safety of care during childbirth. Recommendations may include staff education and training, reinforcement of protocols and guidelines, improved documentation practices, strengthening infection control measures, enhancing fetal monitoring practices, and promoting breastfeeding support and neonatal care.

7. **Implementation and Monitoring**: Implement the recommended interventions and monitor their effectiveness in improving delivery care practices. Monitor adherence levels and evaluate the effects of the changes implemented over a period. Continuously monitor and evaluate delivery care practices to identify any emerging issues or trends.
8. **Reporting**: Prepare a comprehensive report summarizing the audit findings, areas for improvement, and the recommended interventions. Include any relevant data analysis, compliance rates, and specific examples or case studies. Share the report with the healthcare providers involved in delivery care, quality improvement teams, and relevant stakeholders.

Regularly conducting delivery case audits is crucial for ensuring the provision of safe and high-quality care during childbirth. It helps identify gaps in care, promotes adherence to protocols and guidelines, and supports continuous quality improvement efforts to enhance maternal and neonatal outcomes.

Annexure 14A: Delivery Case Audit Checklist

	Medical Audit of Delivery case				
A	**Patient's Personal Particulars**				
1	Name of the Patient:				Age/Sex
2	IP No.				
3	Date of Admission & Time				
4	Name of Ward Department				Name of Treating Physician/Surgeon:
8	Diagnosis on Admission:				
9	Date of Discharge				
B	**Adequacy of documentation**	**No (0)**	**Present but incomplete (1)**	**Yes (2)**	Remarks, if any
1	Was demographic details of the patient complete				
2	Was the case sheet countersigned by the doctor				
3	Was chief complaint of the patient documented				
4	Was diagnosis documented				
5	Whether the diagnosis supported by adequate history notes				
6	Was Whether the adequate clinical notes documented				
7	Whether vitals of the patient documented on arrival in the ward				
8	Does the case contain initial assessment with plan of care				
9	Does the case sheet have consent form of patient				
10	Does the case sheet have Nurse's notes / TPR / Input output charts				
11	Does the case sheet have Discharge summary/note				
12	Is the discharge summary/note countersigned by doctor				
13	Is the handwriting legible				
14	Is partograph present in the case sheet				
15	Is partograph properly filled in real time				

16	Whether appropriate action taken based on the partograph, i.e., Caesarean section performed: Was a difficult delivery detected in time in the referring facility				The audit of the referring facility should include use of partograph, and documented reaction based on partograph findings
15	Caesarean section (CS): Was it referred in time				
16	CS: was the management of the patient started within 10 min of arrival	(>31 min)	(10-30 min)	(<10 min)	
	Max Score=		**Score Obtained =**		
C	**Adequacy of Laboratory and Radiology Investigations**	**No (0)**	**Present but incomplete (1)**	**Yes (2)**	Remarks, if any
1	Are the laboratory and radiographic reports attached and entered in the record at the time of discharge				
2	Was there a prescribed format for reporting and requesting laboratory/radiological investigations				
3	Whether the laboratory findings support the final diagnosis?				
4	Whether the radiological findings support final diagnosis?				
5	Are laboratory investigations sufficient in relation to nature and gradient of illness?				
	Max Score =		**Score Obtained =**		
D	**Adequacy of Clinical Care**	**No (0)**	**Present but incomplete (1)**	**Yes (2)**	Remarks, if any
1	Are medicines prescribed relevant to the diagnosed condition?				
2	Is the diagnosis correct as per the investigation reports?				
3	Are the medicines administered in regular intervals as prescribed?				
4	Any complications during the treatment were treated fairly				
5	Is the dose of the medicines/Injections relevant as per the patient's condition and body weight?				

(continued)

	Medical Audit of Delivery case				
6	Patient was discharged only after symptoms were decreased or ameliorated				
7	Follow-up instructions were given to the patient and documented				
8	Patient was informed about the follow-up visit and same was documented on the case sheet				
	Max Score =		**Score Obtained =**		
	Overall Score				
	Scoring Parameters	**Max score**	**Score Obtained**		**Recommendations**
A	Patient's particulars				
B	Adequacy of documentation				
C	Adequacy of Laboratory and Radiology Investigations				
D	Adequacy of Clinical Care				

Annexure 15: Maternity and Newborn Identification Audit

Objective: The objective of the Maternity and Newborn Identification Audit is to ensure accurate and reliable identification of mothers and newborns throughout the maternity care process. The audit aims to assess compliance with established protocols and guidelines for identification procedures, identify any areas of non-compliance or potential risks, and improve the safety and quality of care for mothers and newborns.

Audit Components

1. **Sample Selection**: Randomly select a sample of maternity and newborn cases from a specific time period (e.g., one month) for review. The sample size should be sufficient to provide a representative overview of the identification practices within the facility.
2. **Audit Criteria**: Define the criteria for evaluating the identification procedures for maternity and newborn care. Some potential criteria include:
   - Adherence to protocols for verifying the identity of the mother upon admission.
   - Verification of the mother's identity before any interventions or procedures.
   - Documentation of the mother's identification details in the medical records.
   - Adherence to protocols for newborn identification immediately after birth
   - Use of identification bands or tags for both the mother and newborn.
   - Verification of the newborn's identity before any interventions or procedures
   - Documentation of the newborn's identification details in the medical records.
   - Proper labelling and matching of specimens (e.g., blood samples, breast milk) to the correct mother or newborn.
   - Training and education provided to healthcare staff regarding identification protocols and procedures.
3. **Data Collection**: Review the selected maternity and newborn case records and assess their compliance with the defined identification criteria. Collect data on each criterion using a standardized data collection form. Record any deviations or non-compliance with the established protocols and guidelines.
4. **Data Analysis**: Analyze the collected data to determine the compliance rates for each identification criterion. Calculate the percentage of records that meet each criterion and identify any patterns or recurring issues.
5. **Identify Areas for Improvement**: Examine the results of the audit and pinpoint areas requiring enhancement. This may include issues such as incomplete or inconsistent documentation of identification details, failure to properly use identification bands or tags, lack of verification procedures, or inadequate staff training on identification protocols. Identify the root causes contributing to these issues.
6. **Develop Recommendations**: Based on the identified areas for improvement, develop recommendations to enhance the accuracy and reliability of maternity and newborn identification. Recommendations may include staff education and training on identification protocols, implementation of standardized identifica-

tion procedures, improved documentation practices, enhanced use of identification bands or tags, and regular auditing and monitoring of identification practices.

7. **Implementation and Monitoring**: Implement the recommended interventions and monitor their effectiveness in improving maternity and newborn identification practices. Track compliance rates and assess the impact of the implemented changes over time. Continuously monitor and evaluate identification practices to identify any emerging issues or trends.
8. **Reporting**: Prepare a comprehensive report summarizing the audit findings, areas for improvement, and the recommended interventions. Include any relevant data analysis, compliance rates, and specific examples or case studies. Share the report with the maternity care providers, quality improvement teams, and relevant stakeholders.

Regularly conducting Maternity and Newborn Identification Audits is crucial for ensuring the safety and well-being of mothers and newborns. Accurate identification procedures minimize the risk of errors, such as misidentification or incorrect interventions, and contribute to improved patient outcomes and satisfaction.

Annexure 15A: Maternity & Newborn Identification Clinical Record Audit Checklist

	Maternity and newborn identification clinical record audit	Yes	No
1.	For Maternity patients, Is there evidence for the last admission of perinatal documentation of Date of rupture of membranes recorded. Time of rupture of membranes recorded. Method of rupture of membranes recorded. Indication for induction/augmentation recorded Type of labor recorded. Date and time of stages of labor. Labor pain relief? Collection of cord blood recorded. Vaginal examination recorded? Second stage observations recorded. Fetal heart rate recorded? Perineal assessment recorded? If a tear, is the degree of tear recorded? If a tear, is the suture material and anesthetic recorded? Indication for operative intervention recorded. Obstetric complications recorded. Skin to skin contact recorded. 2h postpartum observations recorded?		
2.	Is a copy of the perinatal morbidity/statistical form filed in the clinical recorded?		
3.	If yes, is the information recorded on the perinatal morbidity/statistical form consistent with the documentation recorded in the Intrapartum Record notes?		
4.	Is there evidence of assisted birth documentation in the clinical record?		
5.	If yes: Does the assisted birth record include for EACH maneuver: Date & time Fetal heart rate Time of abandonment Printed name, staff category and signature		
6.	Was the newborn "roomed in" on the ward with its mother?		
7.	If yes: Has the newborn's identification been checked and is correct within 24h of birth as per documented hospital policy?		

Annexure 16: Nursing Audit

The purpose of nursing audit is to ensure the provision of high-quality nursing care, improve patient outcomes, enhance patient safety, and continuously monitor and improve nursing practices within the healthcare setting. The main objectives of conducting nursing audits include:

1. Quality Improvement: Nursing audits help identify areas of improvement in the delivery of nursing care. By examining current practices, processes, and outcomes, the audit aims to promote continuous quality improvement and enhance patient care.
2. Compliance with Standards and Policies: Nursing audits ensure that nursing practices align with established standards, guidelines, and policies. They help identify any deviations or non-compliance, allowing for corrective actions to be taken to ensure adherence to best practices.
3. Patient Safety: Nursing audits focus on patient safety measures, such as medication administration, infection control, fall prevention, and patient identification protocols. By evaluating adherence to these safety practices, audits help mitigate risks and enhance patient safety.
4. Documentation Accuracy: Audits assess the accuracy and completeness of nursing documentation, ensuring that all essential information is properly recorded. Accurate documentation is crucial for effective communication, continuity of care, and legal purposes.
5. Staff Performance Evaluation: Nursing audits provide an opportunity to assess the performance of nursing staff. By evaluating their adherence to standards, competencies, and professional practices, audits help identify areas of excellence and areas that require further training or support.
6. Resource Utilization: Nursing audits can also evaluate the appropriate utilization of resources, including supplies, equipment, and staffing. By identifying inefficiencies or discrepancies, audits contribute to the effective and efficient use of resources.
7. Evidence-Based Practice: Nursing audits encourage the integration of evidence-based practices into nursing care. By evaluating the use of evidence-based guidelines and protocols, audits promote the delivery of care based on the best available evidence.

1. Patient Assessment:
   - Adequacy and accuracy of initial patient assessment
   - Timeliness of reassessment based on patient condition or changes.
   - Documentation of assessment findings
2. Care Planning and Documentation:
   - Appropriateness and completeness of nursing care plans
   - Documentation of interventions provided, and outcomes observed
   - Evidence of interdisciplinary collaboration in care planning
3. Medication Management:
   - Accuracy in medication administration and adherence to medication orders

- Documentation of medication administration and any observed adverse reactions
- Adherence to medication safety protocols (e.g., medication reconciliation, double-checking)

4. Infection Control:
   - Compliance with hand hygiene protocols
   - Appropriate utilization and disposal of personal protective gear.
   - Adherence to infection control measures (e.g., isolation precautions, aseptic techniques)
5. Patient Safety:
   - Fall prevention measures implemented
   - Use of patient identifiers and verification processes
   - Reporting and management of incidents or near-miss events
6. Pain Management:
   - Assessment of pain levels and response to pain interventions
   - Administration of analgesics according to prescribed protocols
   - Assessment of the success of pain management approaches
7. Patient Education:
   - Providing patient education materials that are both comprehensive and easy to understand.
   - Assessment of patient understanding and ability to apply self-care instructions.
   - Documentation of patient education provided.
8. Communication and Collaboration:
   - Proficient interaction with patients, families, and members of the interdisciplinary team.
   - Timely and accurate handover or shift-to-shift reports
   - Collaboration and teamwork in patient care delivery
9. Ethical and Legal Considerations:
   - Adherence to nursing code of ethics and professional standards
   - Respect for patient autonomy and confidentiality.
   - Awareness of legal requirements related to nursing practice.
10. Quality Improvement:
    - Participation in quality improvement initiatives and audits
    - Use of evidence-based practices in nursing care delivery
    - Contribution to the identification and implementation of process improvements

Note: The nursing audit can be conducted through direct observation, review of nursing documentation, interviews with nursing staff, and analysis of patient outcomes. The specific criteria and indicators may vary depending on the healthcare setting and nursing practice guidelines. The audit aims to assess the quality of nursing care, identify areas for improvement, and ensure adherence to standards and best practices.

Annexure 16A: Nursing Documentation Audit Checklist

Please mark boxes either with X or a tick				
G1a. Ward / area being audited				
G1b. Specialty (i.e., Med, Surgery, etc.)				
G1c. Were the case notes in a reasonable condition, i.e., not falling apart/tally front cover, etc.				
Yes No				
Section 1: Patient Demographics/Admission Details (100% compliance unless not applicable applies)				
Q1a. Do the case notes contain nursing sheets / notes for the most recent admission?				
If yes, do they contain the following?				
		Yes	**No**	**NA**
Q1b	Patient's surname			
Q1c	Patient's first name - Patient's "preferred" name (i.e., "Shri", "Md" instead of Mohammad")			
Q1d	Hospital registration			
Q1e	Date of Birth			
Q1f	Title (Ms, Mrs, Miss, Mr, Dr, Retd. Etc.)			
Q1g	Patient's address (If it "No fixed abode" documented then select "Yes")			
Q1h	Patient's contact phone number			
Q1i	Patient's religion / spiritual needs			
Q1j	Patient's relationship (i.e., Wife, mother, etc.)			
Q1k	Valuables / property			
Q1l	Diagnosis / reason for admission			
Q1m	Admission date			
Q1n	Admission Time			
Q1o	Information about informing relatives regarding changes to patient's condition			
Section 2a : Baseline observations*(100% compliance with all)* Q2a. Which of the following have been recorded in the nursing documentation? (Please **tick All** that apply) ___ Temperature ___ Pulse ___ BP ___Respiratory rate ___ EWS ___Known allergies				
Section 2b: Allergy screening Q2b. Was the patient Screened for Allergy upon admission? Yes No Q2c. Is allergy information documented at a uniform place in the patient's case note file. Yes No				
Section 3: Risk assessments and maintaining a safe environment				
		Yes	**No**	**NA**
Q3a. Is there evidence of a Falls care plan being undertaken?				
Q3b. Is there evidence of a Bed rails risk assessment being undertaken?				
Q3c. Is there evidence of a nutritional assessment having been undertaken within 24 h of admission? (All inpatients should have nutritional requirements assessed upon admission)				
Q3d. Is there evidence that a pressure ulcer risk assessment has been carried out? (All patients should have a pressure ulcer risk assessment performed within 2 h of admission).				
Section 4: Nursing Record Evaluation		**Yes**	**No**	**NA**

(continued)

Q4a. Are entries made at least daily in the nursing record?			
Q4b. Does the nursing record contain any abbreviations?			
Q4c. Where your ward has a policy allowing nurses to use abbreviations is there an approval list contained within the record explaining the abbreviations?			
Q4d. Are entries recorded in chronological order?			
Q4e. Do recorded entries after an event provide current care and condition of patient *(i.e.,*medication change*)*			
Q4f. Are the text entries in such a manner that they cannot be altered or deleted without evidence of them having been changed *(i.e.,*pencil or large gaps in between entries*)?*			
Q4g. Where your ward has a policy allowing nurses to sign only with initials is there a full identification of the individual or a signature legend contained within the record?			
Q4h. Are there any other forms of "inappropriate filing" contained within the nursing record such as loose documentation or blood results, etc. *(*There should be No loose documentation or inappropriate filing*)*			
Section 5: Nursing pages / Continuation sheets / Progress reports Please Audit the most recent episode of nursing care notes up to a maximum of 10 pages. If there are less than 10 pages, then please audit all of the most recent stay.			
Q5. What is the total number of nursing pages being audit? Q5a. How many of those pages have a patient's name on? Q5b. How many of those pages have a patient's hospital number on?			
Section 6: Nursing Entries			
Q6a. What is the total number of nursing entries being audited?			
Q6b. How many of those nursing entries are dated?			
Q6c. How many of those nursing entries are timed? (For inpatient records only)			
Q6d. How many of those nursing entries are signed?			
Q6e. How many of those signatures also have the nurses' name printed?			
Q6f. How many of those signatures also have the job title or grade documented?			
Q6g. How many of the nursing entries are legible (readable to the auditor)?			
Q6h. How many of the nursing entries are written in **black or dark** ink? (suitable for Photocopying)			
Section 7: Error Corrections			
Q7a. What is the total number of errors / corrections being audited within the nursing notes?			
Q7b. How many of these corrections are crossed through with a single line?			
Q7c. How many of these corrections are dated?			
Q7d. How many of these corrections are timed (only applicable to in-patient records, leave blank if auditing outpatient notes)?			
Q7e. How many of these corrections are signed by the nurse?			

Annexure 17: Laboratory Audit

Laboratory accidents are likely to happen, so checklists are developed to avert them. Laboratories will regularly perform self-inspections to ensure adherence to relevant regulatory standards and to acquaint new staff with the locations of safety equipment. Basic safety rules, hazardous waste, chemical safety, radiation safety, and biological safety will all be covered during these inspections. A laboratory safety audit checklist lays out every detail of the safety processes that must be checked, making the inspection or auditing process easier. The checklist simplifies the process of identifying key elements of a lab that require auditing and ensuring their safety is thoroughly verified. A laboratory audit is a systematic assessment of operations, quality systems, and data compliance with regulatory standards and accreditation regulations, such as ISO 15189:2022 and ISO 17025. It is a key component of continuous quality improvement and clinical governance, aiming to identify areas for enhancement, increase operational efficiency, and ensure the reliability of test results to promote patient safety and quality care.

The scope of a laboratory audit typically includes evaluating the entire laboratory operation, encompassing pre-analytical, analytical, and post-analytical phases. It examines aspects such as equipment calibration, personnel qualifications, quality control measures, documentation practices, and adherence to standard operating procedures (SOPs). The audit may also assess compliance with applicable regulations and guidelines from governing bodies such as the College of American Pathologists (CAP) or the Clinical Laboratory Improvement Amendments (CLIA).

Five key activities under the laboratory audit umbrella include:

- Addressing process or outcome issues.
- Monitoring workload and demand control.
- Overseeing the introduction of new tests or changes in practice.
- Ensuring adherence to best practices and guidelines.
- Evaluating analytical quality.

A laboratory audit involves several key steps to ensure a thorough and effective evaluation of the laboratory's operations. The essential steps involved in conducting a laboratory audit:

1. **Planning and Preparation**:
    - Define the scope and objectives of the audit, considering the specific areas to be evaluated.
    - Schedule the audit, allowing sufficient time for preparation and coordination with laboratory staff.
    - Gather relevant documentation, including standard operating procedures (SOPs), quality control records, regulatory guidelines, and previous audit reports.
2. **Assembling the Audit Team**:
    - Form a multidisciplinary audit team with members who have expertise in laboratory operations, quality assurance, and regulatory compliance. This

team may include laboratory managers, quality assurance personnel, and representatives from relevant departments.
   - Assign specific roles and responsibilities to each team member for the audit process.
3. **Conducting a Pre-Audit Meeting**:
   - Scheduling for laboratory audits should be planned in advance, allowing adequate time for preparation and minimizing disruption to laboratory operations.
   - Audits can be conducted annually or biannually, depending on the laboratory's size, complexity, and regulatory requirements. It's essential to communicate the audit schedule to all staff to ensure their readiness and cooperation.
   - Hold a meeting with laboratory staff to explain the audit process, objectives, and schedule.
   - Address any questions or concerns from the staff to ensure cooperation and transparency.
4. **Data Collection and Observation**:
   - Review relevant documentation, including records related to equipment calibration, personnel training, and quality control measures.
   - Conduct on-site observations of laboratory processes, workflows, and practices to verify compliance with SOPs and regulatory standards.
   - Interview laboratory personnel to assess their understanding of procedures and compliance practices.
5. **Data Analysis and Evaluation**:
   - Analyze the collected data to identify any discrepancies, areas of non-compliance, and opportunities for improvement.
   - Evaluate the effectiveness of quality control measures and the accuracy of test results.
6. **Documenting Findings**:
   - Prepare an audit report that summarizes findings, highlighting strengths and areas for improvement.
   - Include specific examples of non-compliance, along with recommendations for corrective actions.
7. **Feedback and Action Plan**:
   - Share the audit findings with laboratory management and staff in a post-audit meeting.
   - Collaborate with the laboratory team to develop an action plan addressing identified issues, including timelines for implementing corrective actions.
8. **Follow-Up and Monitoring**:
   - Schedule follow-up audits or reviews to ensure that corrective actions have been implemented and sustained over time.
   - Monitor ongoing compliance and effectiveness of the improvements made based on the audit findings.

By following these steps, laboratories can effectively conduct audits that enhance their operations, ensure compliance with regulatory standards, and improve overall quality and patient safety. Annexure 16A illustrates indicators for auditing different areas in the laboratory.

Annexure 17A: Indicators for auditing different areas in the laboratory

Areas of audit in the laboratory	Indicator
Pre-analytical	
Request Forms: Evaluate the usability of request forms. Are all necessary details provided by the user? **Specimen Management**: Ensure that the correct specimen is received at the appropriate time and that laboratory staff selects the relevant investigations. **Phlebotomy and Sample Transport**: Assess the efficiency and effectiveness of phlebotomy services and the transportation of samples to the laboratory.	Ease of use of request forms Completeness of relevant details provided by users Receipt of the right specimen at the right time Appropriate investigations selected by lab staff Specimen container labeling errors Number of specimens rejected Specimens lacking mandatory history Specimens received beyond stability period Inadequate specimen quantity Mismatches between specimen label and TRF details
Analytical	
Appropriateness of Investigations: Evaluate whether the range of available tests meets clinical needs. Audit the number of requests and positivity rates; consider withdrawing tests that are rarely requested or have low positivity rates. **Compliance with SOPs**: Ensure that all test methods are performed in accordance with established standard operating procedures. **Safety Policies**: Each laboratory should maintain a comprehensive safety policy. All accidents must be documented, and necessary improvements implemented. Regular audits should assess the handling of hazardous substances. **Staff Efficiency**: Audit the training and utilization of all staff to ensure optimal performance. **Purchasing Practices**: Review the procurement processes for equipment, reagents, and supplies to ensure efficiency and cost-effectiveness. **Laboratory Reports**: Assess the precision and clarity of laboratory reports to ensure they meet quality standards. **Storage Practices**: Evaluate the conditions under which reagents and specimens are stored to maintain integrity. **Quality Assessment**: Conduct internal and external quality assessments to ensure compliance with regulatory standards. **Test Utilization**: Monitor the appropriateness of test usage in relation to clinical needs and operational efficiency	Appropriateness of available investigations Audit of requests and positivity rates for specific tests Compliance with standard operating procedures (SOPs) Comprehensive safety policies and accident documentation Efficient use and training of staff Purchasing practices for equipment and reagents Clarity and precision of laboratory reports Storage conditions for reagents and specimens Internal and external quality assessment Improvement in IQC and PT performance metrics Equipment downtime and kit availability Trainings conducted
Post-analytical	

(continued)

Areas of audit in the laboratory	Indicator
Turnaround Times: Monitor the turnaround times for each request in all departments to identify areas for improvement. **Reporting Methods**: Evaluate reporting methods, including report types, direct communication, and computer systems. **Reference Ranges**: Review the established reference ranges for accuracy and relevance. **Report Interpretation**: Assess the interpretation, consultation, and comments provided on reports. **Complaints and Corrective Actions**: Monitor received complaints and document the corrective measures implemented in response.	Monitoring of turn-around times for requests Reporting methods used (types of reports, communication) Review of reference ranges Consultation and interpretation comments on reports Inpatient lab results accessible for morning rounds. Timeliness of critical values reporting Clinician/customer satisfaction with lab services Metrics on spills, cost per reportable test, and technician productivity Turnaround time (achieved/defined) Quantity of spills, expense for each reportable test, efficiency per technician, and so on. Number of complaints and corrective actions taken

Annexure 17B: Laboratory Safety—Audit Checklist

General Instructions

- Complete the self-audit and submit it to the Department Safety Champion.
- Download mandatory signage from the OLSEH website or obtain it from the Department Office.
- Always use personal protective equipment (PPE) in the laboratory; ensure staff compliance.
- It is the responsibility of the lab supervisor to ensure safety and quizzes lab users on safety rules during the audit.
- Improper chemical segregation is common; check the inventory to ensure proper separation of solvents, oxidizers, acids, and bases.
- All chemicals must be labelled.
- All chemical waste must be labelled.
- Gas cylinders should be stored correctly and labeled.
- Exit doors, including those leading to stairwells, must not be locked during business hours.
- Replace or repair electrical cords with frayed wires or broken insulation, as they pose significant electrical and fire hazards.
- Install cover plates on all electrical outlets and switches to prevent accidental contact with wires.

Laboratory Safety Checklist

Type	Safety checkpoint	Yes	No	NA	Remarks
General	A working telephone is available. An emergency contact list is posted at eye level near the telephone. The hazard sheet is updated and signed within the last 6 months. All users are wearing closed-toed shoes in the lab. If labs require shoes to be removed, replacement shoes must be provided—no chappals. No evidence of food or drinks in the lab (e.g., teacups, coffee mugs, lunch boxes). The lab is clean and organized, with no clutter like cardboard boxes or trash. The lab in-charge has implemented an orientation program for new users before granting lab access. At least one light is connected to the UPS for backup.				
	Important: Please ensure that the storage of materials in laboratory premises is kept to a minimum. For example, paper goods, plastic containers, materials stored in boxes and empty containers.				
Fire	Maintain two feet of unobstructed walking space around lab areas. Fire extinguishers are of the appropriate type, in working order, and stored in visible, accessible locations. Equipment that generates heat during operation has adequate ventilation. No exposed electrical wires, broken plugs, or faulty switches. Power strips are rated for the appropriate wattage based on the load. Large equipment is properly grounded.				

(continued)

Type	Safety checkpoint	Yes	No	NA	Remarks
Chemicals	All chemical containers must be properly labeled with necessary information—no unidentified bottles. Store chemicals in appropriate cabinets—no open shelves or wooden (flammable) cupboards. Exhausted cabinets are required for large amounts of solvents. All chemical storage must display hazard labels (e.g., corrosive, oxidizer, toxic). Clearly label all chemical waste, and ensure appropriate disposal facilities are available (e.g., for acids). Personal protective equipment (PPE) such as goggles, gloves, aprons, and shoes must be accessible and not locked away. Use splash goggles for liquids, thick nitrile gloves for aggressive chemicals, and avoid latex gloves. Face shields must be worn when handling large amounts of chemicals. Eye wash stations must be nearby when working with corrosive chemicals. Maintain printed Material Safety Data Sheets (MSDS) for hazardous chemicals in an easily accessible file. Segregate acids, bases, oxidizers, and solvents in different cabinets or with secondary containment. Refrigerators used for chemical storage must be marked "No food"—no food should be stored in lab fridges. Mark dustbins clearly to differentiate between hazardous lab waste and general waste. Chemical work must be performed in properly ventilated fume hoods. Sharps should be disposed of in hard containers. Bio-waste must be appropriately disposed of: red bins for soiled items, blood/cultures treated with hypochlorite, and bodies disposed of through Bio departments.				

(continued)

Type	Safety checkpoint	Yes	No	NA	Remarks
High pressure gas cylinders	Gas cylinders must be securely chained to walls at half height or 1/3 + 2/3 height, stored in a tight formation to prevent tipping. Carts are not for permanent storage. Unused or spare cylinders must have valve guards. All cylinders must be tagged with contents, status (empty/in-use/full), and the responsible person. Hazardous gas lines require welded VCR connections; highly dangerous gases need coax connections. Hazardous gas cylinders must be stored in gas cabinets. CVD reactors must be connected to functional exhaust systems. Labs with hazardous gases must have functional gas sensors to detect leaks.				
Cryo	Lab using cryogenic liquids must have cryo-rated personal protective equipment (PPE). Gloves are mandatory for everyone, and large quantities require face shields and aprons. Cryo cylinders must be equipped with safety valves and rupture disks. Cryogenic liquids should only be handled in approved dewars or flasks.				
Electrical hazard	Cover Plates in place for outlets and switches				
	No extension cords used				
	Inspect AC filters periodically				
Exits	Illuminated signs working				
	Paths free from obstructions				
	Emergency exits accessible and marked				
	Other observations/violations/comments				

Additional Checklist for Chemical Storage and Handling

1. Light-duty cabinets should generally not be used. Cabinet units must be securely anchored to the wall.
2. To prevent contamination, food must not be stored in refrigerators or freezers designated for chemical storage.
3. Labelling cabinets by chemical class (e.g., flammable liquids, acids, oxidizers) is essential for segregating chemicals to avoid incompatibilities and for easy identification by emergency response personnel.
4. Highly toxic gases, such as fluorine, phosgene, and certain semiconductor gases, must be stored in ventilated cabinets. These cabinets contain exhaust gas in the event of a leak or fire, protecting lab personnel from exposure.
5. Toxic or flammable substances that can become airborne (e.g., gases, vapors, dusts, fumes, or mists) must not be used in unventilated areas, as contaminants can build up to hazardous levels.
6. Chemical containers must be clearly labelled with at least the chemical name. Ideally, the manufacturer's label should remain intact, as it provides important health and hazard information. When transferring chemicals to a new container, label it accordingly. For small containers, codes or reference systems linked to lab records may be used.
7. To prevent spills or vapor release, containers should remain closed when not in use.
8. For efficient containment, fume hoods should contain minimal chemicals or apparatus during use, ensuring the slots and baffles at the back are not obstructed.
9. Certain chemicals may degrade container materials, e.g., hydrofluoric acid is incompatible with glass, and inorganic hydroxides are best stored in polyethylene containers. Some solvents can soften plastic.
10. Chemicals that can react violently or emit hazardous fumes should not be stored together (e.g., oxidizers and flammables, acids and bases).
11. Corrosive materials pose a high risk of tissue damage, especially to the eyes. Storing them below eye level minimizes this risk.
12. Chemical quantities in storage should match the short-term needs of the lab to avoid excessive stockpiling.
13. When transporting chemicals between rooms or buildings, use secondary containers (e.g., bottle carriers) to contain spills in case of breakage.
14. Certain chemicals, such as ethers, have recommended storage limits to prevent the formation of explosive peroxides. Mark containers with their storage and expiration dates upon arrival.
15. A procedure must be in place to ensure all materials are properly labelled, and unneeded chemicals are disposed of to avoid waste disposal issues.

Annexure 17C: Laboratory Audit Checklist

Category	Checklist Item	Status	Comments
Documentation	The SOPs are updated.	☐ Complete ☐ Incomplete	
	Quality control records are maintained	☐ Complete ☐ Incomplete	
	Equipment maintenance logs are current	☐ Complete ☐ Incomplete	
Personnel	Staff training records are up to date	☐ Complete ☐ Incomplete	
	Competency assessments are performed regularly	☐ Complete ☐ Incomplete	
	Personnel files include required certifications	☐ Complete ☐ Incomplete	
Equipment	Calibration records are maintained	☐ Complete ☐ Incomplete	
	Equipment is labeled with calibration status	☐ Complete ☐ Incomplete	
	Preventive maintenance is performed as scheduled	☐ Complete ☐ Incomplete	
Safety	Safety protocols are followed	☐ Complete ☐ Incomplete	
	Personal protective equipment (PPE) is available and being used	☐ Complete ☐ Incomplete	
	Emergency procedures are posted and known by staff	☐ Complete ☐ Incomplete	
Quality Control	Quality control samples are run and documented	☐ Complete ☐ Incomplete	
	Corrective actions are documented for out-of-range results	☐ Complete ☐ Incomplete	
	Proficiency testing is performed, and results are reviewed	☐ Complete ☐ Incomplete	
Specimen Handling	Specimen collection procedures are followed	☐ Complete ☐ Incomplete	
	Specimen labeling is accurate and complete	☐ Complete ☐ Incomplete	
	Specimen storage conditions are appropriate	☐ Complete ☐ Incomplete	
Data Management	Laboratory information system (LIS) is functioning properly	☐ Complete ☐ Incomplete	
	Data entry is accurate and timely	☐ Complete ☐ Incomplete	
	Backup procedures for data are in place	☐ Complete ☐ Incomplete	

(continued)

Category	Checklist Item	Status	Comments
Customer Service	Turnaround times for test results are within acceptable limits	☐ Complete ☐ Incomplete	
	Customer complaints are documented and addressed	☐ Complete ☐ Incomplete	
	Satisfaction surveys are conducted and reviewed	☐ Complete ☐ Incomplete	

Annexure 17D: Laboratory Audit Checklist

Type	Checkpoint	Yes	No	NA	Remarks
Records	Are records maintained to document the preparation and standardization of volumetric solutions? Is the frequency of reagent standardization clearly defined and in line with current USP/IP/BSP guidelines? Are standardized reagents stored in a way that ensures their integrity? Are there documented procedures outlining the preparation of laboratory reagents and cultures? Are there documented procedures detailing the maintenance of laboratory reagents and cultures? Do the written procedures include measures to prevent contamination of primary standards? Are reference standards stored in a secure area with controlled and monitored environmental conditions? Are there procedures in place to maintain the integrity of reference standards? Are working or in-house standards verified against primary standards at regular intervals? Are stock solutions properly labeled, and are their expiration dates determined based on actual stability? Are there procedures in place for certifying and using non-commercially available reference standards? Are certificates of analysis available for all reference standards, and are they stored as controlled documents? Are provisions in place for the proper handling of controlled substance reference standards? Does the SOP for reference standards address the correct handling of controlled substance reference standards, including a detailed mapping process?				
Laboratory Reference Standards and Solutions	Is there a written procedure in place for ordering and receiving both compendial and non-compendial reference standards? Are the primary standards from the current lot listed in the USP, EP, or JP? Is the receipt of reference standards properly logged? Are all reference standards labeled with the name, source, lot number, and expiration date?				

Annexure 18: Patient Satisfaction Audit

The aim of the Patient Satisfaction Audit is to evaluate how satisfied patients are with the quality of care and services offered by the healthcare facility. The audit aims to identify areas of strength and areas requiring improvement to enhance patient experience and overall satisfaction. Audit Components include:

1. Survey Design: Develop a patient satisfaction survey that includes a comprehensive set of questions covering various aspects of the patient's experience, such as communication with healthcare providers, responsiveness of staff, quality of care, cleanliness of facilities, waiting times, information provided, and overall satisfaction. The survey can be administered using different methods, including paper-based surveys, electronic surveys, or face-to-face interviews.
2. Sampling: Decide on the suitable sample size and choose the sampling technique. The sample should be representative of the facility's patient population. Consider factors such as demographics, types of services provided, and patient volume. Random sampling or systematic sampling can be used to select participants.
3. Data Collection: Administer the patient satisfaction survey to the selected sample of patients. Ensure confidentiality and privacy during the data collection process. Collect both quantitative data (e.g., rating scales, Likert scales) and qualitative data (e.g., open-ended questions, comments) to capture a comprehensive view of patient satisfaction.
4. Data Analysis: Analyze the collected data to determine the level of patient satisfaction across different aspects of care and services. Quantitative data can be analyzed using statistical methods such as calculating means, percentages, and frequency distributions. Qualitative data can be analyzed by categorizing responses into themes or common patterns.
5. Benchmarking: Compare the results of the patient satisfaction audit with relevant benchmarks, such as national or industry standards, or previous audit results. Benchmarking provides context and helps identify areas where the facility performs well or lags behind compared to others.
6. Review the audit results to pinpoint areas requiring enhancement. Look for recurring themes, low satisfaction scores, or specific comments that indicate areas of concern. Common areas for assessment include communication, waiting times, staff attitudes, information provision, and facility cleanliness.
7. Develop Recommendations: Based on the identified areas for improvement, develop recommendations and strategies to enhance patient satisfaction. Recommendations may include improving communication protocols, addressing wait times, enhancing staff training on patient-centered care, implementing patient education initiatives, or enhancing facility amenities. Focus on ranking recommendations according to their likely impact and ease of implementation.

8. Implementation and Monitoring: Implement the recommended interventions and monitor their effectiveness in improving patient satisfaction. Track patient satisfaction scores over time to assess the impact of the implemented changes. Continuously monitor patient feedback and assess patient satisfaction on an ongoing basis.
9. Reporting: Prepare a comprehensive report summarizing the audit findings, areas for improvement, and the recommended interventions. Include data analysis, benchmarking results, and specific examples or quotes from patient feedback. Share the report with relevant stakeholders, such as healthcare providers, administrators, quality improvement teams, and patient representatives.

Patient Satisfaction Survey Parameters

Regularly conducting Patient Satisfaction Audits helps healthcare facilities identify areas of improvement and enhance the patient experience. By addressing patient concerns and improving satisfaction, facilities can improve patient outcomes, increase patient loyalty, and maintain a positive reputation in the community.

1. Personal Information:
   - Name (optional)
   - Age
   - Gender
   - Contact information (optional)
2. Appointment and Registration Process:
   - Ease of scheduling an appointment
   - Wait time for appointment
   - Registration process efficiency
3. Communication:
   - Courtesy and friendliness of healthcare providers
   - Clarity of information provided by healthcare providers
   - Active listening and empathy demonstrated by healthcare providers
4. Waiting Time:
   - Waiting time in the reception area
   - Waiting time in the examination room or consultation area
5. Facilities and Environment:
   - Cleanliness and tidiness of the facility
   - Comfort of waiting areas and examination rooms
   - Privacy and confidentiality measures
6. Quality of Care:
   - Competence and expertise of healthcare providers
   - Effectiveness of treatment or intervention received
   - Clarity of instructions or guidance provided for self-care or follow-up
7. Coordination of Care:
   - Continuity of care across different healthcare providers (if applicable)
   - Timeliness of referrals to specialists (if applicable)
   - Communication between different healthcare providers involved in your care (if applicable)
8. Information and Education:
   - Adequacy and clarity of information provided about diagnosis, treatment options, and potential risks or side effects
   - Availability of educational materials or resources for further understanding
   - Involvement in shared decision-making regarding treatment options
9. Staff Attitude and Professionalism:
   - Respect and dignity shown by all staff members
   - Willingness to address questions or concerns
   - Responsiveness to requests for assistance or support

10. Overall Satisfaction:
    - Overall rating of your experience at the healthcare facility
    - Likelihood of recommending the facility to others
11. Additional Comments:
    - Open-ended space for additional comments, suggestions, or concerns

Note: The survey can be adapted and customized based on the specific healthcare setting and services provided. It can be administered through paper-based forms, electronic surveys, or conducted as face-to-face interviews, depending on the preferences and resources available.

Annexure 18A: Sample Patient Satisfaction Survey

We would like to hear your feedback regarding your recent visit to the provider mentioned above. Please share your thoughts on the service you experienced. This questionnaire will take about 8–10 min. Your responses will help us in identifying gaps in the service provided by us and improving the delivery of services to our patients. Your responses will be kept strictly confidential. In case you have any queries that you wish to be answered quickly about the questionnaire you may contact ………… (Give details of the contact person). Return the completed questionnaire to ……………………. (Give details).

Thanks for your help.

Please rate the following:

	Excellent	Very good	Good	Fair	Poor	Not applicable
A. Your appointment:						
1. Ease of making appointments by phone	5	4	3	2	1	NA
2. Appointments can be scheduled promptly within the time frame.	5	4	3	2	1	NA
3. Receiving treatment for sickness or injury promptly when desired.	5	4	3	2	1	NA
4. Accessing care outside of regular hours when necessary.	5	4	3	2	1	NA
5. The efficiency of the check-in process	5	4	3	2	1	NA
6. Waiting time in the reception area	5	4	3	2	1	NA
7. Waiting time in the examination room	5	4	3	2	1	NA
8. Notifying you about any delays to your appointment time.	5	4	3	2	1	NA
9. The simplicity of obtaining a referral when it was necessary.	5	4	3	2	1	NA
B. Our staff:						
1. The politeness of the individual who answered your call.	5	4	3	2	1	NA
2. The receptionist's friendliness and politeness.	5	4	3	2		NA
3. The caring concern of our nurses/medical assistants	5	4	3	2	1	NA

(continued)

	Excellent	Very good	Good	Fair	Poor	Not applicable
4. Assistance provided by the staff with billing or insurance matters.	5	4	3	2	1	NA
5. The expertise and dedication of our laboratory and radiology personnel.	5	4	3	2	1	NA
C. Our communication with you:						
1. Your phone calls answered promptly	5	4	3	2	1	NA
2. Seeking assistance or guidance as required during business hours.	5	4	3	2	1	NA
3. Explanation of your procedure (if applicable)	5	4	3	2	1	NA
4. Your test results are provided within a reasonable timeframe.	5	4	3	2	1	NA
5. Effectiveness of our health information materials	5	4	3	2	1	NA
6. Our capacity to promptly respond to your calls	5	4	3	2	1	NA
7. Your ability to contact us after hours	5	4	3	2	1	NA
D. Your visit with the provider: (doctor, physician assistant, pharmacist, nurse)						
1. Willingness to listen carefully to you	5	4	3	2	1	NA
2. Taking time to answer your questions	5	4	3	2	1	NA
3. Amount of time spent with you	5	4	3	2	1	NA
4. Clarifying concepts in a manner you can grasp/ understand.	5	4	3	2	1	NA
5. Instructions regarding medication/follow-up care	5	4	3	2	1	NA
6. The thoroughness of the examination	5	4	3	2	1	NA
7. Guidance provided to you on methods to maintain good health.	5	4	3	2	1	NA
E. Our facility:						
1. Hours of operation convenient for you	5	4	3	2	1	NA
2. Overall comfort	5	4	3	2	1	NA
3. Adequate parking	5	4	3	2	1	NA

(continued)

	Excellent	Very good	Good	Fair	Poor	Not applicable
4. The signs and guidance are straightforward to understand and are easy to follow.	5	4	3	2	1	NA
F. Your overall satisfaction with:						
1. Our practice	5	4	3	2	1	NA
2. The quality of your medical care	5	4	3	2	1	NA
3. Overall Evaluation of the care quality from your healthcare provider or nurse.	5	4	3	2	1	NA
Would you recommend the provider to others? Yes No						

If no, please tell us why:

If there is any way we can improve our services to you, please tell us about it:

Some information about you:

1.	Gender	Male	Female
2.	Age (years):		
3.	Are you a new patient or a returning patient?		

Thanks very much for your help!

MIX
Papier aus verantwortungsvollen Quellen
Paper from responsible sources
FSC® C105338

If you have any concerns about our products,
you can contact us on
ProductSafety@springernature.com

In case Publisher is established outside the EU,
the EU authorized representative is:
Springer Nature Customer Service Center GmbH
Europaplatz 3, 69115 Heidelberg, Germany

Printed by Libri Plureos GmbH
in Hamburg, Germany